FOREWORD

GEOTRAP, the OECD/NEA Project on Radionuclide Migration in Geologic, Heterogeneous Media, is devoted to the exchange of information and in-depth discussions on present approaches to acquiring field data, and testing and modelling flow and transport of radionuclides in actual (and therefore heterogeneous) geologic formations for the purpose of site characterisation and evaluation and safety assessment of deep repository systems of long-lived radioactive waste. The project is articulated in a series of structured, forum-style workshops whereby national waste management agencies, regulatory authorities and scientists are able to interact and contribute to the advancement of the state of the art in these areas. Each workshop is also aimed at producing a synthesis report.

Following the success of the first GEOTRAP workshop[1], a second workshop, "Basis for Modelling the Effects of Spatial Variability on Radionuclide Migration", was held in Paris (France) on 9-11 June 1997, and hosted by the *Agence nationale pour la gestion des déchets radioactifs, ANDRA* (French Radioactive Waste Management Agency).

Variability (heterogeneity) in the properties of the rocks over a wide range of spatial scales is a common feature of most geologic media and broad agreement exists that its characterisation and the corresponding analysis of the consequences for groundwater flow and radionuclide transport form an important part of the assessment of the safety of deep geologic repositories.

Twenty national waste management organisations (implementing agencies and regulatory authorities) from ten OECD Member countries were represented at the workshop along with members of nuclear research institutes, the academic community and scientific consulting companies. The workshop provided an overview of current developments in this technical field – both in national waste management programmes and the scientific community. These developments reinforce confidence in the concepts and models used for repository performance assessment.

In addition to the papers presented orally or as posters during the workshop, this publication includes a synthesis of the workshop. The synthesis reflects the materials that were presented, the discussions that took place, and the conclusions and recommendations drawn, notably during the working group sessions. The synthesis also puts these conclusions and recommendations into perspective within the scope of the GEOTRAP project and the state of the art in the technical field.

The opinions, conclusions and recommendations expressed are those of the authors only, and do not necessarily reflect the view of any OECD Member country or international organisation. This report is published on the responsibility of the Secretary-General of the OECD.

---

[1] "Field Tracer Experiments: Role in the Prediction of Radionuclide Migration", Cologne, Germany, 28-30 August 1996 (NEA GEOTRAP Project on Radionuclide Migration in Geologic, Heterogeneous Media), OECD Proceedings, NEA/EC, 1997.

ACKNOWLEDGEMENTS

On behalf of all participants, the NEA wishes to express its gratitude to the *Agence nationale pour la gestion des déchets radioactifs, ANDRA* (France) which hosted the workshop in Paris.

Special thanks are also due to:

- the members of the Programme Committee who structured and conducted the workshop and reviewed the workshop synthesis: Lionel Dewière (ANDRA, France), Björn Dverstorp (SKI, Sweden), Emmanuel Ledoux (École des Mines de Paris, France), John Porter (AEA Technology, United Kingdom), and Paul Smith (Safety Assessment Management Ltd., United Kingdom);
- the NEA consultants: Emmanuel Ledoux (École des Mines de Paris, France) and Paul Smith (Safety Assessment Management Ltd., United Kingdom) who helped the Secretariat in drafting the workshop synthesis, and Rae Mackay (Birmingham University, United Kingdom) who animated the workshop discussions;
- the working groups' chairmen: Cliff Davison (AECL, Canada), Jörg Hadermann (PSI, Switzerland), Rae Mackay (Birmingham University, United Kingdom) and Ivars Neretnieks (Royal Institute of Technology, Sweden), who led and summarised the debates that took place in the four working groups;
- the speakers and the posters' authors for their interesting and stimulating presentations; and
- the participants for their active and constructive contribution.

The Chairmen of the sponsoring NEA groups, Alan Hooper (Co-ordinating Group on Site Evaluation and Design of Experiments for Radioactive Waste Disposal – SEDE) and Piet Zuidema (Performance Assessment Advisory Group – PAAG) made a number of valuable comments and contributions in reviewing the synthesis; their reviews have been addressed in the present publication.

Claudio Pescatore and Philippe Lalieux from the Radiation Protection and Waste Management Division of the OECD Nuclear Energy Agency are responsible for the GEOTRAP Project's scientific secretariat and are members of the workshop Programme Committee.

Anna Beggs from the Radiation Protection and Waste Management Division of the OECD Nuclear Energy Agency was in charge of the compilation and the lay-out of the publication.

TABLE OF CONTENTS

PART A: SYNTHESIS OF THE WORKSHOP

Executive Summary 13

1. **Introduction** 15
1.1. Rationale, Objectives and Structure of the Workshop 15
1.2. Structure of the Synthesis 16

2. **Achievements of the Workshop Objectives** 17

3. **Overall Conclusions** 19
3.1. Building Confidence in Assessment Models 20
3.2. Communication between those Involved in Site Characterisation and
those Involved in Performance Assessment 23
3.3 Effects of Spatial Variability on Geosphere Performance 24

4. **Recommendations and Restrictions** 25

5. **Concluding Observations** 28

Annex 1 Synoptic Table of the Cases Discussed 29

Annex 2 List of Key Questions to be Addressed 40

Annex 3 Conclusions of Working Group A
Approaches to Confidence Building in Models of
Spatial Variability 43

Annex 4 Conclusions Of Working Group B
Definition and Quantification of Effective Parameters
or Flow and Transport 47

Annex 5 Conclusions Of Working Group C
Benefits of Further Site Characterisation for Better
Radionuclide Transport Calculations 49

Annex 6 Conclusions of Working Group D
Future Directions of work in the Treatment of
Spatial Variability for Radionuclide Transport 53

PART B: WORKSHOP PROCEEDINGS

SESSION I

Spatial Variability: Its Definition and Significance to Performance Assessment and Site Characterisation
Chairmen: R. Mackay (Birmingham Univ.,United Kingdom) and L. Dewière (ANDRA, France)

Variability in Natural Systems and Means to Characterise it
E. Ledoux (Ecole des Mines de Paris, France),
and G. de Marsily (Univ. Pierre et Marie Curie, France) 59

Why is Variability Important for Performance Assessment and
What are its Consequences for Site Characterisation and Repository Design?
B. Dverstorp (SKI, Sweden), P.A. Smith (SAM Ltd., United Kingdom),
and P. Zuidema (Nagra, Switzerland) 73

An Example of a Comprehensive Treatment of Spatial Variability
in a Performance Assessment: Nirex-97
C.P. Jackson (AEA Technology, United Kingdom) 83

SESSION II

Experience with the Modelling of Radionuclide Migration in the Presence of Spatial Variability in Various Geological Environments
Chairmen: E. Ledoux (Ecole des Mines de Paris, France) and
J. Geier (Clearwater Hardrock Consulting/Oregon State Univ., United States)

The Geological Basis and the Representation of Spatial Variability in Fractured Media
M. Mazurek (Univ. of Berne, Switzerland),
A. Gautschi and P. Zuidema (Nagra, Switzerland) 101

The Geological Basis and the Representation of Spatial Variability in Sedimentary
Heterogeneous Media
K.A. Cliffe, D.J. Franklin, P.I.R. Jones, E.J. Mcleod,
and J.D. Porter (AEA Technology, United Kingdom) 113

Representation of Spatial Variability for Modelling of Flow and
Transport Processes in the Culebra Dolomite at the WIPP Site
L.C. Meigs and R.L. Beauheim (SNL, United States) 125

The Discrete-Modelling Approach Adopted in SITE-94
J. Geier (Clearwater Hardrock Consulting/Oregon State Univ., United States),
and B. Dverstorp (SKI, Sweden) 135

Approaches Dealing with Fractured Media within the SKB Äspö
Hard Rock Laboratory Project
G. Gustafson (Chalmers University, Sweden),
I. Rhén (VBB Viak AB, Sweden), and A. Ström (SKB, Sweden) 149

Spatial Variability in the Geosphere Models used for AECL's Long Term
Performance Assessment of the Disposal of Canada's Nuclear Fuel Wastes
C.C. Davison, T. Chan, N. Scheier, T. Melnyk,
A. Brown, M. Gascoyne and T.T. Vandergraaf (AECL, Canada) 159

An Integrated UZ Flow and Transport Model for Yucca Mountain, Nevada, USA
G.Y. Bussod, B.A. Robinson, A.V. Wolfsberg, C.W. Gable
and H.S. Viswanathan (LANL, United States) 165

SESSION III

New Areas for Investigation: Two Personal Views
Chairmen: G. Gustafson (Chalmers Univ., Sweden) and E. Fein, (GRS, Germany)

Impact of Uncertainties in Chemical and other Entities on
Radionuclide Migration from a Repository for Spent Nuclear Fuel
I. Neretnieks (Royal Institute of Technology, Sweden) 195

What Can be Learnt from Other Research Fields and Applications
G. de Marsily (Univ. Paris VI, France) 213

SESSION IV

What is Wanted and What is Feasible:
Views and Future Plans in Selected Waste Management Organisations
Chairmen: A. Gautschi (Nagra, Switzerland)

Modelling Strategies and Usage of Site Specific Data for Performance Assessment
A. Hautojärvi (Posiva Oy, Finland), J. Andersson (Golder Associates AB, Sweden),
H. Ahokas (Fintact Oy, Finland), L. Koskinen and A. Poteri (VTT Energy, Finland),
and A. Niemi (Royal Institute of Technology, Sweden) 219

Training for Incorporating Spatial Variability in Performance Assessment Approach
L. Dewière and M. de Franco (ANDRA, France), and E. Mouche (CEA, France) 233

SKB's Approach to Treatment of Spatial Variability in PA Modelling
J.-O. Selroos and A. Ström (SKB, Sweden) 237

Methods to Incorporate Different Data Types in the Characterisation Process
J.J. Gómez-Hernández (UPV, Spain),
J. Carrera and A. Medina (UPC, Spain) 245

POSTER SESSION

Chemical Modelling Studies on the Impact of Small Scale Mineralogical Changes on Radionuclide Migration
A.T. Emrén (Chalmers Univ., Sweden) 259

Upscaling Conductivities and Dispersivities
J.J. Gómez-Hernández and X.-H. Wen (UPV, Spain) 271

Channel Network Model for Flow and Radionuclide Migration
B. Gylling, L. Moreno and I. Neretnieks (Royal Institute of Technology, Sweden) 279

Modelling Transport in Media with Correlated Spatially Variable Physical and Geochemical Properties
M.D. Impey, C.C. Sellar, R.C. Brown and K.J. Clark (QuantSci, United Kingdom),
K. Hatanaka, Y. Iriji and M. Uchida (PNC, Japan) 291

Development of a New Transport Model (PICNIC) which Accounts for the Large- and Small- Scale Heterogeneity
A. Jakob (PSI, Switzerland) 299

Scale-Dependent Heterogeneities in the Gorleben Area and Implications for Model Studies
R. Ludwig, K. Schelkes (BGR, Germany), K.-J. Röhlig (GRS, Germany),
and J. Wollrath (BfS, Germany) 303

A Case Study on the Influence of Sorption Inhomogeneities on the Migration of Contaminants
L. Lührmann, E. Fein (GRS, Germany) and P. Knabner (Univ. Erlangen, Germany) 317

Single Well Harmonic Pulse Testing - A Novel Technique for In-Situ Measurement of the Hydraulic Heterogeneity of a Fracture
A. Selvadurai (McGill Univ., Canada) and P. Flavelle (AECB, Canada) 327

Effect of Sorption Kinetics on the Migration of Radionuclides in Fractured Rock
A. Wörman and S. Xu (Uppsala Univ., Sweden)... 335

LIST OF PARTICIPANTS... 347

PART A:

SYNTHESIS OF THE WORKSHOP

EXECUTIVE SUMMARY

At the completion of the International INTRAVAL Project[1], *the characterisation of the geologic complexity in the field and the theoretical and computational analysis of its impact on the prediction of flow and transport for the space and time scales of relevance in performance assessment* was considered to be by far the greatest issue encountered during the project and the key challenge in future developments. Hence, the decision (i) to launch the GEOTRAP Project with a clear and continuous focus on radionuclide transport in actual (i.e. heterogeneous) geologic formations, and (ii) in the framework of this project, to hold a specific workshop on the modelling of the effects of spatial variability on radionuclide migration.

Twenty national waste management organisations (implementing agencies and regulatory authorities) from ten OECD Member countries were represented at the workshop "Basis for Modelling the Effects of Spatial Variability on Radionuclide Migration", along with members of nuclear research institutes, the academic community and scientific consulting companies. Several techniques were described and discussed during the workshop both to characterise spatial variability in the course of the geologic investigation of a site and to model the consequences for groundwater flow and radionuclide transport. Several types of heterogenities, in different geologic settings and at different spatial scales, were considered. Some new areas for investigation were also discussed along with methods used outside the radioactive waste arena.

The workshop reflected the present status of development in national waste management programmes, and in particular the fact that, on the one hand, detailed treatments of spatial variability have now been successfully performed as part of several performance assessment exercises, and, on the other hand, spatial variability has been more fully treated in connection with groundwater flow than in connection with the transport problem.

In general, the workshop indicated broad agreement both in the importance of considering geologic spatial variability in the development of a deep repository system, and in the need to consider, for performance assessment purposes, geologic variability over a wide range of spatial scales. Progress was demonstrated in the following areas: (i) the development and implementation of alternative models and approaches to represent spatial variability, (ii) the consideration of a broad range of site-characterisation information, and (iii) the enhancement of confidence that can be gained by effective communication between site characterisation and performance assessment specialists[2]. The workshop illustrated advancement in confidence in the concepts and models used for repository performance assessment.

The main conclusions and recommendations that were drawn at the workshop can be summarised as follows:

1. The consideration of a wide range of models, consistent with the available information from site characterisation and model-testing exercises, is important in order to deal with

---

[1] "The International INTRAVAL Project (Developing Groundwater Flow and Transport Models for Radioactive Waste Disposal) – Final Results", NEA/SKI, 1996.

[2] The need for further development in these areas had been identified in the recommendations of previous NEA initiatives, namely the workshop on "The Role of Conceptual Models in Demonstrating Repository Post-Closure Safety" (Paris, 16-18 November 1993; OECD/NEA, 1995), and the Integrated Performance Assessment Group ("Lessons Learnt from Ten Performance Assessment Studies", OECD/NEA, 1997).

conceptual-model uncertainty. Alternative models should include not only models of flow and transport, but also the different possible concepts consistent with existing observations (e.g. the geologic structural models) that form the basis for these other models. In this framework, maximum account should be taken of all available information from site characterisation, and other relevant sources, as it becomes available.

2. The consistency between the model components used to represent a specific system in an assessment (structure, flow, chemistry, transport) should be demonstrated.

3. Additional model testing (e.g. where predictions are made in advance of an experiment or characterisation measurement) should be used, where possible, to narrow the range of conceptual-model uncertainty. Calibration and testing of models should be performed, if possible, at a range of different scales.

4. Analogues, both natural and anthropogenic, provide a possible means to address calibration and testing of models of flow and transport over spatial and temporal scales that are relevant to performance assessment.

5. Models of groundwater flow are currently available that allow, in principle, a realistic representation of the structural variability of a geologic medium. Availability of data is the most important constraint on the use of these models.

6. Simplified models that employ effective parameters are likely to continue to be important, particularly in the modelling of transport. The use and quantification of effective parameters need careful justification. The simplified models resulting from the use of conservative arguments are more valuable when supported by a detailed understanding and, possibly, by more complete (and complex) models to evaluate the degree of conservatism.

7. Progress is being made in unravelling the associations between the different (variable) properties influencing, over a range of scales, radionuclide migration and in determining the potential contribution to bias and uncertainty arising from these associations. Integrated approaches coupling flow, transport and chemical reactions appear very promising. The computational tools are still, however, under development and only a few examples of realistic applications are currently available.

8. A responsive and flexible approach to both performance assessment and site characterisation, with effective inter-communication, has been found useful e.g. for model development and data acquisition. This helps utilise resources in an efficient manner.

Overall the workshop confirmed that a multi-disciplinary approach is necessary in order to address more fully the problems of modelling flow and transport in media characterised by a high degree of spatial variability and in which coupled and, possibly, non-linear processes operate. Effective communication between different groups, including geologists, hydrogeologists, chemists and performance-assessment specialists will therefore continue to be essential. This was also a finding of the first GEOTRAP workshop.

1. INTRODUCTION

1.1. Rationale, Objectives and Structure of the Workshop

At the completion of the International INTRAVAL Project[3], *the characterisation of the geologic complexity in the field and the theoretical and computational analysis of its impact on the prediction of flow and transport for the space and time scales of relevance in performance assessment* was considered to be by far the greatest issue encountered during the project and the key challenge in future developments. After discussions and careful preparation by the organisations represented in the PAAG and SEDE[4], and in the technical community in the field of nuclear waste management, the need for an international forum was identified, in which issues linked with radionuclide transport in actual (i.e. heterogeneous) geologic formations can be discussed in a structured manner between national waste management agencies (implementers), regulatory authorities (regulators) and scientists. Hence, the decision to launch GEOTRAP, the OECD/NEA international project on radionuclide migration in geologic, heterogeneous media. GEOTRAP has a tightly defined and focused scope with an emphasis on the basis for the understanding and modelling of transport in real geologic formations. The project is organised in the form of a sequence of forum-like workshops, each with a specific theme, where intense interaction is necessary between participants. The workshop sequence is fully integrated with the current programmes of work of the PAAG and SEDE.

Variability (heterogeneity) in the properties of the rocks over a wide range of spatial scales is a common feature of most geologic media. It is relevant to the deep geologic disposal of radioactive waste, in that it affects siting feasibility, optimum repository layout and long-term repository safety. A wide variety of techniques exists both to characterise spatial variability in the course of the geologic investigation of a site and to model the consequences of such variability on groundwater flow and radionuclide transport. Developments in these techniques have been rapid in recent years, hence the decision, within the framework of GEOTRAP, to hold a workshop on the basis for modelling the effects of spatial variability on radionuclide migration, keeping in mind the needs of site characterisation and performance assessment.

The workshop objectives were:

- to provide a forum whereby implementers, regulators, and scientists can interact in a structured fashion;
- to learn about, and contribute to, the advancement of the state of the art in the area of the modelling of spatial variability in order to build confidence in predictive modelling of radionuclide transport in geologic, heterogeneous media;
- to discuss the geologic basis, rationale and objectives of current approaches;
- to comment on the approaches used by different programmes;
- to assess the results of current modelling approaches, their uses/relevance for radionuclide transport assessments, and their impact on site characterisation requirements;
- to discuss new/innovative investigation areas, and alternative ways in which to model spatial variability.

---

[3] "The International INTRAVAL Project (Developing Groundwater Flow and Transport Models for Radioactive Waste Disposal) – Final Results", NEA/SKI, 1996.

[4] PAAG: NEA Performance assessment Advisory Group; SEDE: NEA Co-ordinating Group on Site Evaluation and Design of Experiments for Radioactive Waste Disposal.

A final goal of the workshop was the preparation of this synthesis report, which reviews and summarises the lessons learned at the workshop, putting them into perspective within the scope of the GEOTRAP Project and the state of the art in the technical field. This synthesis is also aimed at establishing a link with the outcome of the first GEOTRAP workshop on *"Field Tracer Experiments: Role in the Prediction of Radionuclide Migration"* and emphasising some open questions to be addressed in the third workshop that will be devoted to the *"Characterisation of Water-Conducting Features and their Representation in Models of Radionuclide Migration"*.

The workshop was introduced by three overview papers (Session I) which covered respectively variability in natural systems; the importance of variability for performance assessment; and an example of a comprehensive treatment of variability in a performance assessment. These provided the audience with a common background for the planned discussions. The three following sessions addressed:

- experience gained from transport modelling in the presence of spatial variability in various geologic environments; approaches to the characterisation and representation of the variability were also an important focus of many of the papers (Session II);
- new areas for investigation (Session III); and
- the future plans for dealing with spatial variability in waste management programmes (Session IV).

A poster session dealt with additional test cases and provided technical details on several characterisation and modelling approaches and methods. A key part of the workshop consisted of focused discussions within small, *ad-hoc* working groups, on specific themes, namely:

A. approaches to confidence building in models of spatial variability;
B. definition and quantification of effective parameters for flow and transport;
C. benefit of further site characterisation for better radionuclide transport calculations;
D. future directions of work in the treatment of spatial variability for radionuclide transport.

These discussions were introduced by a presentation by the working groups' chairmen. The outcomes of the working groups provided the basis for a plenary, concluding discussion and for the present synthesis. For each session and for each working group, the Programme Committee had established a series of key questions to be addressed. This proved to be a very effective way of focusing the discussions and reaching practical conclusions.

1.2. Structure of the Synthesis

Following this introductory section, the synthesis provides an assessment as to how each objective specified for the workshop was achieved (Section 2), the overall conclusions (Section 3), the recommendations regarding future work and their potential implementation (Section 4) as well as some concluding observations (Section 5).

The oral and poster presentations discussed at the workshop cover a wide spectrum of aspects ranging from general overviews to case studies, and to technical details. Furthermore, they encompass a broad range with respect to each of the objectives; conceptual and modelling approaches; geologic settings; and scales of interest. Annex 1 provides a synoptic table of the key features of the presentations, that helps set the context in which the workshop achievements, general conclusions and recommendations should be viewed.

Annex 2 details the workshop structure and lists the key questions that were established by the Programme Committee for each session and working group.

In order to further support this synthesis, Annexes 3 to 6 present the main points made during the discussions that took place within the *ad-hoc* working groups. These annexes are made up of the reports that the working groups' chairmen drafted at the end of the workshop. These reports are reproduced here with only minor typographical editing but without further elaboration.

2.   ACHIEVEMENT OF THE WORKSHOP OBJECTIVES

The broad manner in which the objectives of this GEOTRAP workshop, listed in Section 1, were achieved is as follows (technical points are detailed in Sections 3 and 4):

- **To provide a forum whereby implementers, regulators and scientists can interact in a structured fashion.**

The workshop was attended by 44 delegates from 10 countries, with a range of experience including performance assessment, site characterisation, experimental techniques and modelling. Several implementing organisations and a few regulatory bodies were represented, as well as research laboratories, universities and scientific consulting companies. The workshop comprised 13 technical presentations, each allocated a discussion period, and in-depth discussions by 4 *ad-hoc* working groups on a range of detailed topics. The technical presentations triggered discussion as to the level of detail at which it is necessary to consider spatial variability in performance assessment[5], the types/sources of data on variability that should be considered, the modelling techniques that are available for incorporating these data, and the emerging investigation areas. These discussions ultimately led to a clarification of these issues.

- **To learn about and contribute to the advancement of the state of the art in the area of the modelling of spatial variability in order to build confidence in predictive modelling of radionuclide transport in geologic, heterogeneous media.**

Technical presentations gave an overview of the significance of spatial variability to performance assessment and site characterisation and of the experience of various organisations in the modelling of groundwater flow and radionuclide migration in the presence of spatial variability in various geologic environments. The emphasis in most of the presentations was on approaches to the representation of variability in physical/hydrogeologic properties. A separate presentation was, however, made on the impact of uncertainties in chemical and other properties on radionuclide migration. Also, although most presentations drew on experience from the field of radioactive-waste management, a separate presentation was made on what can be learned from other research fields and applications. The discussions of these issues continued in the working-group sessions,

---

[5] In this synthesis, the terms "performance assessment" and "(performance-) assessment models" are taken to include not only those models that are used directly to calculate safety indicators, such as dose, that are often highly simplified, but also the more detailed "process models" that support them (e.g. models of flow that aim at realism).

with the broad topic of confidence building in predictive modelling addressed by Working Group A. The experience gained in the modelling of spatial variability, and the continuing need for improvement, is summarised in the conclusions drawn by the four working groups, appended to this document. The overall contribution to the field can be judged from the conclusions and recommendations of the workshop. As with the first GEOTRAP workshop, the consensus among the participants was that the workshop was successful and useful, both to themselves and, more generally, to organisations working towards the disposal of radioactive waste.

- **To discuss the geologic basis, rationale and objectives of current approaches.**

The oral and poster presentations of the experience of various organisations in the representation of variability and in the modelling of groundwater flow and radionuclide migration in the presence of spatial variability encompass a broad range with respect to each of the rationale; objectives; conceptual and modelling approaches; geologic settings; and scales of interest (see the synoptic table in Annex 1). Specific presentations addressed the geologic basis of current approaches in the cases of fractured crystalline and sedimentary rocks, and unfractured sedimentary media. Common features and differences in the rationale, objectives and approaches of various programmes and in the various geologic media under consideration stimulated interesting discussions.

- **To comment on the approaches used by different programmes.**

While many techniques exist to model the effects of spatial variability on groundwater flow and radionuclide transport, a common feature is that, given the practical limitations on the availability of information and the capability of computer models to handle the full complexity of natural systems, it is necessary to perform a certain amount of simplification and homogenisation of parameters and processes (e.g. in the performance of up-scaling and in the reduction of a 3-D system to a 1-D model). This simplification and homogenisation leads to the introduction of "effective parameters" into the models. The definition and quantification of effective parameters for flow and transport was the topic addressed by Working Group B. While the consensus was that effective parameters are an inevitable feature of practicable calculational approaches, the loss of detail in the description of the system and of the underlying mechanisms, and the bias that these parameters might introduce into modelling results were pointed out and need to be emphasised.

- **To assess the results of current modelling approaches, their uses/relevance for radionuclide transport assessments and their impacts on site characterisation.**

The results of current modelling approaches and their uses/relevance for radionuclide transport assessments were addressed in the discussions that followed the presentations of the experience of various organisations in the modelling of flow and radionuclide migration in the presence of spatial variability in different geologic environments. In the context of performance assessment, there was some debate as to whether simple models are adequate in themselves, or whether the assessor should strive to capture the full complexity of the natural system in his/her model. The consensus was that simple models are useful, particularly as a communication tool, but should at least be supported by more detailed models. Furthermore, advances in computer technology will mean that increasingly complex and detailed process models could, if desired, be incorporated in performance assessment.

It was suggested that the modelling approach used could impact the way in which site characterisation is carried out but the consensus was that site characterisation should not be driven solely by the choice of models. Overall, it was emphasised that the GEOTRAP project contributes to a more effective communication between site characterisation and performance assessment. The benefits of further site characterisation for more realistic radionuclide transport calculations was the topic addressed by Working Group C.

- **To discuss new/innovative investigation areas, and alternative ways in which to model spatial variability.**

The future directions of work in the treatment of spatial variability for radionuclide transport was the topic addressed by Working Group D. It was agreed that, in addition to the identification of major migration pathways, a key question to address in future investigations will be the existence of potential "fast pathways", that may be few in number, but have a dominant effect on radionuclide transport. Although variability in hydrogeologic properties is most important in this regard, the variability of other properties (e.g. sorption, matrix diffusion) should not be overlooked. Complications arise where these properties display non-linear behaviour (e.g. sorption). A further complication, discussed at the GEOTRAP workshop in the context of geochemical behaviour, is that properties may display time-dependence (e.g. porosity distribution may be influenced notably by water flow and reactions with minerals along the flow path), as well as spatial variability. In the development of new/innovative investigation areas, the importance of demonstrating consistency between geologic, hydrogeologic and transport models should not be overlooked.

3.   OVERALL CONCLUSIONS

The following conclusions are based on the presentations made at the second GEOTRAP workshop, the discussions of these presentations and the discussions of the four working groups. Key points are enclosed in boxes.

Overall, the participants of the workshop agreed on the importance of considering spatial variability in the development of deep geologic repositories for radioactive waste. In particular, understanding of spatial variability was seen as important in assessing:

- the feasibility of locating a repository at a particular site;
- the optimum layout of a repository at the selected site;
- the long-term safety of the repository.

The workshop focused on long-term repository safety, but discussions also ranged beyond performance-assessment/site-characterisation issues related directly to spatial variability, and included broad discussions of uncertainty and confidence building in assessment models and of the relationship between site characterisation and performance-assessment modelling. The workshop conclusions reflect the scope of the discussions, and address three broad areas: (i), confidence building in assessment models, (ii), the relationship between site characterisation and performance-assessment modelling, as well as (iii), the effects of spatial variability on geosphere performance.

3.1. Building Confidence in Assessment Models

In view of the existence of uncertainty, it is important to assess and, if necessary, improve confidence in assessment models, particularly if they are to be applied in developing a safety case. Efforts must be made to quantify or bound uncertainty and to understand the sensitivity to uncertainty. The aim is to reach a situation where that which is known is sufficient to evaluate safety with confidence and that which is not known does not undermine the "sufficiency" of safety. Confidence in assessment models, including those that incorporate the spatial variability of the geologic environment of a repository, is built through the application of a number of broad principles. Principles of confidence building that were identified and discussed in the GEOTRAP workshop, and of which the workshop demonstrated an increasing implementation within national waste disposal programmes, include:

> (i) The consideration of alternative models should include not only models of flow and transport, but also the different possible concepts consistent with the observations (e.g. the geologic structural models) that form a basis for these other models.

The GEOTRAP presentations described a number of different techniques currently used to represent spatial variability in models of flow and transport. These include deterministic models (that may include a limited number of discrete features of contrasting properties), stochastic-continuum models, discrete-fracture-network models and channel-network models. In these, as in other areas of performance assessment, it is often possible to identify a range of models that is consistent with the available information. This is referred to as conceptual-model uncertainty. The significance of conceptual-model uncertainty is dependent on the degree to which predictions of the different models diverge when extrapolating from the range over which the models can be tested (and found to be consistent with the available information) to the performance-assessment scale. Importance of uncertainties may be low where there is insignificant divergence in the predictions. The similarity in the predictions may, however, be the result of common basic concepts underlying the different models (e.g. a common geologic structural model underlying different models of flow and transport). In this case there is a potential danger that all the models may be inadequate, and there is thus a need to focus on how to justify the basic common concepts. This is potentially important in the demonstration of repository feasibility and safety, but is rarely done in practice. However, the workshop provided an illustration of the benefit that can be gained by revisiting the geologic structural model and demonstrated that this approach is gaining some acceptance.

> (ii) Broad consideration should be given to a wide range of models consistent with the available observation in order to evaluate conceptual-model uncertainty (only a limited number of models may be selected from this wide range for further quantitative evaluations).

In considering the range of possible models, the danger of entrenchment[6] of prior models should be recognised. This principle applies to all models, including those of geologic structure, of flow and of transport (the characterisation of spatial variability using Gaussian models was cited at GEOTRAP as a possible example of an entrenched model). Having identified that more than one conceptual model is consistent with the available observations, it would, in principle, be desirable to evaluate the consequences of all consistent models, over

---

[6] i.e. the unquestioned acceptance of a model simply due to its repeated usage.

timescales relevant to performance assessment. However, in a repository performance assessment, this is likely to be impractical, and the range of models considered must be narrowed, either by using arguments of conservatism or through further model testing and/or calibration with site-characterisation and other data.

(iii) Conservative arguments represent one method to deal with uncertainty in assessment models. The resulting, simplified models are, however, more valuable when supported by a detailed understanding and, possibly, by more complete (and complex) models to evaluate the degree of conservatism.

There are areas where information is, and will continue to be, sparse and parameter uncertainties are large (e.g. the description of infrequently occurring features that may constitute "fast pathways", and mineralogical and geochemical variability). In such areas, simple, conservative approaches play a role and are likely to continue to do so. Simple approaches are also useful as tools to communicate concepts. The conservatism of these simple approaches must, however, be assessed and, for this reason, simple models are more valuable where supported by a detailed understanding of the site and of the pertinent transport mechanisms and, possibly, by more complex models.

There was some debate at the GEOTRAP workshop regarding the trend towards more complex models of flow and transport. It was noted that there are areas where the available information (e.g. on large-scale geologic structures) exceeds that which can be incorporated into current models. Benefits were identified in developing more complete models to incorporate this information, for example:

- uncertainties related to model abstraction and up-scaling (use of effective parameters), which are difficult to quantify and can lead to a loss of detail in the description of the system and of the underlying mechanisms, can be reduced;
- such models would be more readily calibrated against existing data and may more readily accept new data.

The GEOTRAP presentations indicate that the development of coupled models (e.g. of flow, transport and chemistry) and the integration of detailed process models into the performance-assessment model chain are rapidly progressing areas of activity. Furthermore, with advances in computer technology, the possibility arises to incorporate increasingly complex models in performance assessment. Therefore, limitations in computer resources may, in themselves, provide less motivation to simplify models. It should however be noted that (i) the desirability of the incorporation of detailed models into performance assessment is dependent upon the approach adopted for performance assessment; (ii) it may remain impractical to represent all of the small-scale heterogeneity (in e.g. rock and fracture properties and chemical behaviour) in large-scale radionuclide transport calculations; and (iii) the lack of data to support the complex models currently provides an important motivation for simplification.

(iv) Maximum account should be taken of all available information from site characterisation, and other relevant sources, as it becomes available.

The availability of data increases as a repository programme progresses. In order to build confidence in the selected approaches and models, it is necessary to make full use of this information as it becomes available. It may be necessary to revise the models (including the geologic structural model and other basic conceptual models) as more information is obtained. Consistency between the selected models and the available data should be the goal in the calibration and testing of models. Both "soft" information, such as general geologic experience, and "hard" data, such as hydraulic heads and temperatures, are relevant. Soft data can be helpful in the selection of appropriate models and in the evaluation of the reasonable ranges for parameters that are not fixed by hard measurements. The GEOTRAP workshop highlighted the recent, scientific progress in the techniques available for incorporating different types of soft and hard data into models, but recognised that there is scope for improvement in these techniques. There is currently no single technique to integrate all types of information, and much information is, therefore, not likely to be used in an optimal way.

(v) Model testing (e.g. where predictions are made in advance of an experiment or characterisation measurements) should be used, where possible, to narrow the range of conceptual-model uncertainty. Calibration and testing of models should be performed, if possible, at a range of different scales.

It was pointed out at the GEOTRAP workshop that the identification of properties of the system that can be examined by blind prediction and the comparison of the predictions with actual results is an important aid to the development of science. It is, in particular, a key element of confidence building, maximising the credibility of some models and falsifying others. Extreme care in test design is of the utmost importance if the validity of any rejection of a high-consequence model is to be adequately demonstrated. Such testing may be carried out as part of the site-characterisation/performance-assessment cycle. *Predictive model testing, in the context of field tracer transport experiments, was discussed at length in the first GEOTRAP workshop.* The difference in the scales of space and time between those that are relevant to performance assessment and those that are measurable in field and laboratory experiments is a fundamental limitation of such experiments. It was suggested that the use of natural tracers could, at least in part, overcome this difficulty.

(vi) The consistency between the model components (structure, flow, chemistry, transport) used to represent a system in an assessment should be demonstrated.

Where separate models of structure, flow and transport used to represent a system are applied in a performance assessment, these models should be shown to be consistent with one another. Difficulties in making such a demonstration were pointed out in the workshop. One difficulty concerns terminology. For example, the terms "water age" or "residence time" may have very different significances and representations (e.g. single number vs. wide ranging distribution) depending on the concepts and models (flow, hydrogeochemistry, transport) that are used.

(vii) Transparency in the design, testing, selection and application of assessment models should be sought.

The need for complete openness in the design, testing, selection and application of assessment models was stressed at the workshop. The need for scientific publications to disseminate knowledge of alternative approaches and to provide peer review was deemed essential. Finally, it was noted throughout the workshop that confidence in assessment models, as well as other benefits, can be obtained through effective communication between those involved in site characterisation and those involved in performance assessment, as discussed in 3.2., below.

3.2. Communication between those Involved in Site Characterisation and those Involved in Performance Assessment

The progress of site characterisation should take account of performance-assessment results, the development of a general, realistic understanding of site conditions[7] and the requirements for achieving an optimal design of the repository facility. In this context, effective communication between those involved in site characterisation and those involved in performance assessment is an important aspect in building confidence in the selected approaches and models. *This was also a finding of the first GEOTRAP workshop.* Such communication will also be beneficial in the development of techniques that aim to integrate as wide a range of information as possible.

Those involved in site characterisation need to see a clear relationship between the data that they collect and the way that these data are used in the assessment models, in order to:

- maximise their confidence in the assessment models, e.g. that assessment models do, indeed, include all relevant features, events and processes defined in the site-characterisation process;
- (possibly) stimulate the development of alternative conceptual models.

Those involved in performance assessment should contribute to the strategy for data collection. In particular they can provide guidance on:

- where more or new site characterisation would help reduce uncertainty and discriminate between alternative models, for example through the use of sensitivity analyses;
- which are the necessary data for particular models;
- the likely effectiveness of specific data-collection strategies, for example through the use of scoping calculations.

The modelling approach used could thus impact the way in which site characterisation is carried out. Indeed, some approaches and models are better suited to some types of data and to spatial distributions of data (data points that are uniformly distributed in space vs. focused on specific features) than others. At the workshop, the consensus was that the progress of site characterisation should not be driven solely by the choice and current requirements of assessment models. The interaction between site characterisation and performance assessment is, in any case, constrained by the "time lag" between field measurements and their application in performance assessment.

> A responsive and flexible approach to both performance assessment and site characterisation, with effective inter-communication, has been found useful, e.g. for model development and data acquisition. This helps utilise resources in an efficient manner, e.g. by avoiding the development of models for which the necessary data can never be obtained and the gathering of "interesting" but otherwise irrelevant field data.

---

[7] It should be noted that, depending upon the programme structure, "site understanding" may constitute an integral part of performance assessment.

Overall, it was emphasised that the GEOTRAP project contributes to this effective communication between site characterisation and performance assessment.

3.3.   Effects of Spatial Variability on Geosphere Performance

In the GEOTRAP presentations and discussions that focused on specific aspects of the modelling of spatial variability, the major concern, with respect to long-term repository safety, was the variability in hydrogeologic properties. This variability may lead to discrete pathways through the geosphere with low groundwater travel times. For solute transport, the effect may be accentuated in the presence of fracture coatings, which can reduce the retardation normally attributed to matrix diffusion and sorption. It is concluded that the selected approach to transport modelling should allow for the incorporation, either explicitly or as appropriate effective parameters, of the effects of heterogeneity in detailed pore structure, as well as larger-scale structures and fracture-network effects. The selected approach should thus represent appropriately the observed heterogeneity, taking into account the major migration pathways as well as the potential "fast pathways".

Many GEOTRAP presentations illustrated the need to consider variability of properties over a wide range of spatial scales. This is necessary in order to:

- establish which processes are the most relevant to radionuclide migration in the different parts of the repository system (e.g., in the near field, diffusion from repository system to a discrete, water-conducting feature; in the far field, advection along fast pathways within water-conducting features);
- determine the most appropriate models for representing the key processes of radionuclide migration;
- establish the most appropriate way of deriving effective parameters, if they are used (where effective parameters arise from a "homogenisation of parameters", a simple averaging over inappropriate lengthscale may lead to significant bias in the results);
- evaluate the causes and consequences of uncertainty in the choice of models and in the determination of parameter values;
- determine the most appropriate parameters to measure, drawing on the wide range of methods available for site characterisation.

On this last point, although the majority of presentations in GEOTRAP addressed hydrogeologic properties, variability in other properties may also affect the processes whereby radionuclides are retarded. For example:

- variation in geochemical properties of the groundwater may affect radionuclide sorption;
- variation of mineralogy may also affect sorption and may, through the presence of fracture coatings and infill, affect the process of matrix diffusion.

The modelling of processes affected by spatial variability requires particular care where the processes are non-linear (as exemplified by attempts to incorporate non-linear sorption into models of transport through heterogeneous media); the derivation of effective parameters by a simple averaging of properties is likely to give particularly inaccurate representations in such a case. The mixing processes between pathways are also important if these pathways have very different transport properties. Finally, although it is convenient, in a statistical modelling sense, to consider properties as uncorrelated, the risk of bias introduced by such assumptions in model results is potentially significant, even if only a small correlation exists.

Unravelling the associations between the different (variable) properties influencing radionuclide migration through the geosphere over a range of scales and determining the potential contribution to bias and uncertainty arising from these associations are considered as important areas of development. It was noted that useful work is now being undertaken in these areas.

The characterisation (including the transport properties) of water-conducting features in all types of host rocks and their representation in models of radionuclide migration will be at the core of the 3rd GEOTRAP workshop. The workshop will also address issues linked with the characterisation and representation of fast pathways and the time-dependent properties of water-conducting features.

4. RECOMMENDATIONS AND RESTRICTIONS

The discussion above leads to the following recommendations regarding the modelling of the effects of spatial variability on radionuclide migration:

- It is essential to include a level of description of heterogeneity within the flow and transport models that enables proper account to be taken of the major migration pathways as well as of the potential presence of fast pathways that may dominate the performance of the geosphere as a transport barrier.

- Particular attention must be paid to the avoidance of unquantifiable bias in consequence calculations, which may arise, for example, from the choice of a unique geologic structural model. It is recommended, therefore, that a wide range alternative models, consistent with the available data, is identified and its impact considered. A judgement can then be made as to the way in which this conceptual-model uncertainty is best addressed in performance assessment. It should be noted that such a wide range of alternative models consistent with the available data does not always exist and that only a limited number of models may be selected for further quantitative evaluations.

- Models should be calibrated and tested using experiments and observations covering different spatial scales, in order to narrow the range of conceptual-model uncertainty (*this was also a finding of the first GEOTRAP workshop*).

- Simple models, that compensate for uncertainty through conservative assumptions, are useful in performance assessment, assisting, for example, in the communication of performance-assessment findings. They are, however, more effective in these roles when backed up by more complex models.

- The variability of chemical properties, their possible coupling with other properties and the occurrence of non-linear processes are areas that deserve closer attention in the future.

The current state of the art in modelling techniques, however, imposes some restrictions on the implementation of these recommendations.

(i)	Models of groundwater flow are currently available that allow, in principle, a realistic representation of the structural variability of a geologic medium. An effort still needs to be made, however, to improve our ability to incorporate a wide variety of data (at length scales relevant for the model under consideration) in order to take full advantage of these models.

Different types of groundwater flow model (see 3.1. (i) above) are currently available. Given sufficient and appropriate data, and adequate computational power, they can, in principle, achieve a high level of realism in the conceptualisation of geologic media. The rapidly improving performance of computers means that availability of data, rather than limitations in computational power, is the most important constraint on the realism that can be achieved.

A mix of deterministic and statistical models is proving to be a promising approach to represent the variability at various scales of geologic media in the presence of limited data. For example, a deterministic conceptualisation of the medium, based both on soft and hard geologic data, leads to the partition of the domain under consideration into more or less homogeneous zones, within which the variability of properties can be stochastically represented on the basis of additional (limited) hard hydrogeologic data. In sedimentary media, these homogeneous zones are layers or lenses corresponding to discrete sedimentary bodies; in hard rocks, the zones correspond to major (fractured) volumes of rock. The principal advantage of this approach is that the model can be conditioned on all available local information, including indirect information, such as hydraulic heads, thus increasing confidence in the model. An advantage of the stochastic element in the approach is that, through the generation of a large number of realisations of the medium, the range of uncertainty in parameters required for subsequent transport modelling can be evaluated. The statistical laws must be correctly inferred, which currently requires a large amount of hard data at appropriate scales (though, of course, less than a fully deterministic approach would require). These data can be difficult to collect, particularly bearing in mind that a real repository site should not be jeopardised by the site-characterisation techniques. There is an advantage in developing techniques that use indirect (e.g. geophysical) data, which are non destructively collected, in conditioning the models. An example of such a promising approach was discussed during the workshop. It should be noted that the approach discussed above represents one of the possibilities and that others exist (e.g. indicator geostatistical approach, stochastic continuum model for the entire medium).

(ii)	Simplified models that employ effective parameters are likely to continue to be used, particularly in the modelling of transport. The use and quantification of effective parameters needs careful justification, since these can rarely be measured in the field at the scale they are needed for performance-assessment calculations.

The use of effective parameters is a way of incorporating complexity for modelled lengthscales over which it is not possible to make site-characterisation measurements. The problems in determining effective parameters at an appropriate scale arise from the fact that performance assessment needs to evaluate long-range transport through highly impervious media, in which hydraulic tests and tracer tests can only investigate very short distances. It is, therefore, necessary to infer effective parameter values from the local data that can be obtained in the field, using up-scaling techniques. This approach requires that a model that can be conditioned on these local data is available and that the relevant processes at the local scale have been properly identified. The present state of the art in methods for up-scaling has proved most successful in deriving permeabilities and transmissivities for the modelling of groundwater flow. The methods are less well founded in the case of transport processes, and are still in their infancy in the case of chemical processes. In all cases, the range of uncertainty in the appropriate values for effective parameters may be high. The ideal way to narrow this range would be to calibrate and test the up-scaled model, although this is rarely possible (see (iii), below). In certain cases, however, large-scale field tests (e.g. long-term pumping tests) that affect larger volumes of rock, provide the opportunity to test the up-scaled, effective values, and thus reduce the uncertainty. On the other hand, the relevance that this has to performance assessment depends on the sensitivity of performance-assessment calculations to the uncertainties in effective-parameter values, which, in some cases, is small (e.g., in some cases, geosphere transport of decaying species is found to be insensitive to the value assigned to dispersivity).

| (iii) | Calibration and testing over spatial and temporal scales that are relevant to performance assessment would, in principle, be the best method to build the confidence in models of flow and transport. Such calibration and testing is difficult to achieve, considering notably the very long travel times that are expected through the rock formations that are potentially acceptable to host a repository and the possible transient nature of the groundwater flow system. Analogues, both natural and anthropogenic, provide a possible means to address this difficulty. |

It is rare for any experiment, carried out using artificial tracers over spatial scales that are relevant to performance assessment, to yield useful information for model calibration and testing. *As discussed in the first GEOTRAP workshop, this is due both to experimental difficulties (duration required) and difficulties in interpreting the results (non-uniqueness where the experimental system is complex and incompletely characterised).* Indeed, such experiments become impractical, in terms of the experimental duration required, especially where the tracers interact chemically with the rock. A possible means of overcoming these difficulties is through the use of long-term natural and anthropogenic phenomena, particularly where these involve several independent migration processes, which reduces the range of possible interpretations of the observations. Successful examples of such an approach were given in the course of the GEOTRAP presentations. It was shown, in the hydrogeologic system of the Paris Basin, how the same model was able to account simultaneously for the observed hydraulic heads, temperatures and the concentrations of salt and rare gases. Another case study, where the groundwater flow model was calibrated against site data for temperatures, heads and salinities, was presented.

| (iv) | Integrated approaches coupling flow, transport and chemical reactions appear very promising. The computational tools are still, however, under development and only a few examples of realistic applications are currently available. |

A difficulty in facing those developing coupled models for flow, transport and chemical reactions is that the variability of water chemistry may be influenced by several factors, including the local heterogeneity of the host rock (mineralogical composition, local porosity and permeability). Effort is required both in the development of suitable computational techniques and in the development of conceptual methods for averaging, or up-scaling, the interaction between rock, fluid and trace elements in heterogeneous media. The capacity to model the variability of groundwater composition is important if one wishes to acquire an understanding of *in situ* observations of natural tracers and to predict the consequences of chemical perturbations caused by the presence of a repository.

5.   CONCLUDING OBSERVATIONS

The workshop reflected the present status of development in national waste management programmes, and in particular the fact that, on the one hand, detailed treatments of spatial variability have now been successfully performed as part of several performance assessment exercises, and, on the other hand, spatial variability has been more fully treated in connection with groundwater flow than in connection with the transport problem.

Vis-à-vis the state of the art at the end of the International INTRAVAL Project[8], progress was demonstrated in the following areas: (i) the development and implementation of alternative models and approaches to represent spatial variability, (ii) the consideration of a broad range of site-characterisation information, and (iii) the enhancement of confidence that can be gained by effective communication between site characterisation and performance assessment specialists. It should be noted that the need for further development in these areas had been identified in the conclusions and recommendations of previous NEA initiatives, namely the workshop on conceptual models[9], and the Integrated Performance Assessment Group[10].

Overall the workshop illustrated advancement in confidence in the concepts and models used for repository performance assessment, and confirmed that a multi-disciplinary approach is necessary in order to address more fully the problems of modelling flow and transport in media characterised by a high degree of spatial variability and in which coupled and, possibly, non-linear processes operate. Effective communication between different groups, including geologists, hydrogeologists, chemists and performance assessment specialists will therefore continue to be essential. This was also a finding of the first GEOTRAP workshop.

---

[8]   "The International INTRAVAL Project (Developing Groundwater Flow and Transport Models for Radioactive Waste Disposal) – Final Results", NEA/SKI, 1996.

[9]   "The Role of Conceptual Models in Demonstrating Repository Post-Closure Safety", Proceedings of an NEA Workshop, Paris, 16-18 November 1993, OECD/NEA, 1995.

[10]   "Lessons Learnt from Ten Performance Assessment Studies", OECD/NEA, 1997.

ANNEX 1

SYNOPTIC TABLE OF THE CASES DISCUSSED

The oral and poster presentations that were given and discussed at the workshop cover a wide spectrum of aspects (e.g. general overviews, case studies, technical details). They encompass a broad range with respect to each of conceptual and modelling approaches; geologic settings; and scales of interest. Furthermore, the rationale and objectives of the presented studies differ from one programme to another, notably according to their stages of development and documentation. In order to give an overview of the presentations, and to set the context in which the achievements, general conclusions, limitations and recommendations of the workshop should be viewed, Table 1 summarises the main characteristics of these studies, including their current status and principal aims. This table is not intended to be comprehensive and is to be considered as a guide to see where the effort has been focused and what sort of progress has been communicated.

Table 1 comprises a breakdown of the workshop papers in terms of:

- geologic settings and location;
- scale(s) of interest;
- aims of the study (what to model and for which purposes);
- conceptual approach to flow and transport;
- approach to PA (up-scaling, abstraction, simplification);
- computer codes; and
- status of the study (at the moment of the paper).

In order to provide the reader with an entry to the proceedings, Table 1 has been structured according to the chronological order of the sessions, and oral and poster presentations.

Title	Geologic Settings	Scale of Interest	Aims of the Study	Conceptual Approach	Relevance to PA	Computer Codes	Status of the Study
Session I *Variability in Natural Systems and Means to Characterise it* Ledoux, de Marsily	Fractured and porous medium. Crystalline and sedimentary rocks.	Macroscopic: rock sample. Megascopic: rock formation.	General methodology. Modelling methods.	CPM and DF Medium. Deterministic and Stochastic.	Up-scaling of parameters and model validation.	not applicable	State of the art.
Session I *Why Is Variability Important for PA and What Are Its Consequences for Site Characterisation and Repository Design?* Dverstorp et al.	Fractured crystalline medium. Äspö HRL (Sweden) and crystalline basement of Northern Switzerland.	Near- and far-field around a potential repository site.	General understanding of a site for PA investigations.	CPM and DF Medium. Deterministic and Stochastic.	Role of the variability in the knowledge of the medium. Consequences to PA.	not applicable	Used in the completed PAs: Site-94 (SKI) and Kristallin-I (Nagra).
Session I *An Example of a Comprehensive Treatment of Spatial Variability in a PA:* Nirex 97 Jackson	Quarternary deposits, channel sandstones and fractured volcanic rocks. Sellafield (UK).	Features at different length scales (cm to km) in a regional flow and transport model.	Integration of variability at different scales in various units for PA calculations.	Explicit modelling of variability in each unit. Stochastic. Simple analytical and numerical models used to up-scale.	Up-scaling to derive effective parameters for regional scale models. Systematic treatment of uncertainties.	Up-scaled parameters used in 2D and 3D regional CPM for flow and transport. Local-scale DFN.	Part of Nirex 97 PA study (Nirex).

Table 1:   Overview of the studies presented at the workshop

               (DF: Discrete Fracture, DFN: Discrete Fracture Network, CPM: Continuum Porous Medium)

Title	Geologic Settings	Scale of Interest	Aims of the Study	Conceptual Approach	Relevance to PA	Computer Codes	Status of the Study
Session II *The Geologic Basis and the Representation of Spatial Variability in Fractured Media* Mazurek et al.	Fractured crystalline and sedimentary media. Fractures in Äspö Granite (Sweden) and in limestone beds within Palfris Marls (Switzerland).	Macroscopic: single fracture, drill-core scale. Megascopic: fracture network, outcrop scale.	Description of single fracture heterogeneity due to structural conditions. Methodology for up-scaling to a network for flow and transport modelling.	Use of structural and mechanistic principles of fracture development for conceptualisation and up-scaling. Deterministic conceptualisation of the geologic medium with stochastic description of parameter values.	Deterministic modelling of transport through fractures; flowpath divided into legs with constant properties. Sensitivity analysis of geologic conceptual models and parameter values.	PICNIC: geosphere transport code coupling advection, sorption, matrix diffusion and decay, with explicit representation of longitudinal and transversal heterogeneity.	State of the art of the Swiss approach for modelling fracture network.
Session II *The Geological Basis and the Representation of Spatial Variability in Sedimentary Heterogeneous Media* Cliffe et al.	Channel sandstones. Sellafield (UK).	Representation of sub-facies (metre scale) in a regional model (hectometre scale)	Investigate the impact of various conceptual models of the variability of sandstone sub-facies on calculations of flow and transport.	Explicit representation of the channels and of their internal sub-facies. Stochastic description of parameter values.	Use of detailed calculations to support up-scaling for PA.	STORM 3D Stochastic Reservoir Modelling. SPVMG: multigrid flow and particle-tracking code.	Used to support approach to up-scaling and the values of up-scaled parameters in Nirex 97 PA (Nirex).

Table 1: **Continued from previous page**
(DF: Discrete Fracture, DFN: Discrete Fracture Network, CPM: Continuum Porous Medium)

31

Title	Geologic Settings	Scale of Interest	Aims of the Study	Conceptual Approach	Relevance to PA	Computer Codes	Status of the Study
Session II *Representation of Spatial Variability for Modelling of Flow and Transport Processes in the Culebra Dolomite at the WIPP Site* Meigs, Beauheim	Culebra Dolomite: fractured and porous aquifer above bedded rock salt. WIPP (US).	Megascopic: aquifer scale	Evaluation of the effects of variability in aquifer transmissivity on the distribution of flow pathways. Evaluation of the effects of small-scale heterogeneity on diffusion rates.	2D CPM with stochastic distribution of flow parameters. Dual porosity, multi-rate diffusions for transport.	Explicit representation of variability for flow models (T-fields). Monte-Carlo simulations for transport.	Inverse methods for generating T-fields. Particle tracking model for pathways identification. Double porosity model for tracer tests.	Confirmed conceptual and numerical approach. Used in the completed WIPP PA (USDOE)
Session II *The Discrete-Modelling Approach Adopted in SITE-94* Geier, Dverstorp	Fractured granite of Äspö HRL (Sweden).	Megascopic : site scale including the vicinity of the repository and the regional fracture zone.	Prediction of hydrogeologic parameters for RN migration on the basis of all Äspö data. Development of a methodology for integration of various types of field data, at various scales.	3D DF multi-scale flow model (deterministic DF semi-regional model and stochastic DFN local model). 1D transport model.	Methodology taking in account local- and regional-scale hydrogeologic and structural data. Evaluation of flow and transport effective parameters from Monte-Carlo simulations.	DF and DFN codes for flow and transport	Confirmed conceptual and numerical approach used in the completed Site-94 PA exercise (SKI).
Session II *Approaches Dealing with Fractured Media within the SKB Äspö HRL Project* Gustafson et al.	Fractured granite of Äspö HRL (Sweden).	Site scale (100-1000m).	Testing of flow models at different scales (e.g. pre-excavation predictions vs. observations).	Site-scale, stochastic CPM in between a network of deterministic large-scale DF.	Incorporation of all available data in predictive flow modelling (before excavation) and testing with post-excavation measurements.	PHOENICS: finite-volume, stochastic CPM, 3D, flow code (with 2D large-scale DF).	State of the art. R&D at Äspö HRL (SKB).

Table 1: **Continued from previous page**
(DF: Discrete Fracture, DFN: Discrete Fracture Network, CPM: Continuum Porous Medium)

Title	Geologic Settings	Scale of Interest	Aims of the Study	Conceptual Approach	Relevance to PA	Computer Codes	Status of the Study
Session II *Spatial Variability in the Geosphere Models used for AECL's Long Term PA of the Disposal of Canada's Nuclear Fuel Wastes* Davison et al.	Fractured granite of the URL site in Manitoba (Canada).	Megascopic: regional scale around an hypothetical repository.	To perform a site characterisation and PA exercise on the basis of site-specific data (URL).	Deterministic Equivalent CPM for flow and solute transport using 2D finite elements for fractures and 3D elements for matrix.	Vault located near a major fracture zone. Particle tracking to construct a 3D network of 1D paths on the basis of the 3D flow results.	MOTIF 2D and 3D finite element code for flow and transport. GEONET 1D code for transport to the biosphere.	Confirmed conceptual and numerical approach used in the completed EIS (AECL).
Session II *An Integrated UZ Flow and Transport Model for Yucca Mountain, Nevada, USA* Bussod et al.	Unsaturated and fractured tuffs, Yucca Mountain (Nevada, US).	Site scale: near-field and far-field unsaturated and saturated zones of a geologic repository system (10-75000 m).	Prediction of migration of key RN to the accessible environment (considering heat effects coupled to hydrologic, geochemical and geologic databases and parameters).	1D, 2D and 3D stochastic and deterministic dual-permeability, dual-porosity DF flow and transport simulations in unsaturated and saturated rocks.	Integration of regional-scale, site-scale, field and lab measurements into 2D and 3D process flow and transport models for abstraction to dose calculations using particle tracking.	FEHMN: multiphase, fully coupled finite element/finite volume flow and reactive transport (+geostatistical package). GEOMESH: griding tool kit linked to industry geologic framework models (STRATAMODEL, LYNX, Earth Vision).	Codes being validated/calibrated through saturated and unsaturated field transport tests. (USDOE).

Table 1: Continued from previous page
(DF: Discrete Fracture, DFN: Discrete Fracture Network, CPM: Continuum Porous Medium)

Title	Geologic Settings	Scale of Interest	Aims of the Study	Conceptual Approach	Relevance to PA	Computer Codes	Status of the Study
Session III *Impact of Uncertainties in Chemical and other Entities on Radionuclide Migration from a Repository for Spent Nuclear Fuel* Neretnieks	Single fracture in hard rocks.	Macroscopic : flow, transport and sorption processes within a single fracture close to a repository.	Identification of important parameters for sorption linked to matrix diffusion. Role of the flow-wetted surface. Compilation of sorption and matrix diffusion data.	Equivalent CPM within a fracture and within its walls following a streamline.	Examples of calculation of uncertainties in sorption, diffusion and flow-wetted surface close to the RN source.	Not applicable.	Present state of an ongoing research work.
Session III *What Can Be Learnt from other Research Fields and Applications* de Marsily	Several cases: Rhine alluvial aquifer, Paris Basin (France), …	Regional (aquifer) to local (fracture) scales.	Current development in methods (geophysical, hydraulic testing, natural tracer) to detect anomalies in the continuity of confining strata.	Not applicable.	Not applicable.	Not applicable.	Present state of ongoing research works.

Table 1: Continued from previous page

(DF: Discrete Fracture, DFN: Discrete Fracture Network, CPM: Continuum Porous Medium)

34

Title	Geologic Settings	Scale of Interest	Aims of the Study	Conceptual Approach	Relevance to PA	Computer Codes	Status of the Study
Session IV *Modelling Strategies and Usage of Site Specific Data for Performance Assessment* Hautojärvi et al.	Fractured crystalline basement (Finland).	Macroscopical (near-field) to megascopical (far-field).	Strategy for flow and transport modelling at different scales for site-specific PA.	Nested models -flow at the scale of deposition hole -groundwater flow: equivalent CPM, DFN, stochastic continuum. Deterministic, stochastic.	Adoption of different conceptual models for the various scales studied.	not applicable	Ongoing reflection for a PA strategy in Finland.
Session IV *Training for Incorporating Spatial Variability in Performance Assessment Approach* Dewiére et.al.	Fractured media.	Megascopical (far-field)	Development of a modelling method.	Stochastic approach.	Help in focusing the whole site characterisation and performance assessment work by uncertainty assessment.	Not applicable.	Tested on a sub-surface low-level waste disposal site.
Session IV *SKB's Approach to Treatment of Spatial Variability in PA Modelling* Selroos, Ström	Fractured crystalline basement (Sweden).	Macroscopical (near-field) to megascopical (far-field).	Strategy for flow and transport modelling for site-specific PA.	Stochastic continuum model; DFN model; and channel network model.	Use of various conceptual models of the spatially variable geosphere to address conceptual model uncertainty.	HYDRASTAR (stochastic CPM); FracMan/ MAFIC (DFN); CHAN3D (channel network).	Ongoing PA exercise SR 97 (SKB).

Table 1: **Continued from previous page**
(DF: Discrete Fracture, DFN: Discrete Fracture Network, CPM: Continuum Porous Medium)

Title	Geologic Settings	Scale of Interest	Aims of the Study	Conceptual Approach	Relevance to PA	Computer Codes	Status of the Study
Session IV *Methods to Incorporate Different Data Types in the Characterisation Process* Gómez-Hernández et al.	Various natural and synthetic examples of aquifers.	Macroscopical (local scale measurements) to megascopical (far-field).	Development of methods to extract as much information as possible from the available site-characterisation data.	Random function model.	Confidence building in the models and reduction of uncertainty by incorporating different types of data.	not applicable	Present state of ongoing developments.
Poster *Chemical Modelling Studies on the Impact of Small Scale Mineralogical Changes on Radionuclide Migration* Emrén	Fractured granite of Äspö HRL (Sweden).	Macroscopic : variability of the chemical composition of groundwater in a single fracture due to small-scale mineralogical changes.	Investigation of several redox control models in groundwater reacting with minerals in fracture walls in equilibrium or non-equilibrium situation.	Coupled chemical and transport simulation of the propagation of water in a fracture using diffusion cells. Water composition computation in cells.	Evaluation of the changes in hydrochemistry on engineered and natural barrier performances.	CRACKER for coupled transport and chemistry. PHREEQUE for the computation of speciation in water.	Present state of an ongoing research.

Table 1: Continued from previous page

(DF: Discrete Fracture, DFN: Discrete Fracture Network, CPM: Continuum Porous Medium)

36

Title	Geologic Settings	Scale of Interest	Aims of the Study	Conceptual Approach	Relevance to PA	Computer Codes	Status of the Study
Poster *Upscaling Conductivities and Dispersivities* Gómez-Hernández, Wen	Theoretical heterogeneous 2D continuous medium.	Macroscopic (local scale measurements) to megascopic (reservoir scale simulation).	Improving up-scaling techniques to evaluate transmissivity and dispersivity on a coarse grid.	Simulation of a 2D continuous medium using stochastic generation methods for heterogeneous parameters.	Evaluation of effective parameters	Not applicable	Present state of an ongoing research.
Poster *Channel Network Model for Flow and Radionuclide Migration* Gylling et al.	Fractured granitic medium (e.g. Äspö HRL, Sweden).	Macroscopic (near field) to megascopic (far field) around a repository site.	Coupling near-field and far-field simulation tools for PA using local field data.	Near-field of a canister with a compartment model. Coupling in the far-field to a 3D channel network model with matrix diffusion and sorption.	Evaluation of the released activity from a repository.	NUCTRAN for canister release computation. CHAN3D for flow and transport.	State of the art of a PA method.
Poster *Modelling Transport in Media with Correlated Spatially Variable Physical and Geochemical Properties* Impey et al.	Synthetic medium made up of a mixture of sorbing and non-sorbing beads.	Microscopic (beads properties) to macroscopic (transmissivity and sorption properties of blocks of beads).	Design of an experimental program aiming to evaluate the correlation between spatial variability of flow properties and sorption properties.	Fractal Model Description of variability of flow properties. Flow and transport experiments within a synthetic medium.	Identification of preferential pathways taking into account flow and sorption variability.	Coupled experimental and computing program MACRO II.	Preliminary design of a research program.

Table 1: **Continued from previous page**
(DF: Discrete Fracture, DFN: Discrete Fracture Network, CPM: Continuum Porous Medium)

Title	Geologic Settings	Scale of Interest	Aims of the Study	Conceptual Approach	Relevance to PA	Computer Codes	Status of the Study
Poster *Development of a New Transport Model (PICNIC) which Accounts for Large- and Small-Scale Heterogeneity* Jakob	Fractured medium such as crystalline rock bodies.	Macroscopic (scale of a local channel) to megascopic (scale of the geologic barrier around a repository).	To describe the flow and transport through the geologic barrier using local fracture properties.	Deterministic simulation of advection, dispersion, matrix-diffusion and linear sorption in a 3D network.	Consideration of large- and small-scale heterogeneities in transport modelling.	PICNIC to be coupled later with the flow code NAPSAC	Code under development, to be used in Nagra's future PAs. Tested on the Kristallin-I cases.
Poster *Scale-Dependent Heterogeneities in the Gorleben Area and Implications for Model Studies* Ludwig et al.	Sedimentary aquifer and aquitard structure around the Gorleben salt-dome (Germany).	Microscopic to megascopic (site characterisation); macroscopic to megascopic (model studies).	Characterisation of flow and transport around Gorleben site. Study of the influence of geologic heterogeneity on the geometry of pathways and on transfer-time.	Deterministic 2D and 3D simulations of flow and transport with variable density. Stochastic treatment of parameters and model variabilities (geostatistics and Monte-Carlo simulations).	Evaluation of the pathways and transfer times to the human environment.	Not indicated (finite element flow and transport codes, geostatistical and uncertainty/sensitivity codes).	Present state of several site characterisation and safety analysis studies.

Table 1: Continued from previous page

(DF: Discrete Fracture, DFN: Discrete Fracture Network, CPM: Continuum Porous Medium)

38

Title	Geologic Settings	Scale of Interest	Aims of the Study	Conceptual Approach	Relevance to PA	Computer Codes	Status of the Study
Poster *A Case Study on the Influence of Sorption Inhomogeneities on the Migration of Contaminants* Lührmann et al.	Theoretical continuous medium including megascopic scale heterogeneities.	Macroscopic to megascopic.	Benchmarking of a transport computer code accounting macroscopical sorption processes, applied to heterogeneous medium.	Deterministic simulation of advection, macro-dispersion and sorption.	Not applicable	TRAPIC finite volumes	Present state of an ongoing research.
Poster *Single Well Harmonic Pulse Testing - A Novel Technique for In-Situ Measurement of the Hydraulic Heterogeneity of a Fracture* Selvadurai, Flavelle	Fractured medium.	Macroscopical flow properties of a single fracture.	Determination of transmissivity and storativity of a single fracture assuming a sinusoidal flow rate.	Deterministic modelling using equivalent CPM.	Field method for local scale site characterisation of fractured medium.	Analytical solution for solving the direct and inverse problem.	Field equipment and measurement procedure under development
Poster *Effect of Sorption Kinetics on the Migration of Radionuclides in Fractured Rock* Wörman, Xu	Fractured crystalline medium (based on Äspö HRL data, Sweden).	Macroscopical: single fracture.	Study of the effects of sorption kinetics on RN migration.	2D transport with matrix diffusion, instantaneous matrix sorption and sorption kinetics.	Theoretical model tested with batch sorption tests.	Migration, flow.	Present state of an ongoing research (SKI).

Table 1: Continued from previous page

(DF: Discrete Fracture, DFN: Discrete Fracture Network, CPM: Continuum Porous Medium)

39

ANNEX 2

LIST OF KEY QUESTIONS TO BE ADDRESSED
(per session and working group)

SESSION I : **SPATIAL VARIABILITY: ITS DEFINITION AND SIGNIFICANCE TO PERFORMANCE ASSESSMENT AND SITE CHARACTERISATION**

1. What are the types of variability relevant for transport at different scales?
2. What is the geologic basis (quantitative and qualitative) for the representation and conceptualisation of spatial variability?
3. What are the consequences of spatial variability on physical and chemical processes and the relevant properties for transport of solutes in geologic media?
4. What are practical limitations in the estimation of spatial variability?
5. Why is up-scaling needed for performance assessment purposes?
6. What are the modelling tools currently used? What are their strengths and limitations?
7. Limitations in predictions of safety arising from the intrinsic variability of the system and from the resulting uncertainty.
8. Is further site characterisation beneficial for better treatment of variability (in terms of performance assessment)?
9. What is the geologic setting? Which information is available?
10. How is variability in each unit modelled?
11. How is up-scaling performed? How is uncertainty treated?
12. What are the consequences of variability on the PA results?
13. What are desirable follow-ups?

SESSION II: **EXPERIENCE WITH THE MODELLING OF RADIONUCLIDE MIGRATION IN THE PRESENCE OF SPATIAL VARIABILITY IN VARIOUS GEOLOGIC ENVIRONMENTS**

1. What were the objectives of the modelling?
2. What was the contribution of these models to PA, and in particular to the evaluation of the effect of spatial variability on radionuclide migration?
3. How were the models of variability selected?
4. What field data were available? What were the types of information on which the models of spatial variability are based?
5. How were parameters derived?
6. How was up-scaling performed? How were the various levels of up-scaling (from mineralogy and porosity to basin) integrated?
7. How feasible was the testing of models of spatial variability?
8. How were early choices relating to models, system understanding, site characterisation, ... refined?

9. Where was the greatest success? The most significant failure? How do you assess the results of the modelling exercise in the light of its objectives? Can they be improved?

SESSION III: NEW AREAS FOR INVESTIGATION: TWO PERSONAL VIEWS

(no key questions were defined for this session)

SESSION IV: WHAT IS WANTED AND WHAT IS FEASIBLE: VIEWS AND FUTURE PLANS IN SELECTED WASTE MANAGEMENT ORGANISATIONS

1. What is your future strategy for dealing with spatial variability at different scales? Are there suitable robust and simple ways of dealing with variability?
2. On what grounds is this strategy based? How are lessons from previous exercises integrated in the new strategy?
3. What do you expect to gain (in terms of radionuclide transport) from a more sophisticated treatment of variability?
4. To what extent is spatial variability perceived as an issue that can be solved (for PA purposes)?
5. What are future priorities for resource allocation?
6. What can be expected of further developments (what is possible - what is not)? What are typical mistakes that need to be avoided?

SESSION V : IN-DEPTH DISCUSSIONS BY WORKING GROUPS

Working Group A : **APPROACHES TO CONFIDENCE BUILDING IN MODELS OF SPATIAL VARIABILITY**

- How certain are you about your uncertainty?
- Spatial correlation
- Model bias
- How are conceptualisation(s) and models chosen?
- Importance of the extrapolation up to the PA scale and of the uncertainty introduced during that process
- How is "soft" (qualitative) geologic information to be used to support the whole process of conceptualisation/modelling/up-scaling. Multiple lines of reasoning to support this process
- Confidence building and decision making

Working Group B : **DEFINITION AND QUANTIFICATION OF EFFECTIVE PARAMETERS FOR FLOW AND TRANSPORT**

- What are the consequences of geometrical simplifications?
- Can we produce all the sensible effective parameters we need (e.g. dispersion length)? Do they have a physical significance?
- What is the bias introduced by using the effective parameter approach?
- Current methods for separating the effects of variability and uncertainty.

Working Group C : **BENEFITS OF FURTHER SITE CHARACTERISATION FOR BETTER RADIONUCLIDE TRANSPORT CALCULATIONS**

- What data are/are not measurable in the field?
- What level of detail is useful/beneficial to PA and radionuclide transport calculations? Is this achievable?
- Are we satisfied with the effective parameter approach? If not, what are the consequences on site characterisation and modelling?
- How can "soft" (qualitative) geologic information be used to support and improve the whole process of conceptualisation, modelling, up-scaling?
- Better justification of the models versus further site characterisation

Working Group D : **FUTURE DIRECTIONS OF WORK IN THE TREATMENT OF SPATIAL VARIABILITY FOR RADIONUCLIDE TRANSPORT**

- What are the processes which could be enhanced by spatial variability, and have an impact on radionuclide transport?
- Treatment of geochemical variability
- Interactions between density driven flow and spatial heterogeneity
- Consistency between geologic and flow models, and transport models
- Relation between current scientific knowledge and PAs

POSTER SESSION

(the posters dealt with additional test cases and provided technical details on several characterisation and modelling approaches and methods)

ANNEX 3

CONCLUSIONS OF THE WORKING GROUP A

Approaches to Confidence Building in Models of Spatial Variability

Chairman: R. Mackay (United Kingdom)
Members: P. Smith (United Kingdom), G. de Marsily (France), G. Gustafson (Sweden), E Frank (Switzerland), P. Jackson (United Kingdom), K-J. Röhlig (Germany), K. Clark (United Kingdom), J. Astudillo (Spain), C. Pescatore (NEA)

1. Introduction

The topics posed by the programme committee for discussion were as follows:

a) How certain are you about your uncertainty?
b) Spatial correlation
c) Model bias
d) How are model conceptualisation(s) and models chosen?
e) Importance of the extrapolation up to the performance assessment scale and of the uncertainty introduced during that process.
f) How is "soft" qualitative geologic information to be used to support the whole process of conceptualisation/modelling/up-scaling. Multiple lines of reasoning to support this process.
g) Confidence building and decision making.

To focus the discussions on the broad range of issues posed by these topics, a brief review of the performance assessment process and the actions of the process in which spatial variability issues are relevant were discussed and agreed. The discussions arising from this review then followed a natural pattern of development without the need to address specifically each of the topics above. The consequence of this approach to organising the meeting was that the group addressed topics a), c), d), f) and g) in some detail; summarised the major issues under topic b) and omitted from the discussion topic e). It was assumed that topic e) would receive greater attention by Working Group B.

The group's starting point was to consider the meaning of confidence building and this discussion is reported in the next section. The group then considered a range of issues related to conceptual model building and the testing of conceptual models (reported in Section 3). Finally, broader based discussions took place to address future directions for improving the confidence in models (Section 4).

2. What Does "Confidence Building" Involve?

Confidence building at its simplest is a demonstration, to an individual or group, that all reasonable care has been exercised in the development and application of the performance assessment. At its most complex, it is the validation of the approaches used in a performance assessment and of the accuracy of the output of the assessment. The latter was accepted as an intractable problem for radioactive waste disposal given the lack of any equivalent problem for which results are already known and on which the performance assessment approach has been successfully demonstrated. Consequently, it was necessary to limit the discussions to the lesser aim of demonstrating the reasonableness of the approaches used.

The early discussion addressed "Who is the target audience for confidence building?". The general consensus was that *all* groups expressing a stake or an interest in the disposal of radioactive waste and the potential long-term consequences of disposal are included. The audience was thus defined to include the performance assessment team, the scientists and engineers contributing to the supporting research, the management team controlling the disposal design and assessment, the regulator, relevant government departments, recognised interest groups as well as members of the public.

The later discussion considered "How can confidence building be achieved?". As presented above, it was agreed that the nature of radioactive waste disposal only permits confidence being raised by the development and application of the components contributing to a performance assessment and not by the whole assessment. The intended (rather than guaranteed) result should be to produce confidence in the whole procedure. The analysis and modelling of the spatial variability of geologic properties and its consequences can be considered to be a component of the assessment procedure.

The group adopted as a useful example, the public confidence that a car manufacturer had achieved in its products. The group used this example to compare and contrast the problems that are faced in preparing a performance assessment for radioactive waste disposal. Two points arising from the discussions are worth noting in this summary:

– a product is considered good if the manufacturer has a demonstrated track record of high quality production;

  The suggestion was made that it might be possible to exploit the prior experience of those concerned with spatially heterogeneous geologic problems, notably the oil industry, to demonstrate quality and expertise. The problem is that the exploitation of stochastic modelling in the oil industry is relatively recent and previous difficulties in using deterministic modelling approaches to provide accurate forecasts of reservoir behaviour over time-scales of more than one year do not provide convincing evidence of specific or unsullied expertise.

– a car manufacturer uses the components of a car when advertising the car's quality but they seldom sell the car on the component quality, rather they sell it on the component's innovation.

  Quality is apparently implicit in the manufacturer's reputation. Exploitation of component quality for confidence building in performance assessment is not therefore a tested procedure. Although, innovation is necessarily important in the performance assessment process it appears that innovation alone, without full testing is not important on its own for confidence building in the car industry.

There appears to be few parallels that can be usefully developed to construct a confidence building strategy for performance assessment.

3. The Conceptual Basis

The group spent a great deal of time discussing the relationship between the conceptual framework for modelling a hydrogeologic system, the modelling of the spatial variability of the system and the conditioning of the model using the available hard and soft data.

Model choice:

In general it is not practical to define only one model of the spatial statistical structure of any property of the system using the available data, and it is necessary, therefore, to consider all models that satisfy the constraints defined by the data.

If a model is to be removed from the set of alternatives then it is necessary to show by a suitable test that such a model is incorrect. Only those models that present the greatest risk in the performance assessment can be explored as candidates for testing on the grounds of economics. However, such an approach implies a desire to reduce perceived risk by rejecting high-risk cases where these arise. The need for complete openness in the design of the testing programme is therefore essential. This requires full traceability of decisions, procedures and data use through the documentation of all actions that influence the calculation of risk and the rejection of high risk hypotheses.

Extreme care in test design is of the utmost importance if the validity of any rejection of a high-risk model is to be adequately demonstrated. Such testing must be carried out as part of a site investigation/performance assessment cycle.

In performance assessment it is easy to confuse confidence in the site defined by the calculated risk and confidence in the calculated risk required for acceptance of the performance assessment. By mixing the requirements of testing models dependent on their contribution to risk the group felt that a possibility for confusion between these two was possible and that this would be to the detriment of confidence building. However, no suggestions were made to overcome this possibility.

Historical Precedence:

Improvements in knowledge are commonly hampered by the tendency for Paradigms (prior models) to become entrenched. The reliance on particular numerical models and the preference for particular approaches to the characterisation of spatial variability using Gaussian models were seen as potential problems for proving the acceptability of the assessment procedure. The need for scientific publications to disseminate knowledge of alternative approaches and to provide peer review of approaches was deemed essential.

If significant changes of understanding arise then these should be put forward not as an expression of a failure of the assessment programme but as a confirmation of new understanding and new insights obtained by the programme.

Conditioning:

One important concern of the group was the possibility for unwanted biases associated with the conditioning of a spatial variability model in which the underlying concepts were in error. As an example, the possibility of defining a two dimensional model in which vertical fluxes were assumed negligible would cause transmissivities to be modified incorrectly by any conditioning of the transmissivity fields to the observed head distribution.

The purpose of conditioning is to reduce uncertainty by calibration using data obtained from field observations. Reducing uncertainty by conditioning is desirable as long as bias is not increased. The use of conditioning must, therefore, be linked to procedures to explore the bias in the conditional model results.

One approach is to explore the predictive accuracy of any new model by identifying properties of the system that can be examined by blind prediction and comparing the predictions with actual results. As uncertainty decreases the predictive accuracy improves or remains the same for a good model. [A cautionary note was raised about the use of blind predictions in the assessment process since the science is in some instances very young and failures to predict could be construed as incompetence. Whilst, this point was accepted it was felt that the discipline of testing our understanding is an important aid to the development of science. In such cases any downside due to bad press should be compensated by the greater awareness of the limits of our present knowledge.]

Correlations:

Although it is desirable in a statistical modelling sense to consider properties as uncorrelated, the risk of bias introduced by such assumptions in model results is potentially significant even if only a small correlation does exist. Consequently, it was felt that much greater attention should be paid to unravelling the associations between the different parameters influencing radionuclide migration and in determining the contribution to bias and uncertainty arising from these associations. [It is worthy of note that useful work is now being undertaken in this area.]

4. Future Directions

For confidence building new approaches need to be found. Most concerns in performance assessment can be reduced to "How do we quantify what we do not know?". Thus techniques are needed in spatial variability assessment which either demonstrate that what we do not know is not important or that what we know is sufficient. The approach of bounding what we do know about the spatial variability of the system and imposing an additional uncertainty that provides a safety factor for what we do not know may be one solution to the problem. However, is it compatible with the desire to use all available data and to avoid contradictions in data interpretation?

The prior definition of the regulation problem, in a way that does not exaggerate the importance of minor features of the system, may prove beneficial. The creation of a regulatory target for risk and the method of its identification *a priori* could help greatly in improving confidence. At present it remains unclear, for example, as to the level of aggregation of radiation doses both spatially and temporally that is permissible for the calculation of risk to a particular critical group. Consequently, no averaging of spatial properties is being explicitly included in the analysis and this leads to an over-emphasis on spatial heterogeneity as a problem for carrying out the assessment calculations.

As noted in this summary, the main emphasis technically must be on methods to demonstrate the validity of the conceptual models used to describe the system and on test procedures to validate the acceptability of different models applied to the data.

ANNEX 4

CONCLUSIONS OF WORKING GROUP B

Definition and Quantification of Effective Parameters for Flow and Transport

Chairman: J. Hadermann (Switzerland)
Members: J. Porter (United Kingdom), O. Jacquet (Switzerland), B. Bonin (France), A. Hautojärvi (Finland), E. Mouche (France), K. Schelkes (Germany), J. Wollrath (Germany), J. Gómez-Hernández (Spain), B. Dverstorp (Sweden), T. Clemo (United States)

The participants of Working Group B first discussed the question: **What are effective parameters?**

There was agreement that flow and transport cannot be considered by going back to the fundamental physical laws and a microscopic description of the system on atomic scale. In that sense, all parameters of all models are effective parameters. The participants discussed a classification of effective parameters for practical purposes. They agreed that such a classification is dependent on scale and on aspect to be considered. Considering models for flow and transport, effective parameters can be divided into two broad classes:

1.   Those which arise from homogenisation in the process of up-scaling. These can be subdivided into two classes:

    1.1   Homogenisation of processes. A typical example is dispersivity. These parameters are not locally measurable.
    1.2   Homogenisation of parameters. A typical example is porosity. These parameters are – loosely speaking - locally measurable.

2.   Those which arise from simplifications. A typical example is the reduction from 3D to 1D. Clearly, these parameters are not measurable.

With respect to the **physical significance** and the physical meaning of effective parameters, the opinions were diverse. Consider for example the triple chain of up-scaling:

Brownian motion $\rightarrow$ Diffusion $\rightarrow$ Transfer coefficients.

A model for transfer coefficients would attribute the physical significance to diffusion, a model for diffusion to Brownian motion. This shows the impact of the scale under consideration. The Group stressed that a physical understanding of the processes and mechanisms is the most important point. Based on such an understanding, the effective parameters can be evaluated and, in principle, calculated.

The **possibilities of calculation** are dependent on the context, e.g. single versus multiphase flow, unfractured sedimentary versus fractured medium. The situation is generally clearer for local parameters. For non-local parameters also boundary conditions (not always well known) are of importance. For practical applications holds: "We use them, even if when we cannot produce them".

The **consequences** of using effective parameters at a given scale are both positive and negative. They allow for a judgement of importance of processes and for a ranking among them. They permit the identification where further experiments and model concept development are needed and allow for large amounts of calculations to be done in a safety/performance assessment. The negative side is that introduction of effective parameters leads to a loss of details in the system description and to a loss of understanding of the underlying mechanisms. Introduction of effective parameters might also lead to a bias depending on the aim of the system description/investigation.

The discussion on **separating variability and uncertainty** confirmed the known facts: Variability of parameters leads to uncertainty. Uncertainties in model concepts and their effective parameters arise from uncertainties in boundary conditions and uncertainty in process description. A means to evaluate the last one, is down-scaling. At any rate, the way to investigate impact of model uncertainty is to use different models.

During the Working Group's meeting the issue of time dependence of effective parameters, and how to best evaluate the impact of choice of effective parameters came up. These were not discussed due to lack of time.

ANNEX 5

CONCLUSIONS OF WORKING GROUP C

Benefits of Further Site Characterisation for Better Radionuclide Transport Calculations

Chairman: C.C. Davison (Canada)
Members: J.-C. Mayor (Spain), P. Flavelle (Canada), J.-O. Selroos (Sweden), M. Mazurek (Switzerland), J. Geier (United States), K. Jakobsson (Finland), G. Bussod (United States), S. Norris (United Kingdom), J. Carrera (Spain), P. Lalieux (NEA)

1. Introduction

Our working group spent a considerable time initially discussing and debating terminology. What was meant by "performance assessment" and should we restrict ourselves to the "safety analyses" that often derive from the final (simplified) abstracted models of the site characteristics? We finally agreed that we ought to consider this topic as encompassing the full range of performance assessment model development. We should consider it to include the site "process models" that currently form the main building blocks for most country's performance assessment models. We felt the detailed process models could/will soon become integral components of the performance assessment models. We were confident that increased computational power and mesh generation capabilities were rapidly leading us to the situation where there is becoming less need to simplify the process models in order to incorporate them into performance assessment. Therefore, we should not limit ourselves in our thinking about abstracting site characterisation data into performance assessment models because of current past/computational inadequacies.

2. Question #1. What data are/are not measurable in the field?

Are: Intrinsic properties (fundamental physical, chemical, mechanical, thermal properties or characteristics).
  – permeability, pore/crack structure, thermal conductivity, compressibility/stiffness, strength, etc.
  – extrinsic properties – influences that can cause a difference response (properties that vary spatially and temporally).
  – hydraulic head, infiltration, temperature.
  – need to emphasise that spatial and temporal variability can be (are) intimately linked.
  – ultimately the group conceded that we could not produce a comprehensive list of what could be measured in the field, but rather site characterisation staff needed some guidance on what needed to be measured in the field for the process models. Although each country may need to determine this for its own purposes (depending on disposal concept and geologic media) there appeared to be agreement that using some sort of FEP's list was an important starting point for both the development of the process modelling approach that should be used as well as for the development of the list of processes that require parameterisation for the modelling.

– there appear to be some very useful examples of how some advanced programs have developed lists of site characterisation data to feed into the performance assessment process (Yucca Mountain and WIPP are good examples which are at relatively advanced stages of developing performance assessments and safety assessments with the objective of site/facility licensing and regulatory approval).

3. **Question #2. What level of detail is needed for performance assessment? Is this achievable?**

In general, the level of site characterisation or site description that can be performed (and usually is done) is done at a level of detail that is greater than that needed for the development of most process models and certainly much more detailed than is usually required for the simplified performance/safety assessment models. Exceptions might be in the quantification of connectivity of flow paths or dimensional descriptions of channelled flow paths for those models that use these conceptualisations. In general, most other characteristics are measured in the field at measurement scales that are much more detailed than the scale they are incorporated into performance/safety assessment and process models. Models of up-scaling and other overlapping models or conceptualisation are currently used to derive effective parameters to incorporate them into process models or performance/safety assessment models.

– It is particularly valuable to use hydraulic head and groundwater chemistry measurements during model development to integrate process/effects over long times and large space.

– There is a need to strive for site characterisation measurements at the modelling scale, although good up-scaling approaches are being developed.

– We see no reason why greater detailed understanding of the spatial and temporal variability of site characteristics cannot eventually be directly incorporated in performance/safety assessment models and this will serve to make the performance/safety assessment models more realistic (because at least one step of abstraction will be eliminated) and also make the performance/safety assessment models more responsive to accept new site characterisation data or to be directly calibrated against existing data.

– Furthermore, we felt that it was important to emphasise that the development and sensitivity analyses of the process models and performance/safety assessment models could/should be useful tools in aiding site characterisation. These models can be manipulated to show where new or additional data would help in reducing uncertainty or aid in confidence building (i.e. there needs to be strong feedback from the teams developing the process/performance/safety assessment models and the teams involved in site characterisation to allow proper integration, refinement and improvement). Finally, the site characterisation teams need to provide their expertise in determining how their data should be treated or assessed in the process models or performance/safety assessment models. There was a concern that the hand-off of data from site characterisation to performance/safety assessment could result in a lack of ownership by the site characterisation team of the final performance/safety assessment models. This should not happen and steps need to be taken to ensure it doesn't. Otherwise the scientific/technical support for the process and performance/safety assessment models that are finally developed will be weakened. The best way of ensuring this will happen is to incorporate as much site characterisation realism as possible into the site models that comprise the final performance/safety assessment models.

4. **Question #3. How can "soft" information on spatial variability be used to support (or improve) the conceptualisation and modelling for performance assessment?**

   – We began discussing this question by trying to agree on the definition of soft information. There were many views on what was meant by soft versus hard information. We decided the distinction between these two was mainly model dependent, however some examples of site specific soft information were: geophysical measurements (quantitative) and geologic interpretations that were often stylised representations of conceptual models. (This was especially the case for the structured geologists' interpretations of fracture zone geometry.) Hard information could be "direct" measurements of a property (such as permeability) or "indirect" measurements that provide property information (such as hydraulic head).

   – Non site-specific soft information came from expert opinion (experience) and scientific literature (including other scientific disciplines or areas of study). Non site-specific hard information can be obtained from natural analogue studies, especially for understanding events and processes that cannot be observed at the site under current conditions.

   – There was agreement that formalised, well-documented approaches were needed to show how both soft and hard information were incorporated into the conceptual models, process models and abstracted performance/safety assessment models. It was also emphasised that the model development process should be structured to ensure continuous integration of the experimentalists, site characterisation groups and the modelling teams. There were numerous examples given of situations where even various site characterisation teams (often discipline based) did not communicate effectively with each other. If poor communication amongst the information gatherers and the model developers exists, situations can develop whereby the groups responsible for site characterisation are unable to recognise their understanding in the models that are being using for performance/safety assessment.

   – A formalised approach would help to ensure the proper use of the site characterisation/ experimental information in developing the process and performance assessment models as well as help to ensure there was proper feedback from the models to the supporting groups. For instance, this could allow the process models/performance/safety assessment models to be refined or tested (confirmed) by obtaining more or new site specific information.

5. **Question #4. How to decide if further site characterisation is (will be) helpful to performance assessment?**

   – While we did not have much time to discuss this question directly, we felt that our discussions of the earlier three questions had often included this topic.

   – In general, we believed that if there was a formalised, clearly-understood, and well-documented approach to the model abstraction exercise, there would be proper linkages between the site characterisation groups and the modelling groups. This would ensure that the modelling groups could provide feedback to the site characterisation teams to show where more or new site characterisation would help remove uncertainty, choose between alternative conceptual models or provide a needed parameter value or process understanding.

– Usually the application of the process models or performance/safety assessment models will determine if further site characterisation is needed. The need must be assessed in the context of whether or not performance/safety assessment will be improved.

6. General Comments

Several themes emerged throughout the various discussions by members of Working Group C that related to the overall topic of the workshop but were not specific to the questions we were asked to address.

One recurring theme was the general concern that site characterisation should not be undertaken solely to meet performance/safety assessment requirements. Firstly, site characterisation must provide information for engineering design and optimisation which can require different information than that required for performance/safety assessment. Second, site characterisation can often reveal information that leads to completely new or unexpected conceptual models of processes. In these cases, if site characterisation had not collected the information (despite the fact it was not clear at the time it was collected how the information would be used), incorrect conceptual models of site processes would have been used for the performance/safety assessment. (Examples from USDOE's YMP and AECL's regional characterisation of the rocks at the Whiteshell Research Area were cited.) Site characterisation needs to be undertaken in such a way that it can evolve a realistic understanding of the site conditions. Such an approach requires the process model development teams and performance/safety assessment abstraction modelling teams to be both flexible and responsive as well as interactive with the site characterisation teams. This approach allows a design-as-you-go philosophy to be used during the site characterisation/performance/safety assessment model development stage that should also follow through the construction/design and operating phases of the disposal facility.

ANNEX 6

CONCLUSIONS OF WORKING GROUP D

Future Directions of Work in the Treatment of Spatial Variability for Radionuclide Transport

Chairman: I. Neretnieks (Sweden)
Members: E. Ledoux (France), A. Poteri (Finland), A. Emrén (Sweden), L. Dewière (France),
A. Gautschi (Switzerland), C. H. Kang (Korea), L. Meigs (United States), A. Jakob (Switzerland),
E. Fein (Germany), A. Wörman (Sweden)

1. Q1 : "What processes can be enhanced by spatial variability that impact radionuclide transport"

The group found that the major concern relates to identifying the major pathways for the escaping nuclides. Of special concern is to identify what is termed potentially unexpected short circuiting paths. The water pathways and how they are influenced by the variability of the hydraulic properties are the foremost entities to determine. This will make it possible to assess the major flowpaths. The variability of sorption properties, matrix diffusion, specific flow wetted surface, block size distributions and mixing processes along the major flowpaths are then important to determine. The mixing processes between pathways are of special importance if adjacent paths can have strongly different radionuclide transport properties

Any non-linear processes that can be expected are especially important to identify because they need special attention when following the nuclides along the pathways.

2. Q2: "Treatment of geochemical variability"

The variability of the rock mineral composition and its correlation to flowpaths can have a strong impact on radionuclide transport and should thus be investigated. Fracture alteration and fracture gouge material, if present, can also have a large impact both as additional sorption mass and potentially as a barrier to diffusion into the matrix.

Variability of the water chemistry is different in the sense that it changes over time because it is influenced by the flow of water and by the reaction with the minerals along the flowpath. Mixing processes can also influence the local water composition. Potentially the variations of the water chemistry can be used to test the flow and transport models that are to be used for the site. This can lead to a better understanding of the radionuclide transport properties of the site.

The pH, Eh, concentration of complexing agents such as carbonate and the ionic strength influences the sorption and diffusion properties of some nuclides. The variations of these entities over space in time should thus be considered.

3. Q3: "Consistency between geologic, flow, transport and geo-hydrologic models"

The conceptualisation of the site and processes must be consistent for all the models.

In performance assessment reports a chapter should be dedicated to show how the different models fit together and that they are not contradictory. Also discussions on how relevant observations have been used to derive understanding of processes and to interpret basic data into parameters should be shown in this chapter.

An effort must be made to develop better ways for the different groups to communicate and to learn to understand each other.

An example of discrepancies is the use of the term water age. This conveys the notion that an identifiable packet of water has travelled for a certain time. This contradicts the model of radionuclide migration and its mediation by sorption and matrix diffusion mechanisms. The residence time of the radionuclide as for any molecule, including the water molecules cannot be characterised in a meaningful way by a single number. The distribution of residence times is extremely large and so called age dating methods can give practically any number depending, on the tracer or technique used. We rely on this distribution when trying to prove that possibly only the faster radionuclides will reach the biosphere before they have decayed.

4. Q4: "Relation between scientific knowledge and performance assessment"

An efficient communication between the different representatives must be assured from the beginning. This applies to planning experiments, developing conceptual models, interpreting experiments, and documenting the findings.

Attempts at alternative conceptualisation of the observation should be encouraged. Devils advocates should be used. Close contacts with regulators should be upheld.

Publication of findings in the open literature should be strongly encouraged.

PART B:

WORKSHOP PROCEEDINGS

SESSION I

Spatial Variability: Its Definition and Significance to Performance Assessment and Site Characterisation

Chairmen: R. Mackay (Birmingham Univ., United Kingdom)
and L. Dewière (ANDRA, France)

Variability in Natural Systems and Means to Characterise it

E. Ledoux
Ecole des Mines de Paris, France

G. de Marsily
University Pierre et Marie Curie, France

Abstract

A common characteristic of natural geological media is that they display an immense spatial variability with respect to their structural, physical and chemical properties. This is hardly surprising when one remembers the complexity of formation and evolution undergone by rock masses within the earth. It is, however, extremely difficult to describe this variability in space, given the amount of effort needed to explore directly natural media and the shortcomings of indirect geophysical methods as they are very often insufficiently sensitive to the variations of relevant parameters.

The purpose of this paper is to give a general review of the existing approaches to describe variability and quantify its consequence with regards to migration processes through the geological medium.

A number of key questions arise because of this variability:

- what are the different scales of variability and their impact on characterisation of relevant processes for the quantification of migration;
- are we able to identify these different scales from the field;
- do we have the proper tools to model this variability once it is identified;
- are we able to obtain the parameters involved in the tools describing the variability;
- are we able to practically run the computer models including a realistic level of heterogeneity;
- how can we evaluate the uncertainty of model prediction due to the spatial variability.

Present knowledge and state of the art on these issues will be reviewed here.

1. DIFFERENT SCALES OF VARIABILITY

Variability in natural geological media applies to the structure of the medium and to the physical processes. Table 1 shows the different scales of variability that are relevant to hydrogeological processes.

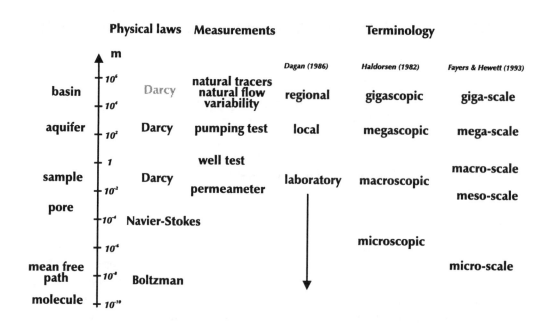

Table 1. Scales of variability for flow conceptualisation

In safety analysis for radioactive waste disposal, the prediction of the phenomena must be given at the gigascopic scale which is the one to which it can be expected that the geological barrier would be efficient to protect the human environment from radioactivity. The issue is that at the gigascopic scale, very few measurement methods are available in the field and only models of spatial variability, fitted on discrete or approximate integral observations can be used.

Macroscopic measurements are the most commonly used, coming from laboratory or field tests, but they must be upscaled before being properly introduced in the models. In the representation of flow, the microscopic scale is practically never used. This could however be relevant in the case of very impervious media like clays in which the interaction between water molecules and mineral surfaces are known to be important. Table 2 describes the different scales of variability in geochemical processes.

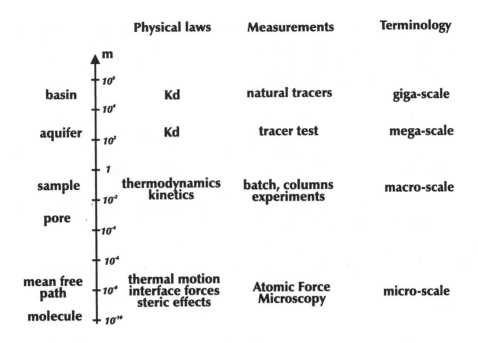

	Physical laws	Measurements	Terminology
basin	Kd	natural tracers	giga-scale
aquifer	Kd	tracer test	mega-scale
sample	thermodynamics kinetics	batch, columns experiments	macro-scale
pore			
mean free path	thermal motion interface forces steric effects	Atomic Force Microscopy	micro-scale
molecule			

Table 2. Scales of variability in geochemical processes

In safety analysis, the geochemical processes are commonly represented in a semi-empirical way by a global distribution coefficient Kd which is supposed to be an effective parameter mostly obtained from laboratory tests. At the macroscopic scale, a thermodynamic and kinetic representation of the geochemical processes has been shown to better represent fluid-rock interaction. In such a case, the parameters become intrinsic and the predictive power of the models is enhanced. The major problem is the computational complexity of the codes which make them hardly usable in reality for performance assessment. Upscaling methods to derive effective thermodynamic and kinetic parameters for average rock properties in spatially variable media have not yet been developed, even if preliminary tests have shown the difficulty and the potential bias of such an averaging [1], [2].

The microscopic scale for the description of geochemical processes is rarely investigated in performance assessment due to the fact that the upscaling methods are still in their infancy. The study of the behaviour of processes at the microscopic scale is however fruitful and cannot be avoided when handling with complex phenomena such as surface complexation. An example is given by the role of colloids on radionuclide migration, which involves physico-chemical processes controlled by electrostatic phenomena [3].

2. GEOLOGICAL BASIS FOR THE GENERATION OF VARIABILITY

Variability at different scales is the result of the generation processes which took place during the formation and evolution of rocks. At the megascopic and gigascopic scales, the sequences of sedimentation are responsible for the initial heterogeneity of the geological structure. Sedimentary media are thus organised in a series of layers each showing a more or less homogeneous lithology. The magmatic differentiation produces somehow similar lithological effects for crystalline rocks during the cooling phase of a magma. Erosion processes and tectonics can then modify the geometry and the properties of the initial rock bodies by carving, folding and faulting. Flow pathways that can play a dominant role in migration are thus the consequence of all these processes. The knowledge of the geological history of the rock mass is therefore of major importance to establish the structural model at the megascopic and gigascopic scales.

At the macroscopic scale, variability is mainly due to micro-sequences of sedimentation for sedimentary media and crystallisation for crystalline media. Compaction, diagenesis, fracture generation and weathering are processes playing a role in the evolution of the rock mass that can generate preferential flow pathways or, on the contrary, close pre-existing ones.

Petrophysical properties of the rocks are relevant at the microscopic scale. The initial properties can be strongly modified through the whole history of the rock formation due to diagenis and weathering.

It is thus clear that the causes of variability of geological medium are very numerous and can take place at various steps of the geological history. The result can be a very complex structure in which the prediction of the heterogeneity is hardly possible.

Table 3 summarises the different processes relevant for the generation of variability at different scales in natural media.

	GENESIS	**EVOLUTION WITH TIME**
mega/giga scale	**sequences of sedimentation magmatic differentiation**	**tectonics erosion**
macro-scale	**micro-sequences of sedimentation crystallisation**	**compaction diagenesis diaclasis weathering**
micro-scale	**petrographic and mineralogic features**	**diagenesis weathering**

Table 3. Geological basis of the genesis of variability

3. MIGRATION PROCESSES TO BE CONSIDERED AT DIFFERENT SCALES

The relevant processes to be considered in transport modelling are largely dependent on the scale at which the geological medium is described.

At the megascopic scale, the advective transport trough pathways induced e.g. by faults or sedimentary channels, is dominant. The key question is therefore the good knowledge of the discontinuities and the characterisation of their connectivity and hydraulic properties. At that scale the retardation processes which are the consequences of fluid-rock interaction will only be represented by using effective parameters like distribution coefficients (Kd).

At the macroscopic scale, diffusion processes become more important since they can interact with the advective transport. Dispersion, matrix diffusion, chemical interaction with the rock matrix or fracture coating should be addressed.

At the microscopic scale the pertinent mechanisms become more complicated. However, taking them in consideration seems necessary if the detailed geochemical processes are to be accurately represented. For instance, the microscopic distribution of minerals influences local precipitation or dissolution and thus the way in which permeability can be reduced or enhanced. As an other example, the electrostatic properties of minerals surfaces determine surface complexation reactions and the way in which colloids will migrate through the medium and incorporate or not radionuclides.

In conclusion, the complexity of the mechanisms that should be involved in the models is strongly dependent on the scale of representation.

4. GEOLOGICAL BASIS FOR THE CONCEPTUALISATION OF VARIABILITY

In sedimentary basins, the classical method of representing a complex geological formation in order to model its hydrogeologic behaviour is to first decompose it into aquifers (layers of high permeability) and aquitards (layers of low permeability). A multi-layered schematic representation of the medium is then made. One or the other of these layers may sometimes be discontinuous, in case for instance of the interruption of an aquitard resulting in the creation of a single aquifer from the ones of each side. This schematic structural model is built by the geologists who, with the help of field measurements and geophysics, establish the continuity of the layers, while, at this stage, disregarding the local petrophysical properties of the rock bodies and extrapolating in space the borehole and geophysical information. Fig.1 shows an example of such a multilayered representation of Paris basin.

Once this geometrical work has been done, a mesh can be generated in 2 or 3-D to describe this geometry. The picture must be then completed by parameters values in each one of the meshes. This is usually done with the help of pumping test results from the wells. Based on these local values, the unknown parameters are then initially estimated either by "homogeneous zones" or preferably by geostatistical methods. The last stage in this process is the fitting of the model on hydraulic head data: this can be done either manually or by the "inverse" method, to improve the values from the initial estimates.

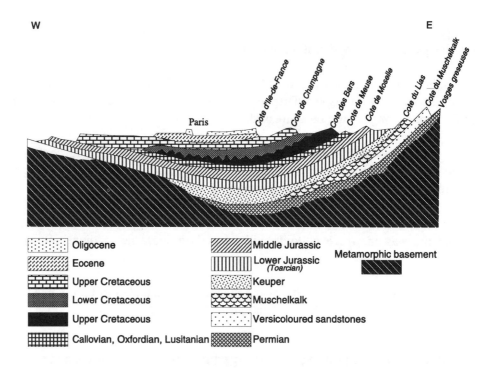

Figure 1. Multilayer representation of the Paris basin showing aquifers in grey and aquitards in black

This approach is not entirely satisfactory for several reasons. The major objection is that the initial geometric decomposition is rather arbitrary and never questioned. Any cross-section of a fairly thick sedimentary basin shows that facies variations inside the same unit are constant because of the sedimentation processes and that the definition of aquifers and aquitards based purely on lithostratigraphic descriptions is imprecise. It thus remains possible that discontinuities at smaller scale might be ignored, leading to the risk that major pathways like faults or thin continuous permeable sedimentary layers can jeopardise the continuity of the geological unit.

The second objection is that inside a layer the global permeability is not only determined by the distribution of local permeabilities but also by the connectivity of high permeability units. The geostatistical tools make, by construction, the hypothesis of the continuity of the medium, with parameter values only varying around an average assumed to be constant or slowly varying. It thus happens a tendency to exaggerate the permeability of a layer which, for instance, would consist on isolated sandy lenses inside a clay matrix. As opposed to this, the global permeability of an impervious layer which may contain thin continuous permeable sections would be underestimated.

The third objection is that, in general, only one structural model of reality is produced by this approach and that the residual uncertainty of this model cannot be quantified.

In recent years, several overviews of how to describe sediment spatial variability have been presented, by e.g. Haldorsen and Damsleth [4], Fayers and Hewett [5], Yarus and Chambers [6] in the oil industry, or Marsily [7], Kolterman and Gorelick [8], Anderson [9], in groundwater hydrology. Causes for sedimentation variation with time can be traced to the sequential stratigraphy concepts developed by Vail [10]. We will try to summarise some of the major findings that may improve the way to represent heterogeneity in models.

4.1 Geostatistical simulation of parameters

The difference between an estimation by kriging and a simulation is the following. An estimation is by definition unique and it gives at all points the optimal value of the parameter that can be obtained from available data. However, this estimation is never meant to be an image of reality because it is by definition smoother than reality.

As opposed to this, simulations are possible realisations of the unknown magnitude, all equally probable, but none of which can presume to be the closest to reality. The simulations are said to be conditional if they respect the measurement points. The simulations can be conditioned by the mean of direct values of the permeability but also by the hydraulic heads through the inverse method.

Compared to the estimations the simulations have the advantage of producing the right spatial variability of the parameters and not a smooth field.

Moreover, with the simulations it is always possible to quantify the residual uncertainty of the model prediction after a fitting by using the simulations as input to the Monte Carlo method. To do this, one generates as many predictions with the model as one has parameter simulations and one then examines the statistics of these predictions.

Two categories of geostatistical methods are presently available for the conditional simulation of geological facies: the truncated Gaussian method and the indicator approach

The gaussian thresholds method [11], [12], is a technique for simulating the geometry of facies in 3-D. The variation of a random function in space with a Gaussian distribution is divided into intervals defining geological facies. The cumulated probability of the normal distribution function between thresholds corresponds to the percentage of distinct lithofacies in the geological formation. In order to recreate the stratification of facies a spatial covariance of the Gaussian function is prescribed. The fitting of the model consists in first choosing the thresholds that define the facies in order to obtain the right proportion of each facies in the basin; one then fits the horizontal and vertical covariance functions from field observations.

The method can easily be conditioned on observations in the wells and on the outcrops. The generation of facies in space generally uses very fine grids including several millions of meshes, which can go far beyond the capacity of the computer codes. It is therefore necessary to combine a large number of meshes to create a suitable grid for a hydraulic or transport model.

Figure 2 shows an example of a model using three facies.

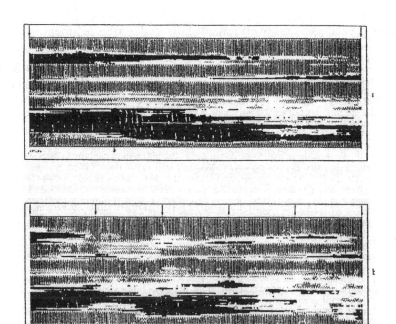

Figure 2. Example of a model of three facies using threshold Gaussian approach [12]

The indicator approach [13] leads to similar results but using different hypothesis. Each facies is defined by a class. Binary (indicator) variables indicate if a given point in space belongs to a given facies class. The probability of the parameter belonging to a given class is estimated by kriging. The results can be easily conditioned to the data observed from boreholes and can take into account the information from outcrops to determine the covariance functions of each facies indicator.

4.2 Boolean methods

In the petroleum literature, the Boolean method was made popular in 1986 by Haldorsen and Chang [14] and is known as "stochastic shales" method. It entails randomly generated in space objects whose position and dimension are randomly selected whereas their form is chosen *a priori*. A parameter value is then simply attributed to each object and another one to the homogeneous matrix containing the objects, so that a 3-D description of the medium is obtained. This method produces an number of simulations of the medium and never an estimate. However, there are many drawbacks. First, it is necessary to select *a priori* the geometry of the objects which opens up infinite possibilities. In fractured medium, the traditional choice is a circular disk (Fig 3). In an alluvial medium one suggestion has been made to choose half-moon cross-sections with zigzag longitudinal extension. Second, one has to decide which value of parameter to attribute to each facies. Third, an unsolved problem is that of the conditioning of the simulation on the data. This is usually done in a non-random manner; the objects observed in a borehole are added to the model and the random objects that have not been observed are eliminated.

Figure 3. Example of Boolean model in a fractured medium

4.3 Simulation of sedimentary processes

This method is still in infancy. Its aim is to generate facies, not by simulating statistical processes that have a resemblance to the observed distributions, but as a result of a physical sedimentation process. The effort is concentrated on the mechanisms that create the deposits in the course of time as function of the climate, topography, erosion and solid transport, etc... One of the most remarkable achievements is perhaps the work by Koltermann and Gorelick [15] on the sedimentation of the San Francisco Bay. The authors have simulated the processes over 600,000 years starting with a reconstruction of the climate, of the variations in the sea level and of the solid load of the rivers as a function of the climate, based on present day observations of these variations at different latitudes. The method is highly computer intensive, but the result gives a very realistic representation of the sedimentary structure as it is known from the bore holes (Fig 4). However this method cannot be presently conditioned on observations and only one simulation was generated as opposed to the preceding methods which systematically provide a great number of simulations.

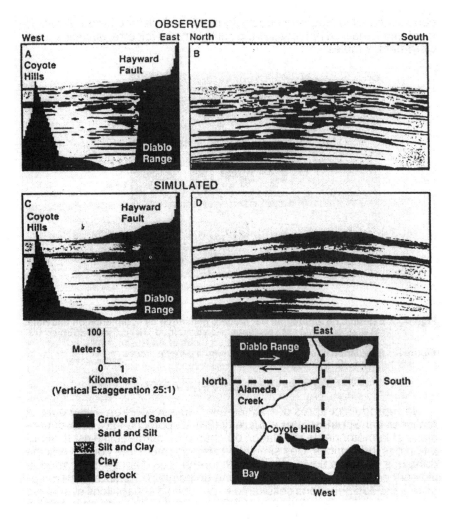

Figure 4. Simulation of the deposit of San Francisco Bay using a genetic model [15]

In conclusion of this chapter, it seems that the geostatistical simulations (Gaussian threshold or indicators for facies) are readily available tools. They have the immense advantage that they can be fully conditioned on all the available local information, even indirect such as hydraulic heads, and that they are fully objective, which means that no empirical decision is left to the user once the statistical parameters have been chosen.

Boolean methods are best suited to objects whose shape is clearly defined, such as fractures. For sediments their role is less clear, because they are very hard to condition on the data.

Sedimentation models are presently only research tools but they provide a mean of introducing a genetic perspective into the description of heterogeneity. It is conceivable that in the future the geostatistical and the genetic approach could be combined.

5. LIMITATIONS IN THE CONCEPTUALISATION OF HETEROGENEOUS MEDIUM

The handling of variability introduces some limitations to the modelling of geological medium. These limitations concern, for instance, the determination of effective parameters and the bias of the computer models that can results as a consequence of heterogeneity.

5.1 Determination of effective parameters

At the scale needed for performance assessment which is basically the megascopic ,or even the gigascopic one, model parameters can rarely be identified from field tests. It is thus necessary either to upscale effective parameters at the proper scale from data obtained at a smaller scale (macroscopic), or to calibrate the model over spatial and time scales that are relevant to performance assessment. Such calibration exercises are very difficult to achieve considering the very long travel times that are expected through the rock formations that are potentially acceptable for waste disposal. A possible means to overcome this difficulty is through the use of long term natural or anthropogenic phenomena involving flow and migration processes.

An other method under development consists on theoretically upscaling the local parameters using mathematical algorithms. Facies simulation models generate such a number of meshes that flow and transport models cannot use the realisations directly, due to the restrictions imposed by the computing power. To overcome such restrictions averaged hydraulic parameters are needed at scales appropriate to the numerical resolution of the problem. The necessary requirement of the upscaled parameters is that the model reproduces on average the flow patterns which would be observed in the full fine scale simulation. The purpose of the upscaling is thus to support stochastic analysis of flow and transport. The present tendency is to derive quick upscaling techniques. A review of existing methods for upscaling permeability is presented by Renard and Marsily [16].

5.2 Modelling bias induced by heterogeneity

An attempt of computing the effective permeability was made by Lachassagne [17] using mathematical groundwater simulation codes in finite differences and finite elements applied on a 2-D heterogeneous medium within which the permeability was randomly generated from a log-normal distribution. This distribution is defined by its mean and its standard deviation. In that case, it can be shown that the actual effective permeability is the geometrical mean. The effective permeability of the medium can also be numerically inferred by computing the value of the flow through a square grid in which a prescribed head conditions has been settled along two opposite boundaries. It appears (Fig. 5) that the numerical models deviates significantly from the theory for high values of standard deviation of the permeability; the finite element code tends to overestimate the effective permeability while the finite differences one shows a tendency to underestimate it. This phenomenon could be explained by the chanelling effect, for high values of standard deviation, which leads to the fact that the majority of flow would take place in particular pathways.

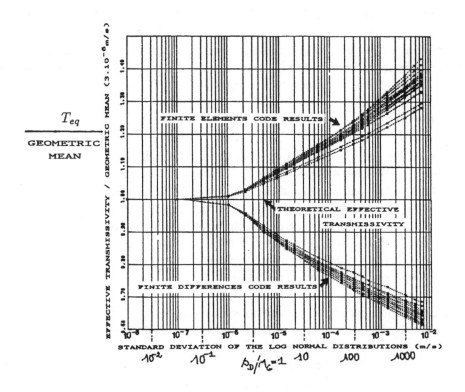

Figure 5. Values of the effective permeability computed with numerical models versus the standard deviation of the permeability distribution [17]

6. CONCLUSION

For present-day applications, methods are existing which allow a realistic representation of the heterogeneity of the geological medium. Geostatistical tools are available ones that offer the advantage that they can be conditioned on field data and that they can propose several realisation of the possible reality, which enables to associate to the model an approach of the uncertainty of the simulation results. Sedimentation models provide a means of introducing a genetic perspective into the description of heterogeneity.

A key question is that the computer tools are not readily able to handle with such a level of heterogeneity due to the fact that geological models generate a great number of meshes. Upscaling the model parameter values is thus a necessity which can be obtained through calibration of the models against field data involving phenomena at the proper scale or through theoretical upscaling techniques.

7. REFERENCES

[1] Emren, A.T., The influence of heterogeneous rock chemistry on the sorption of radionuclides in flowing groundwater, Journal of Contaminant Hydrology, 1993, 13, 131-141.
[2] Pelletier, I., Traitement des données de réservoir en vue d'une simulation de la diagenèse. Application à l'illitisation du Brent de la Mer du Nord, Doctoral Dissertation, Institut de Physique du Globe, Paris, (to appear).

[3] Van der Lee, J., Modélisation du comportement géochimique et du transport des radionucléides en présence de colloides, Doctoral Dissertation, Ecole des Mines de Paris, 1997.

[4] Haldorsen, H.H. and Damsleth, E., Stochastic modelling, Journal of Petroleum Technology, 1990, 42, 404-412, discussion by Saleri, N., 929-930.

[5] Fayers, F.J. and Hewett, T.A., A review of current trends in petroleum reservoir description and assessing the impacts on oil recovery, Mathematical Modelling in Water Resources, 1992, 2, 3-33 (Russel, T.F. et al. Eds, Computational Mechanics Publications and Elsevier).

[6] Yarus, J.M. and Chambers, R.L., Stochastic modelling and geostatistics : AAPG Computer Application in Geology, 1994, 3.

[7] Marsily, G. de, Quelques méthodes d'approches de la variabilité spatiale des réservoirs souterrains : Hydrogéologie, 1993, 4, 259-268.

[8] Kolterman, C.E. and Gorelick, S.M., Heterogeneity in sedimentary deposits : a review of structure-imitating, process-imitating and descriptive approaches, Water Resources Research, 1996, 32(9), 2617-2658.

[9] Anderson, M.P., Characterization of geological heterogeneity, Stochastic Subsurface Hydrology, 1996 (Neuman, S.P. and Dagan G. Eds, Cambridge University Press), in press.

[10]Vail, P.R., Audemard, F., Bowman, S.A., Eisner, P.N., and Perez-Cruz, C., The stratigraphic signatures of tectonics, eustacy, and sedimentology -an overview, Cycles and Events in Stratigraphy, 1991, 617-657 (Eisele, G., Ricken, W., Seilacher A. Eds, Springer Verlag, New York).

[11]Matheron, G., Beucher, H., Fouquet, C. de, Galli, A., Guerillot, D., and Ravenne, C., Conditional simulation of the geometry of fluvio-deltaic reservoirs, 1987, Society of Petroleum Engineers SPE 16753.

[12]Matheron, G., Beucher, H., Fouquet, C. de, Galli, A., and Ravenne, C., Simulation conditionnelle à trois faciès dans une falaise de la formation du Brent, Sciences de la Terre, Sér. Inf., Nancy, 1988, 213-249.

[13]Journel, A.G., Fundamentals of Geostatistics in Five Lessons, Short Course in Geology, American Geophysical Union, Washington DC, 1989, 40.

[14]Haldorsen, H.H., and Chang, D.M., Notes on stochastic shales from outcrop to simulation models, Reservoir Characterization, New York, 1986 (Academic Press).

[15]Kolterman, C.E., and Gorelick, S.M., Paleoclimatic signature in terrestrial flood deposits : Science, 1992, 256, 1775-1782.

[16]Renard, Ph. and Marsily G. de, Upscaling the permeability : a review, Advances in Water Resources, (to be printed).

[17]Lachassagne, P., Estimation des perméabilités moyennes dans les milieux poreux fortement non-uniformes étudiés sous l'angle stochastique. Application aux essais de débit en aquifère captif, Doctoral Dissertation, Ecole des Mines de Paris, 1989.

Why is Variability Important for Performance Assessment and What are its Consequences for Site Characterisation and Repository Design?

B. Dverstorp
SKI, Sweden

P. A. Smith
Safety Assessment Management Ltd., UK

P. Zuidema
Nagra, Switzerland

Abstract

The importance of spatial variability is discussed in terms of its consequences for site characterisation and for repository design and safety. Variability is described in terms of various scales of discrete structural features and a pragmatic classification is proposed according to whether the features are:

- feasibility-determining (i.e. features within which repository construction and operation is not practical and which preclude long-term safety);
- layout-determining (i.e. features which, if avoided, would enhance long-term safety);
- safety-determining features (i.e. features which cannot be shown to be avoidable and which strongly influence the calculated long-term safety of the repository system).

The significance with respect to the geosphere-transport barrier of small-scale pore structure within the various classes of feature is also discussed.

The practical problems of characterising variability and modelling its effects on radionuclide transport are described. Key factors affecting groundwater flow and radionuclide transport are identified, models that incorporate spatial variability are described and the estimation of appropriate parameters for these models is discussed.

1.   Importance of Variability

Spatial variability over a wide range of scales is a characteristic of most geological media and its presence must be taken into account in order to:

- assess the feasibility of siting a repository in a particular area or region;

- determine the optimum layout for a repository at a particular site;

- assess the safety afforded by the host rock at the site, both in terms of (a), satisfactory and prolonged performance of the engineered barriers and (b), retardation and decay in the geological barrier of any radionuclides released from the engineered barriers.

Figure 1 illustrates schematically the scales at which structural features typically exist and their significance in terms of repository design and safety.

(i)   Feasibility-determining features.

At the largest scale are major structural features, which are mechanically weak and are often associated with high fluxes of groundwater. Within these features, repository construction and operation are not (economically) feasible and, where there are high groundwater fluxes, long-term safety is not considered achievable. The features thus define the boundaries of discrete volumes of rock that might host all, or parts of, the repository. Early within a repository programme, the broad characteristics and spatial frequency of such features must be defined in order to establish the suitability of the siting area or region under investigation. Economic viability may not be achievable if the repository is partitioned between too many discrete volumes. A key task is to establish a preliminary boundary, in terms of properties, between these and the smaller-scale features that permeate the remainder of the rock.

(ii)   Layout-determining features.

Whereas the properties and frequency of the largest-scale features determine the feasibility of siting a repository in a particular area or region, smaller-scale features determine the optimum use of the site from the point of view of repository performance. Features that do not preclude repository construction, operation and long-term safety, but are nevertheless both less effective at retarding radionuclide transport and are avoidable during construction, may be referred to as "layout determining". In performance assessment, it may be possible to take considerably more credit for the geological barrier if it is demonstrated that emplacement tunnels and/or deposition holes can be located away from such features. In this respect, performance assessment is an important tool for developing relevant criteria for the adaptation of repository layout to local geological conditions. Since they are "avoidable", layout-determining features are, by definition, detectable at some stage during site characterisation. In some instances, however, they may be detected only in the later stages of a repository programme (e.g. near-vertical features are difficult to detect using vertical boreholes). In such cases, until it is demonstrated that the features are avoidable, they need to be considered as a part of geological barrier and hence there is likely to be an emphasis on the near field in performance assessment.

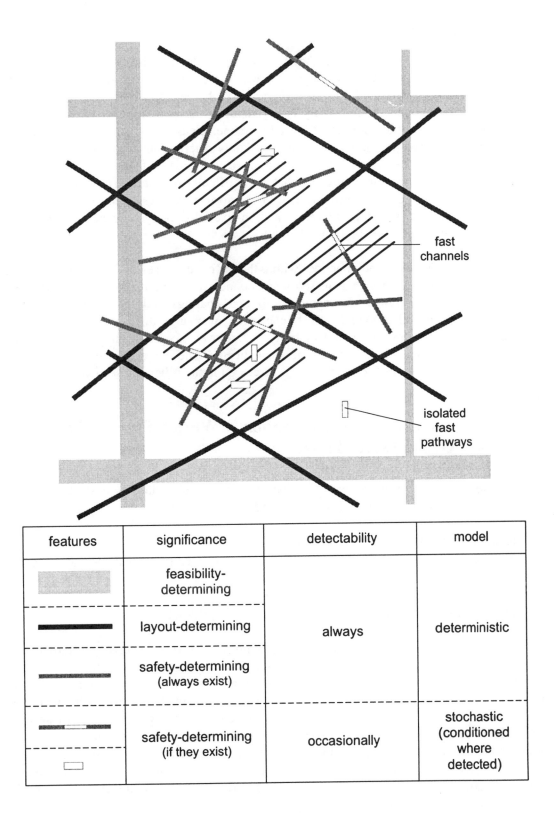

features	significance	detectability	model
![feasibility]	feasibility-determining		
![layout]	layout-determining	always	deterministic
![safety]	safety-determining (always exist)		
![safety2] ![isolated]	safety-determining (if they exist)	occasionally	stochastic (conditioned where detected)

Figure 1. The features that determine (i), the feasibility of repository construction and operation, (ii), its layout and (iii), its long-term safety

(iii) Safety-determining features.

Smaller-scale features that cannot be shown to be avoidable are safety determining in that their individual properties, and also the degree to which they are connected, controls the spatial distribution of groundwater flow in the vicinity of the repository. The distribution of groundwater flow around the repository is key input to the evaluation of the longevity of the engineered barriers, to the assessment of radionuclide releases from the near-field and to the assessment of transport through the geosphere. Safety-determining features may either (a), be detected, but found to be too frequent to avoid, or (b), be isolated features (e.g. "fast channels" within a larger-scale fracture network) that are difficult to detect. Those features that are detected may be represented deterministically in performance-assessment models. Features that are not detected, but whose presence cannot be excluded on geological grounds, must be represented stochastically (see section 2, below).

The transport times of radionuclides within the different classes of features described above depends not only on their average transmissivity and connectivity, but also on their small-scale pore structure, including the presence of fracture infill and coatings and the pore structure of the surrounding rock matrix, all of which affect retardation by matrix diffusion and/or sorption. An understanding of this smallest scale of variability is thus also necessary in order to quantify the performance of the geosphere transport barrier realistically. The characterisation of variability at different scales, and the modelling of its effects, is discussed in the following sections.

2. Techniques for the Characterisation of Variability and Modelling its Effects of Groundwater Flow

Figure 2 presents an overview of the evaluation of spatial variability in site characterisation and performance assessment. The broad steps involved are:

- accumulation of site data by a range of techniques;
- characterisation of spatial variability in terms of one or more geostatistical- and/or geological-structure models;
- flow modelling;
- transport modelling (discussed, in more detail, in section 3, below).

Several complementary techniques are typically used for characterising heterogeneity and estimating the spatially varying properties of geological media. Large-scale features, such as major fracture and fault zones, and major lithological units can often be identified deterministically through geophysical measurements, borehole data and hydraulic testing in a site investigation programme. Such features can thus be incorporated explicitly in models of groundwater flow and transport. For smaller-scale features, complete characterisation of a site may never be possible. It is, however, necessary to take account of the impact of incompletely characterised parts of the rock, in particular in the case of the rock close to the repository, when assessing the performance of the geological barrier. Geostatistical models of heterogeneity as input to stochastic model simulations of groundwater flow is one way to address this problem.

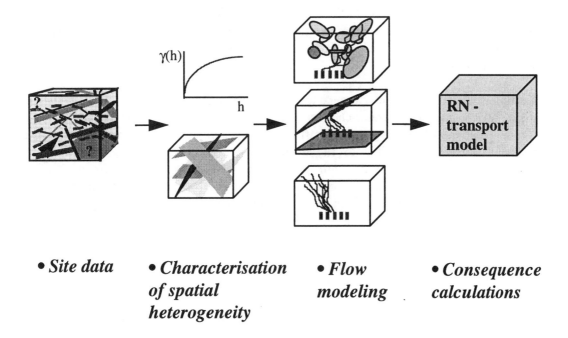

• *Site data*　　• *Characterisation*　　• *Flow*　　　• *Consequence*
　　　　　　　　　　of spatial　　　　*modeling*　　*calculations*
　　　　　　　　　　heterogeneity

Figure 2.　　**Evaluation of spatial variability in site characterisation and performance assessment**

Stochastic discrete-fracture or channel-network models of groundwater flow build on statistical distributions of fracture parameters (for example, fracture frequency, transmissivity, orientations), many of which can be estimated from borehole data. Stochastic continuum models build primarily on distributions of hydraulic conductivities estimated from small-scale hydraulic tests performed in boreholes. Common to both concepts are the often highly uncertain assumptions on connectivity properties or correlation structures, i.e. the spatial arrangement of hydraulically conductive elements between measurement points.

Most site-investigation methods, such as the borehole measurements, result in "point values" of the parameters defining hydrogeological models, for example hydraulic conductivities estimated from packer tests and geophysical evidence of fracture locations along boreholes. Up-scaling to a site-scale hydrogeological model will build on conceptual assumptions that often are difficult to test directly using site data. In a semi-stochastic discrete fracture network model, geological evidence on long-range correlation, hence location and orientation of major fracture zones, may be incorporated explicitly, whereas small-scale fractures and connectivity between major features are represented stochastically. In a stochastic continuum model, up-scaling is treated by means of geostatistical models describing the autocorrelation of hydraulic conductivities, although recent developments of indicator simulation techniques also allow incorporation of soft geological information. Such information can also be used to test the consistency of a model ("validation").

Experience from SKI's SITE-94 performance assessment [1], where both a discrete-feature model [2, 3] and a stochastic-continuum model [4] were applied, indicates that both concepts have strengths and weaknesses. In particular, each is better suited to the incorporation of some types of information than others. An important strength of discrete models is their ability to incorporate readily much of the geological information (and direct observations) concerning structures. The explicit representation of structures also makes it possible to evaluate connectivity properties and structural parameters such as flow wetted surface area [5]. The stochastic continuum model incorporates less of the fundamental site-specific data on hydraulic structure, but different realisations can be conditioned more easily on measured "point" hydraulic data than was found to be the case for a discrete-feature model, as implemented in SITE-94.

Multiple-scale cross-hole interference tests, tracer tests and consistency checks between the hydrogelogical models and groundwater-chemistry distributions may help to discriminate between alternative conceptual models for groundwater flow but, in a highly heterogeneous medium, it is likely that it will always be possible to fit several alternative conceptual models to the data. Furthermore, while consistency checks with groundwater-chemistry distributions are generally possible, cross-hole interference tests are not possible in tight rock formations. In performance assessment, the consequences of the resulting uncertainty may be bounded by applying several alternative conceptual models of the rock heterogeneity.

3.   Geosphere Transport Modelling

The most influential factors associated with the migration of radionuclides in fractured rocks are:

- the spatial distribution of groundwater flow - i.e. the presence of transport pathways with a range of hydraulic properties;
- the properties that affect radionuclide retardation along the paths and, in particular:
  - flow-wetted surface area;
  - sorption properties of fracture surfaces and the adjacent porous matrix and, to a lesser extent, matrix diffusivity and porosity.

In terms of the spatial distribution of groundwater flow, models of groundwater flow tend to be geometrically more complex than models of radionuclide transport. Generally in transport studies, due to limitations in the available geological data and to limitations in the available computational tools, a small number transport-pathway groups (frequently one), with uniform properties, is taken to represent transport through the geosphere as a whole. This high degree of simplification of a natural system that, in reality, would be expected to comprise many pathways with a wide range of spatially variable physical and chemical properties, inevitably introduces some bias into the results. The properties assigned to the groups of pathways must be carefully considered in order to ensure that the bias is conservative (while, if possible, avoiding excessive conservatism). As illustrated schematically in Figure 3, the bias introduced by a coarse grouping may nevertheless be acceptably low because of the phenomenon of radioactive decay. Due to decay during transport, the slower (low transmissivity) pathways tend to be of low importance with respect to radionuclide release to the biosphere and a large number of such pathways may either be grouped together, or neglected entirely in transport calculations (even if they are significant in the calculation of flow). Conversely, it is the relatively small number of fast (high transmissivity) pathways that tend to dominate release to the biosphere and these must be represented with greater differentiation and accuracy. The fastest pathways may also have a low spatial frequency of occurrence, but, as discussed below in section 4, this frequency of occurrence may be difficult to assess.

In terms of the properties that affect radionuclide transport times along the paths, the greatest uncertainties are generally those associated with the evaluation of groundwater flux and flow-wetted surface area. Figure 4, taken from SKI's SITE-94 performance assessment [1], shows estimates of the groundwater flux on the scale of single canister deposition holes in a hypothetical

Swedish repository, based on different conceptual models of groundwater flow, that range over several orders of magnitude. The flow-wetted surface area may be similarly uncertain. Figure 5, taken from Nagra's Kristallin-I performance assessment [6], shows bore-core observations, and a simplified conceptualisation for transport modelling, of cataclastic zones - one of the types of feature that convey groundwater through the crystalline basement of Northern Switzerland. In Kristallin-I, the flow-wetted surface area is estimated from the observed spatial frequency and width of open channels within fractures, both of which show considerable variability. The transport time along a path is a function of the ratio of flow-wetted surface to groundwater flux. As shown in Figure 6, which is taken from SITE 94, the current uncertainty in this ratio is such that the calculated performance of the geosphere transport barrier ranges from "good" (substantial attenuation of near-field releases during geosphere transport) to "poor" (little substantial attenuation of near-field releases during geosphere transport). A similar conclusion is drawn in Kristallin-I, which results in an emphasis on the performance of the engineered barriers at the current stage of site characterisation.

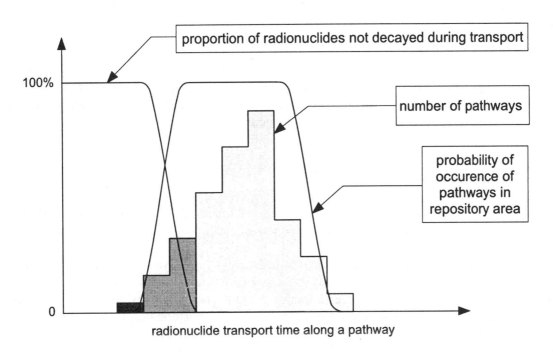

group	radionuclide transport time / half life	probability of occurence	relevance to flow	relevance to transport
■	low (highest transmissivity)	low	low	high - if paths exist
▨	similar (high transmissivity)	moderate - high	moderate - high	moderate - high
☐	high (low transmissivity)	high	high	low - due to decay

Figure 3. **Schematic illustration of the grouping of transport pathways (3 groups in this example) in order to facilitate transport modelling.**

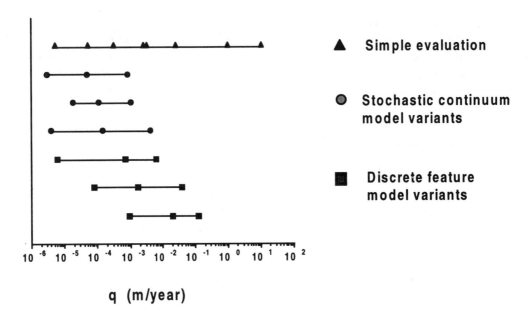

Figure 4. Predicted ranges of near-field flux, q, based on (i) a simple-evaluation model, (ii), stochastic continuum model variants and (iii), discrete-feature model variants, from the SITE-94 performance assessment

5. Discussion and Conclusions

This paper has proposed a pragmatic division of structural features within a repository host rock according to their relevance to repository design and safety.

Whereas major structural features (the feasibility- and layout-determining features) are relatively simple to detect and characterise, greater difficulties are faced during site characterisation in determining the properties of the features that give rise to the most transmissive pathways within the undisturbed host rock and that, to a large extent, control the performance that can be attributed to geological barrier (the safety-determining features). Problems arise in that:

- the features may be aligned such that they are not easily detected by boreholes;

- it is difficult to determine the extent to which any features that are detected are interconnected (it can, however, be conservatively assumed that all detected channels form a connected network);

- highly transmissive channels may not always be associated with particular geological structures.

This paper has also discussed the characterisation of spatial variability and the models of groundwater and transport that are used to assess the consequences of this variability in performance assessment.

It may be that, even at a late stage within site characterisation (e.g. during the operation of a rock laboratory), the location and properties of the features can only be defined in a statistical sense – i.e. in terms of probability distributions. From the point of view of performance assessment, it is the high-consequence tails of these distributions that are of key importance and a priority is to define "cut-off" values so that inappropriate parameter values, that would otherwise dominate the consequence analysis, are not sampled by the performance assessor. It is necessary to draw upon the basic geological understanding of the host-rock type to fill gaps left by site characterisation and to set "cut-offs" and a concern of repository programmes is to confirm that basic geological understanding is adequate for this task, both for relatively homogeneous sedimentary rocks and, more problematically, for highly heterogeneous and fractured hard rocks.

Acknowledgements

The authors would like to thank Dr. J. Schneider (Nagra) for useful technical discussions and would also like to acknowledge Drs. M. Mazurek (University of Bern) and A. Gautschi (Nagra), who provided Figure 5 of this paper.

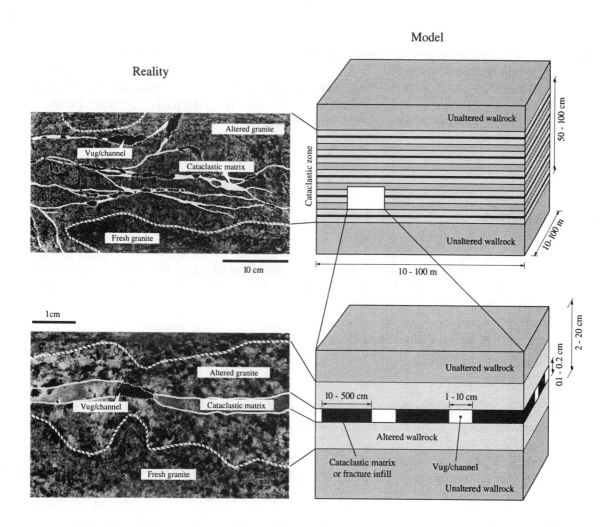

Figure 5. **Cataclastic zones in the crystalline basement of Northern Switzerland, as observed in bore cores and as simplified for the purposes of transport modelling**

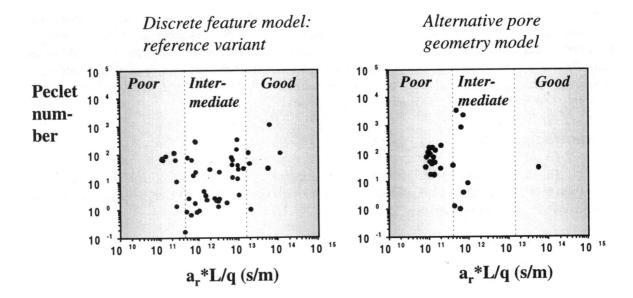

Figure 6. The "geosphere retardation potential" - i.e. the product of flow-wetted surface, a_r, and transport-path length, L, divided by near-field flowrate, q, for different canister positions in the hypothetical SITE-94 repository. Values are derived using a discrete-feature hydrogeological model, with two alternative concepts for the internal structure of water-conducting features.

References

[1] SKI, SITE 94, Deep Repository Performance Assessment Project, SKI Report 96:36, Vols. 1&2, Swedish Nuclear Power Inspectorate, Stockholm, 1996.

[2] Geier, J. E. and Thomas, A. L., Discrete-feature Modelling of the Äspö Site: 1. Discrete-fracture Network Models for the Repository Scale (SITE-94), SKI Report 96:5, Swedish Nuclear Power Inspectorate, Stockholm, 1996.

[3] Geier, J. E., Discrete-feature Modelling of the Äspö Site: 2. Development of the Integrated Site-scale Model (SITE-94), SKI Report 96:6, Swedish Nuclear Power Inspectorate, Stockholm, 1996.

[4] Tsang, Y. W., Stochastic Continuum Hydrological Model of Äspö for the SITE-94 Performance Assessment Project, SKI Report 96:9, Swedish Nuclear Power Inspectorate, Stockholm, 1996.

[5] Geier, J. E., Discrete-feature Modelling of the Äspö Site: 3. Predictions of Hydrogeological Parameters for Performance Assessment (SITE-94), SKI Report 96:7, Swedish Nuclear Power Inspectorate, Stockholm, 1996.

[6] NAGRA, Kristallin-I Safety Assessment Report, Technical Report Series NTB 93-22, Nagra, Wettingen, Switzerland, 1994.

An Example of a Comprehensive Treatment of Spatial Variability in a Performance Assessment: Nirex 97

C P Jackson
AEA Technology, UK

Abstract

The treatment of spatial variability and uncertainty adopted for the Nirex 97 performance assessment for a repository at Sellafield is discussed. The geosphere was classified into hydrogeological units. Within each unit, the variability of the hydrogeological properties on all length scales was recognised and appropriate conceptual models for the variability, based on geological understanding, were developed. On the basis of these models, a process of upscaling was undertaken to derive appropriate effective parameters for each hydrogeological unit suitable to represent the groundwater flow system on different length scales. The aim was to develop a consistent picture of the system across all length scales, and ultimately to derive effective parameters for regional models.

The process is discussed for three of the units at Sellafield: the Quaternary, the St Bees Sandstone and the Borrowdale Volcanic Group (BVG, the potential repository host rock). The Quaternary was subdivided into several different hydrogeological domains. Each domain has a layered structure, possibly with 'windows' through low permeability layers. The St Bees Sandstone is predominantly a fluvial channel sandstone. The deposits within a channel have structure, tending to fine upwards, and later channels have tended to erode earlier channels. In the BVG, groundwater is considered to flow mainly in a subset of the total set of discontinuities, which can be identified in cores by the presence of recent (in geological terms) calcite. In boreholes, these features are strongly clustered, and the clusters appear to have hydrogeological significance. A key question is then whether the clusters themselves are connected. In order to address the uncertainty about the connectivity of the clusters, effective parameters were determined for unconnected and connected models, and the uncertainty about the connectivity incorporated into the overall uncertainty.

Simple analytical models (models in which the effective parameters can be expressed in terms of the parameters using simple formulae) that represent the main features of the conceptual models for each unit were used to calculate the effective parameters (and associated uncertainties). This helped to provide understanding of the main factors affecting each effective parameter. Upscaling was generally undertaken in several steps, addressing variability on successively larger length scales. A pragmatic subdivision of the variability into three broad ranges of length scale (small, medium and large) was adopted. Large-length-scale variability was effectively included in the uncertainty. Approximations in the simple models were also taken into account in the uncertainty.

The work described has shown that heterogeneity can be systematically taken into account in calculations for a performance assessment. The systematic treatment of uncertainty enables the key uncertainties to be identified to serve as a potential focus for future experimental work.

1. Introduction

Nirex has been undertaking an assessment (Nirex 97) of the performance of a repository at Sellafield, updating the preliminary Nirex 95 assessment [1]. In this paper, the treatment of spatial variability and uncertainty adopted for Nirex 97 is discussed. The treatment is described in detail in Reference [2].

Sellafield is near the North West coast of England, on the margin of the East Irish Sea Basin. The potential repository host rock was the Borrowdale Volcanic Group (BVG) which is of Ordovician age. In the Potential Repository Zone (PRZ) the BVG comprises mainly welded tuffs, which have been interpreted as ignimbrites. In the vicinity of the PRZ, the BVG is overlain by Carboniferous and Permo-Triassic sedimentary rocks, but it outcrops a few kilometres to the North East of the PRZ. At the base of the Permian rocks is a breccia (the Brockram) which interdigitates with shales and evaporites. These are overlain by a predominantly fluvial sandstone (the St Bees Sandstone), which in turn is overlain by predominantly aeolian sandstones (the Calder and Ormskirk Sandstones). Over most of the site, the sandstones are overlain by a thin, highly-variable Quaternary cover.

An extensive investigation of a region around the PRZ, focused on the PRZ itself, has been carried out over many years, using both surface geophysical techniques and a number of deep boreholes. The available data [3] include the interpretation of the geological structure (based mainly on seismic surveys) and measurements of hydrogeological properties on a range of length scales. The hydrogeological measurements range from porosity and permeability measurements on core samples (with a length scale of a few centimetres) through single borehole tests on intervals with lengths of the order of tens of metres and with time scales of order an hour through to larger length-scale inter-borehole tests for periods of order months.

The starting point for the Nirex 97 assessment was a conceptual model of groundwater flow in the region of Sellafield [3]. The conceptual model was developed from interpretation of the available data. It identifies the driving forces for groundwater flow at the site and provides a framework for characterising the hydrogeological behaviour of the rocks there. At Sellafield, one of the main driving forces is the so-called 'topographic driving force' resulting from gravity acting on an elevated water table, and groundwater flow is strongly coupled to transport of salinity and heat, particularly the former.

At the highest level of description, the variability in the hydrogeological properties of the rocks was characterised by classifying the rocks in terms of a number of hydrogeological units, within each of which the hydrogeological properties are sufficiently similar to justify the use of the term 'unit', and are sufficiently different from the properties of the other hydrogeological units that the units can be distinguished. The conceptual model identifies the nature of the dominant flow channels in each hydrogeological unit. The hydrogeological units were closely related to the geological structure of the site. This provided a means whereby geological understanding could be taken into account in the characterisation of the hydrogeological properties.

2. Upscaling

Within each of the units the variability of the hydrogeological properties on all length scales was recognised and appropriate conceptual models for the variability, based on geological understanding, were developed. On the basis of these conceptual models and the available data, a process of upscaling was undertaken to derive appropriate effective parameters for each hydrogeological unit suitable to represent the groundwater flow system on different length scales. The aim was to develop a consistent picture of the system across all length scales, in which the variability on each length scale was identified and related to data, and ultimately to derive effective parameters, and associated uncertainties, for use in regional-scale models. A consistent picture can only be achieved by relating the experimental observations on all length scales to the variability on the smallest length scale.

The variability was addressed on three broad ranges of length scales: small (lengths of order centimetres to metres), medium (lengths of order metres to hundreds of metres) and large (lengths of order hundreds of metres to tens of kilometres). This subdivision was a pragmatic treatment of the available data, and reflects the fact that the length scales of geological features at the site tend to fall into one of the categories. For example, the channels in the St Bees Sandstone, which are discussed below, are medium length scale. It is considered that the overall results of the analyses would not be greatly changed by a more refined subdivision.

It is generally accepted that variability on a significantly smaller scale than the scale represented in a model effectively averages out to give appropriate effective parameters. It must be stressed that the averaging process is in general not a simple arithmetic average, but depends on the organisation of the properties within the domain, so that different ways in which the same overall distributions of permeability or porosity are organised within a domain can lead to very different effective parameters. The effective parameters on a given scale are not necessarily homogeneous, but have the variability appropriate to the scale. There are uncertainties associated with the effective parameters. These uncertainties depend on the available data.

Most of the upscaling calculations were made using simple analytical models, that is models in which the quantities of interest are approximated by simple analytical expressions, usually simple algebraic formulae in the parameters of the models. The relevant parameters are the parameters that characterise the variability on the various length scales. The approach of using simple analytical models is very powerful:

(a)   It enabled approximations to the effective parameters to be evaluated fairly rapidly;

(b)   It helped to provide an understanding of the most important factors affecting each effective parameter, and of potential correlations between effective parameters. This helped to identify key parameters. More detailed numerical studies to support the simple analytical models could then be undertaken.

(c)   It enabled the uncertainties in the calculated effective parameters to be estimated in a practicable way. This was done by propagating the uncertainties in the parameters that characterise the variability through the calculations of the effective parameters, using the First-Order Second-Moment Method [4].

The fact that approximations were made was explicitly taken into account by including a 'modelling uncertainty factor' in each analysis. Realistic estimates of the uncertainties in the modelling uncertainty factors, that is the errors introduced by the approximations, were made. These uncertainties were then taken into account in the analyses leading to the uncertainties in the calculated effective parameters. This approach of formally recognising the uncertainties introduced by the approximations made has two benefits. It leads to more realistic estimates of the uncertainties in the effective parameters, and it enables the relative benefits for reducing uncertainty of gathering more data and of undertaking more detailed analyses to be assessed.

The hierarchy of models used for upscaling is shown in Figure 1. The process of upscaling is illustrated by three examples: the Quaternary, the St Bees Sandstone and the BVG.

3.   Quaternary

On the basis of analysis of borehole logs, the Quaternary was subdivided into a number of hydrogeological domains that have different vertical lithological profiles [3]. Each domain consists of a layered sequence of materials with different permeabilities. In some cases, there are effectively 'windows' through low permeability layers. Cross sections through some of the domains are shown in Figure 2.

The models used to upscale the permeability of the various hydrogeological domains were based on the models shown in Figure 3, which capture the main features of the domains (the layering and the possible windows). As for all the hydrogeological units, the upscaling was undertaken in several steps, addressing variability on successively larger length scales. The first step was to upscale from the small length scale to the length scale of the thickness of the layers, which is at the lower end of the medium length scale. In this step, the variability on the small length scale within a single layer, which was taken to be unstructured, was addressed. The next step was to take the layering into account. The third step was to upscale to the large length scale. In this step, the component of variability on the medium length scale was addressed. This leads to effective permeabilities that still have variability on the large length scale. In the regional-scale groundwater flow and transport calculations carried out for Nirex 97, homogeneous effective parameters were used for each hydrogeological unit, and the large-length-scale variability was combined with the uncertainty. This tends to slightly overestimate the uncertainty because, in calculations in which the variability on the large length scale was explicitly represented, there would effectively be some averaging over the variability on the large length scale.

The end result of the upscaling for the hydrogeological domains represented by the model of Figure 3(a) is that the effective permeabilities in the directions parallel and normal to the surface are

$$k_{ea} = k_{eb} = C_a \frac{\sum_i \hat{t}_i \hat{k}_i C_{hsi} 10^{x_{li}}}{\sum_i \hat{t}_i} C_{hma} \quad , \quad k_{ec} = C_c \frac{\sum_i \hat{t}_i}{\sum_i \dfrac{\hat{t}_i}{\hat{k}_i C_{hsi} 10^{x_{li}}}} C_{hmc}$$

where

a, b, c are coordinates with c normal to the surface;

$\wedge$ denotes geometric mean;

k_i is the permeability of layer i;

t_i is the thickness of layer i;

C_{hsi} is a factor that represents the effects of upscaling small-length-scale heterogeneity;

C_{hma}, C_{hmc} (which may be different) are factors that represent the effects of upscaling medium-length-scale heterogeneity;

x_{li} is the component of variability on the large length scale for the logarithm of the permeability of layer i;

C_a, C_c (which may be different) are modelling uncertainty factors (which were judged to be between 0.5 and 2).

The heterogeneity factors C_{hsi} depend on the variances σ_{si}^2 that characterise the variability of the logarithm of the permeability of layer i on the small length scale ($C_{hsi} = 10^{\frac{1}{6}\sigma_{si}^2 \log_e 10}$). Similarly, C_{hma} and C_{hmc} depend on the variance σ_m^2 that characterises the variability on the medium length scale for the logarithm of the permeability. They also reflect the fact that the Quaternary is a relatively thin layer compared to its lateral extent. Thus parallel to the surface, the Quaternary is effectively two-dimensional on the medium length scale, which leads to $C_{hma} = 1$. In the direction normal to the surface, the effective permeability is given by the arithmetic average of the effective permeabilities of blocks with dimensions comparable to the thickness of the Quaternary, which leads to $C_{hmc} = 10^{\frac{1}{2}\sigma_m^2 \log_e 10}$. Thus, if the variability on the medium length scale is large, the effective permeability in the normal direction is significantly greater than the geometric mean, as result of the presence of local regions with above average effective permeability on the medium length scale.

A similar, but more complicated, analysis was carried out for the hydrogeological units represented by the model shown in Figure 3(b).

4.   St Bees Sandstone

The St Bees Sandstone is predominantly a channel sandstone, although there is a basal layer comprising laterally extensive sheetflood sandstones and siltstones interbedded with mudstones [3]. The channels are sinuous and braided. An important aspect of the system is that later channels have eroded into earlier channels, so that the present structure of the channel sandstone consists of the preserved remains of the original channels as shown in Figure 4(a). Within a channel the deposits tend to fine upwards, so that a simplified generic description of the original structure within a channel might be as shown in Figure 4(b). At the base of the channel is a discontinuous basal lag containing mudstone clasts. Above this are trough- and planar-cross-bedded fine- to coarse-grained sandstones that form the bulk of the channel. There may be a layer of ripple-laminated fine-grained sandstone above this and even a mudstone plug above this. Not all of the upper parts of this structure are preserved in many channels. This is a simplified idealisation. In practice, more complicated structures will be present.

This structure forms the framework for characterising the variability of the hydrogeological properties. There is variability on the small length scale within each of the sub-facies within a channel discussed above, and variability between channels. Overall, the view of experts on sandstone architecture is that most of the variability is internal to a channel, rather than between channels.

The effective parameters were calculated using a simplified generic model of a channel sandstone that represented the main features described above. Figure 5(a) shows the generic model of a single channel, and Figure 5(b) shows the way the generic channels fit together. As for the Quaternary, the permeability was upscaled to the regional-scale in several steps. The permeability was upscaled from the core scale to the scale of a sub-facies within a channel; then the sub-facies within a channel were combined to determine the effective parameters of a channel using simple models of networks of 'flow resistances' as shown in Figure 5(c); and then the permeability was upscaled to the regional scale.

The results of the calculations showed that the system behaves in a very similar manner to a simple layered model, even though the low permeability channel plugs and lags are not continuous. The plugs and lags are sufficiently thin relative to their lateral extent and their permeability is close enough to that of the cross-bedded sandstones, which form of the bulk of the channels, that it is easier for groundwater to flow through them than around them.

The simple network calculations were supported by detailed finite-element calculations for the generic channel. Calculations in which the heterogeneity in each facies within a channel and between the channels is explicitly represented have also been undertaken within the Nirex programme, and are reported in another presentation to GEOTRAP [5].

In the near-surface sandstones, open near-horizontal bedding-plane features, possibly resulting from stress relief are considered to play an important role. The effects of these features were taken into account using a simple model.

5. BVG

In the BVG, groundwater is considered to flow mainly in discrete features. Recent work [3] has shown that the flow is predominantly in a subset of the total set of discontinuities. This subset, the Flowing Features (FFs), can be identified in cores by the nature of their surface mineralogy - essentially by the presence of calcite of a certain crystal form that is considered to be precipitating from recent (in geological terms) groundwaters - combined with open porosity. The corresponding features in the boreholes have been classified as Potential Flowing Features (PFFs). The PFFs in boreholes are strongly clustered, implying that the FFs are clustered, at least local to the boreholes.

Measurements of the hydrogeological properties of the BVG have been made on a range of length scales. Core measurements have been made. Borehole tests have also been carried out for intervals of order 1m, 20m and 50m in length. The measurements are consistent with the view that FFs and especially FF clusters are important hydrogeologically, with FF clusters being associated with high effective permeability in borehole tests. Figure 6 presents a comparison of the effective hydraulic conductivity derived from Environmental Pressure Measurements (EPMs) for intervals (of length about 50m) crossed by identified PFF clusters and for intervals crossed only by background PFFs that do not form part of PFF clusters. The effective permeability of intervals crossed by PFF clusters is generally higher than the effective permeability of intervals crossed only by background PFFs, although the distributions have significant overlap.

A key question is then how the FF clusters are connected. It was considered that a range of models is possible (see Figure 7) from clusters that form isolated 'blobs' (which are nevertheless connected by background FFs) to clusters that form connected 'sheets'. Connected 'tubes' of FF clusters are also possible. These form different conceptual models for the BVG.

The available data were analysed to determine the distribution of transmissivity of FF clusters (with each cluster considered as a single feature) and the distribution of transmissivity of background FFs. As part of the analysis, the large-length-scale variability was also evaluated. The distribution of the orientation of the FFs was determined by analysis of the borehole data. The FF clusters were taken to have the same orientation distribution. The distribution of the length of the FF clusters and the distribution of the length of background FFs were taken to be truncated power-law (fractal) distributions based on the observed distribution of the lengths of features in the BVG on different length scales. (These observations are mainly for surface lineaments.)

The effective hydrogeological parameters of the BVG were then calculated for the two extreme conceptual models: the model in which the FF clusters form a connected network in their own right; and the model in which the FF clusters are not connected and the large-length-scale flow is predominantly through the background FFs.

These calculations were initially carried out using the following very simple analytical model for the effective permeability in three orthogonal directions:

$$k_{ei} = \frac{C_{Gi} G_i \hat{T}}{b}$$

where

b is the average spacing of the transmissive features (for convenience, as measured in a vertical borehole);

$\hat{T}$ is the geometric mean transmissivity of the transmissive features;

G_i is a geometric factor, which accounts for the orientation of the features relative to the direction of interest;

C_{Gi} is a factor that represents the combined effects of modelling uncertainty and the connectivity and variability of the features.

This expression was derived by considering a network of several sets of infinitely long fractures with constant transmissivity. The transmissivity within a feature was upscaled to give the effective transmissivity of the feature and then the features were combined. The values of C_{Gi} were judged to be within half an order of magnitude of 1 on the basis of previous experience of the results of calculations for numerical fracture network models.

The upscaling analysis was supported by numerical calculations of the large-length-scale properties. Realisations of networks of FF clusters and networks of background FFs in suitably large cubes were generated. The individual features in the networks were tessellated into small domains within which the transmissivity was taken to be constant. The effective large-length-scale permeabilities obtained from the numerical calculations were in good agreement with the values calculated using the simple analytical formula. This provides support for the simple analytical model. The results of the numerical calculations were then used to calibrate C_{Gi}.

As discussed above, there is uncertainty about whether or not the FF clusters form a connected network in their own right. Therefore, an overall estimate of the uncertainty about the effective regional-scale permeability of the BVG was obtained by combining the uncertainties for the two extreme conceptual models considered above with the uncertainty about the conceptual model. It was found that the main contribution to the uncertainty about the effective permeability of the BVG is the uncertainty about the conceptual model, and the next most important contribution comes from large-length-scale variability.

6. Discussion

For Nirex 97, effective parameters were calculated for regional-scale models in which homogeneous effective parameters are used for each hydrogeological unit. Local-scale variability of the BVG was represented in calibrated models. The effect of the small-length-scale variability not explicitly represented in the models was taken into account through dispersion in the radionuclide transport calculations, both in detailed multi-dimensional calculations and in one-dimensional PSA calculations. It was anticipated that in future studies, some calculations would be undertaken in which heterogeneity within the hydrogeological units would be explicitly represented. The analyses discussed above also provided the parameters required for such models.

The models used represented a combination of structured and unstructured variability as appropriate to each hydrogeological unit. The work described has shown that heterogeneity can be systematically taken into account in calculations for a performance assessment. The systematic treatment of uncertainty enables the key uncertainties to be identified to serve as a potential focus for future experimental work.

Acknowledgment

This work was funded by United Kingdom Nirex Limited.

References

1.   Nirex 95 A Preliminary Analysis of the Groundwater Pathway for a Deep Repository at Sellafield, Nirex Science Report S/95/012 (3 Volumes), 1995.

2.   An Assessment of the Post-closure Performance of a Deep Waste Repository at Sellafield. Volume 2: Hydrogeological Model Development - Effective Parameters and Calibration, Nirex Science Report S/97/012, 1997.

3.   An Assessment of the Post-closure Performance of a Deep Waste Repository at Sellafield. Volume 1: Hydrogeological Model Development - Conceptual Basis and Data, Nirex Science Report S/97/012, 1997.

4.   Dettinger M.D., and Wilson J.L., First-order Analysis of Uncertainty in Numerical Models of Groundwater Flow. 1 Mathematical Development, Water Resour. Res. 17, 149-161, 1981.

5.   Cliffe K.A., Franklin D.J., McLeod E.J. Jones P.I.R. and Porter J.D., The Geological Basis and the Representation of Spatial Variability in Sedimentary Heterogeneous Media, Proceedings of the second GEOTRAP workshop, "Modelling the Effects of Spatial Variability on Radionuclide Migration", Paris, France June '97, OECD/NEA, 1998 (This volume).

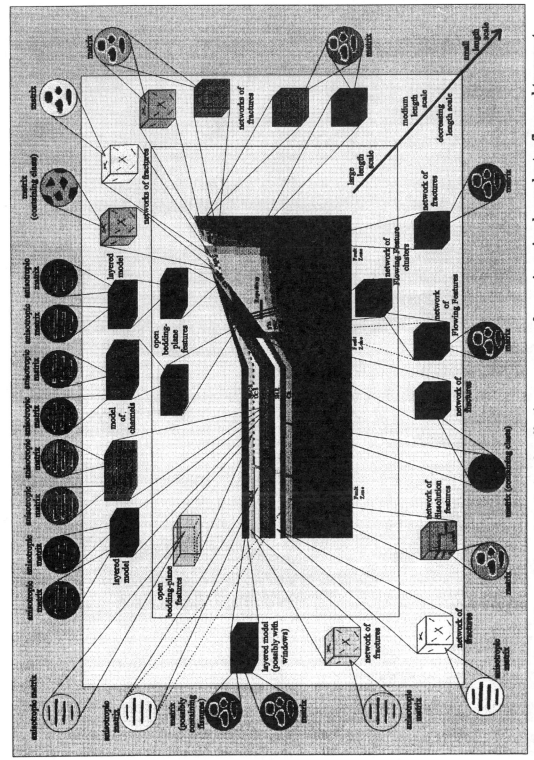

Figure 1. Overview of the models used do derive the effective parameters for use in regional goundwater flow and transport modelling. At the centre the figure shows schematically the organisation of the hydrogeological units at Sellafield. Reading outwards the figure shows the models used to represent the large-length-scale, medium-length-scale and small-length-scale features of the conceptual model for flow in each unit.

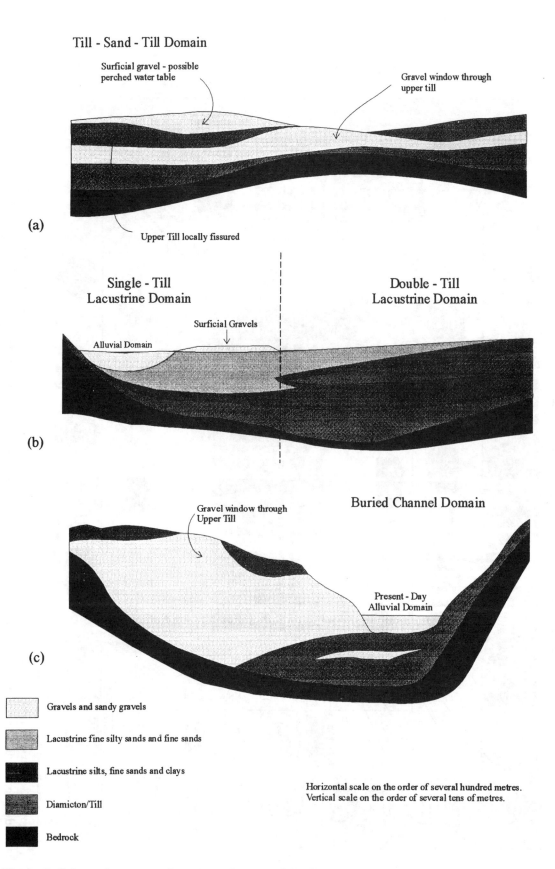

Figure 2. Schematic cross sections through some of the Quarternary hydrogeological domains.

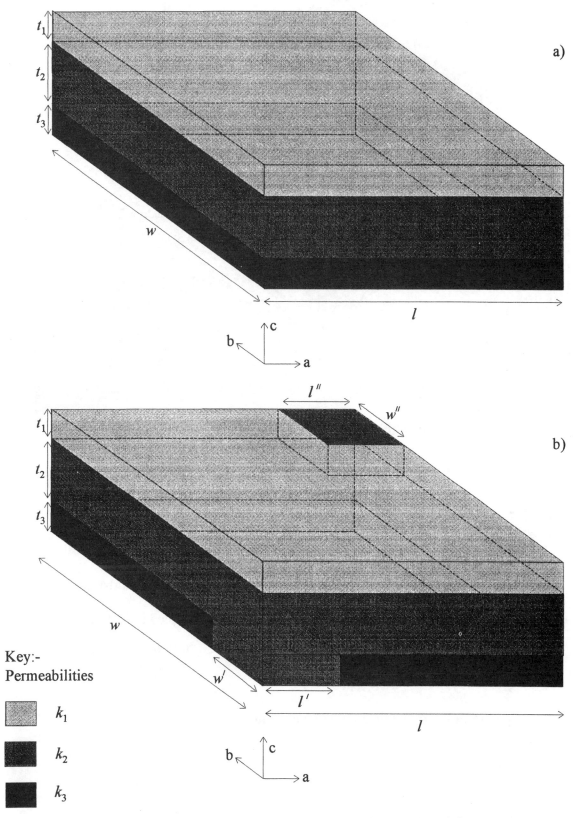

Figure 3. The simple models used for upscaling the hydrogeological properties of the Quartenary hydrogeological domains: (a) Simple alyered models (b) Model with windows through low permeability layers.

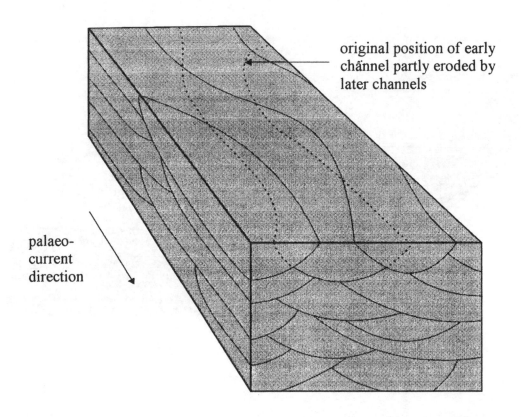

Structure within channel sandstone (St. Bees Sandstone)

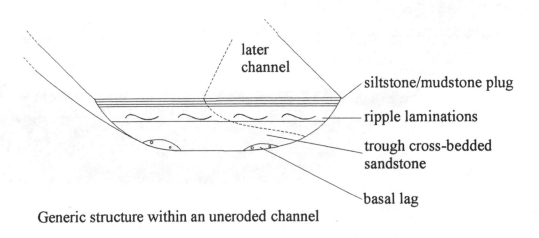

Generic structure within an uneroded channel

Figure 4. Schematic representation of the sandstone channel architecture and structure within a channel.

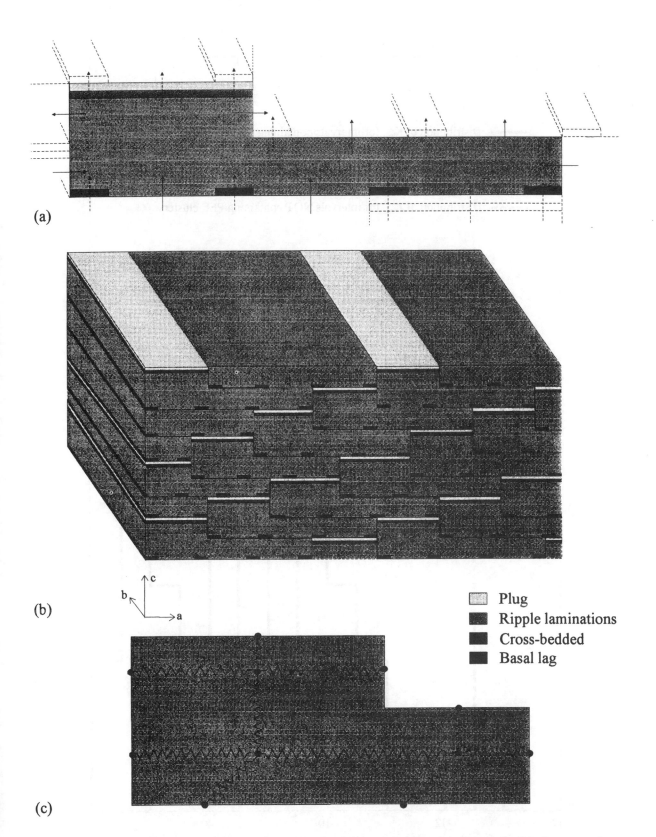

(a)

(b)

▲c
b↖   →a

Plug
Ripple laminations
Cross-bedded
Basal lag

(c)

Figure 5. A simplified generic model of a channel : (a) a single channel; (b) the way the channels fit together to form the overall representation of the channel Sandstone; (c) An effective `flow resistance' network used for upscaling the permeability normal to bedding.

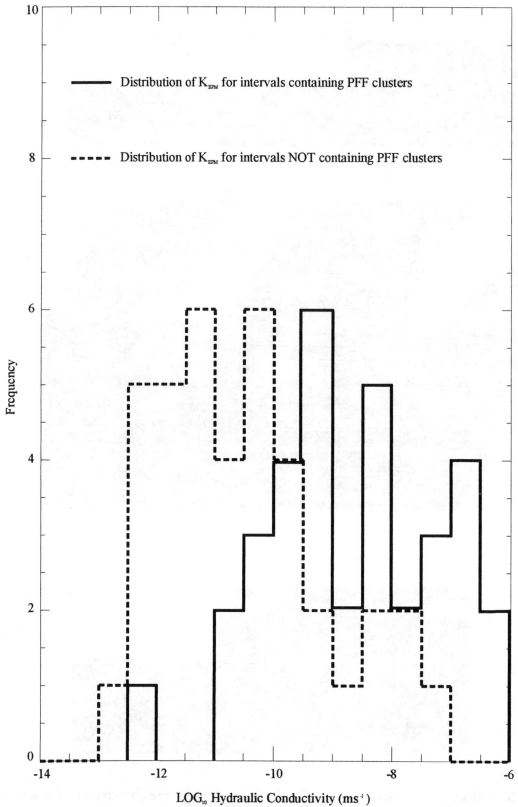

Figure 6. Comparison of the distributions of the logarithm to base 10 of the hydraulic conductivity determined from EPMs for intervals crossed by PFFs clusters and for intervals crossed by only by background PFFS.

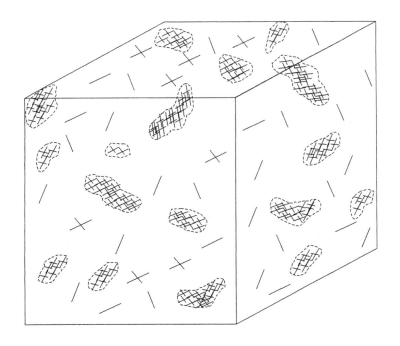

(a)   Unconnected FF clusters

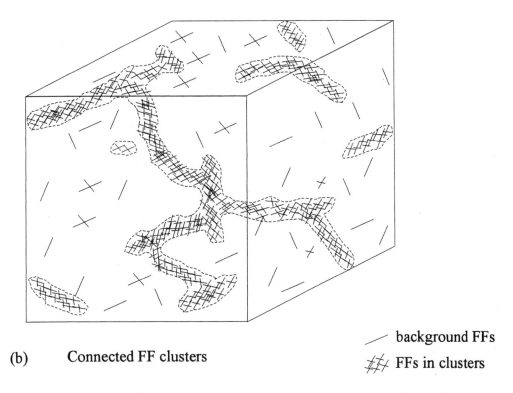

/ background FFs

⌗ FFs in clusters

(b)   Connected FF clusters

Figure 7. Alternative models for the connectivity of FF clusters in the BVG: (a) The 'unconnected' model in which flow is predominantly in the background of PFFs; (b) The 'connected' model in which the FF clusters, considered as features in their own right, form a well-connected network.

SESSION II

Experience with the Modelling of Radionuclide Migration in the Presence of Spatial Variability in Various Geological Environments

Chairmen: E. Ledoux (Ecole des Mines de Paris, France) and
J. Geier (Clearwater Hardrock Consulting/Oregon State Univ., United States)

The Geological Basis and the Representation of Spatial Variability in Fractured Media

Martin Mazurek
Rock/Water Interaction Group (GGWW), University of Bern, Switzerland

Andreas Gautschi & Piet Zuidema
Nagra, Switzerland

Abstract

Spatial variability of features and parameters relevant for contaminant transport modelling occurs on all scales of interest for the quantification of processes that govern solute migration, typically decimeters to hundreds of meters. Two types of spatial variability are distinguished, namely the internal heterogeneity of each individual water-conducting feature (e.g. the complex architecture of a fault) and the larger-scale heterogeneity that results from groundwater flow through different types of water-conducting features along the flowpath from the repository to the discharge areas. An upscaling procedure is required to obtain hydraulic parameters and the properties of the overall flowpath, whereas the heterogeneity of many other geologic features (geometry of flow and matrix porosity, mineralogy, etc.) can be fed directly into coupled codes that quantify radionuclide transport. The procedures needed to derive conceptual models integrating geological and hydraulic field measurements and observations at a given site are illustrated by examples from both crystalline and sedimentary rock formations.

1. TYPES OF SPATIAL VARIABILITY

Many planned repository sites are situated in crystalline rocks, indurated shales or other hard sedimentary rocks. In spite of the diversity of the geologic situation in various geotectonic environments (e.g. pre-Cambrian shield, orogenic belts, sedimentary basins), a large number of structural and hydrogeological properties of the host formations appear to be common to most sites investigated to date. In all these environments, any radionuclides that are released from a repository would be transported through the host rock by advection in discrete water-conducting features, i.e. in fractures which intersect a low-permeability rock body (which itself is accessible by diffusion).

In general, the geologic history of a site is complex and multi-phase, such that water-conducting features with different characteristics were generated and overprinted in the course of different geological events. Thus the present-day situation, which is the basis for the characterization and the prediction of groundwater flow and transport of radionuclides, is the result of the overlay and interference of a number of events of brittle rock deformation with or without ductile deformation precursors. The resulting flowpaths from the repository to the biosphere are heterogeneous, i.e. they have spatially variable characteristics.

Site investigations indicate two types of spatial variability of water-conducting features, relevant both for geologic and for hydraulic characteristics:

- A heterogeneous *internal* structure is characteristic of each water-conducting feature because of the multi-phase history of deformations and mineralizations

- A larger-scale variability is due to the fact that a typical flowpath from the repository to the biosphere is composed of several types of water-conducting features with different properties, e.g. joints, mineralized veins, complex shear-zones, regional faults.

2. HETEROGENEITY WITHIN SINGLE WATER-CONDUCTING FEATURES: EXAMPLE 1 - FAULTS

2.1 STRUCTURAL HETEROGENEITY

At many sites, both in crystalline and sedimentary rock, faults are the dominating type of water-conducting feature. Surface or tunnel observations indicate that each fault is structurally heterogeneous. A fault typically consists of a variable number of discrete shear planes (Figure 1a), with or without fault gouge, surrounded by a zone of damaged rock (increased fracture frequency, see Figure 1c). The character of faults changes along strike over short distances, e.g. evolving from one single fault plane into a complex network of shear surfaces and joints.

This kind of structural heterogeneity is explained by the generally accepted mechanistic principles of brittle fault growth by the linkage of smaller segments (e.g. [1], [2], [3]). Faults with shear displacement in both crystalline and sedimentary rocks preferably develop along arrays of pre-existing discontinuities, such as joints, lithologic boundaries, cleavage planes etc. With increasing shear strain, the individual, initially spatially isolated fault planes grow and eventually link with their neighbours via *fault steps* (Figure 1a,b; extensive discussion and examples in [4]). In three dimensions, faults steps are chimney-like, largely one-dimensional features, and their orientation is a function of the fault orientation and of the deformation mechanism. In the case of normal and thrust

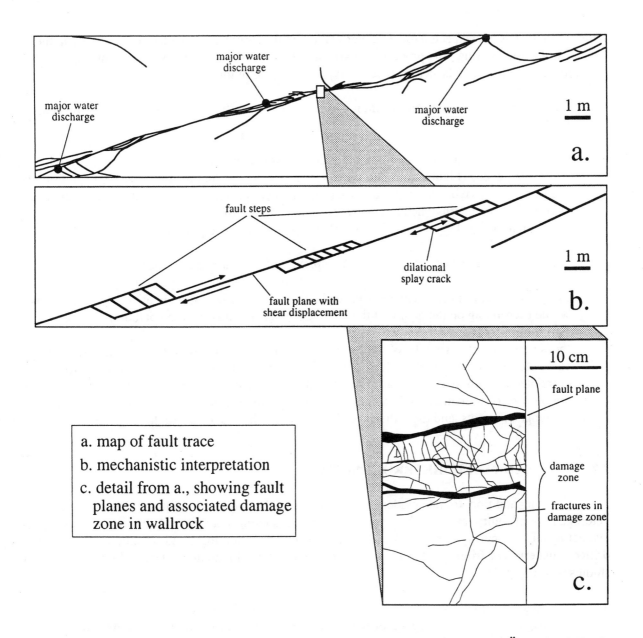

Figure 1.   Anatomy and mechanistic interpretation of a fault in granitic rocks (Äspö Hard-Rock
Laboratory, Sweden), from [4].

faulting, fault steps are expected to be subhorizontal, while subvertical orientations are likely in the case of strike-slip faulting. The process of segment linkage results in geometrically complex structures that contain two basically different types of discontinuities:

-   *Fault planes* accomodate shear strain and so may contain fault rocks, such as fault breccias or fault gouges

-   *Splay cracks* ("dilational jogs") link individual fault planes and are dilational features without a shear component (Figure 1b). The open space created by dilation may represent preferential fluid pathways. Hydrothermal water/rock interactions postdating the deformation events may also affect the internal heterogeneity of faults, either by creating new flow channels (by dissolution) or by clogging existing flowpaths (by mineral precipitation).

The consequences of the natural complexity of fault architecture are that:

-   the point observations obtained from drillcores may not be representative of the fault as a whole (depending on the nature of the fault segment penetrated by the borehole)

-   the structural heterogeneity of faults implies a heterogeneity of transmissivity, resulting in channeled flow within each fault structure. The length scale characterizing this heterogeneity is correlated to fault size. In small faults, significant variations in geometric properties (such as the number of individual fault planes, frequency of splay cracks, etc.) occur over distances in the range of meters, and these distances increase with the size of the faults.

2.2   HYDRAULIC HETEROGENEITY

In addition to the evidence derived from the study of fault architecture, hydraulic heterogeneity of faults (and of other types of water-conducting features) can be studied directly by evaluating and comparing in-situ hydraulic tests (packer tests, fluid logs) and structural evidence from drillcore materials or from structural and hydrogeological tunnel mapping. The results of several case studies (e.g. [4], [5]) show that:

1. At any given depth level below surface, transmissivity of faults can vary considerably, and this is thought to be a consequence of the structural heterogeneity within the faults. For example, this variability is ±0.7 log transmissivity units (1σ standard deviation) in the marls of the Wellenberg (Central Swiss Alps), and is independent of depth at this site.

2. The structural inventory (i.e. the number) of faults exceeds the number of hydraulically identified permeable zones (i.e. inflow points into boreholes). This means that the transmissivity of a part of the structurally identified faults is at least locally (i.e. within the tested rock volume, typically decimeters to dekameters) below the detection limit of the hydraulic investigation method.

3. No geological criteria, such as mineralogy, fault gouge thickness, etc., have been observed that would correlate with the measured transmissivities of the faults. Transmissivity is primarily a function of the fracture-aperture distribution within a fault, but apertures generally cannot be measured in core materials. Core damage due to stress release, drilling operations and sample preparation enhances the in-situ fracture apertures and overprints the natural fracture patterns (typical transmissivities measured at many sites of 10^{-7} - 10^{-12} m²/s correspond, according to

the cubic law ([6]), to in-situ apertures in the range of micrometers). Moreover, in boreholes the scales of observation are different between geology (core diameter, ca. 10 cm) and hydraulics (tested rock volumes with sizes of decimeters to dekameters).

4. Fracture orientation may, at least in some cases, be a parameter that allows a discrimination of water-conducting and tight fractures. Within the total fracture inventory, fractures oriented perpendicularly to the smallest compressive stress trajectory of the present-day stress field tend to be more transmissive than fractures with other orientations (e.g. [4]).

2.3   REPRESENTATION OF INTERNAL FAULT HETEROGENEITY IN FRACTURE NETWORK FLOW MODELS

In fracture network models, faults (and other water-conducting features) are often represented as planes with rectangular or elliptic shapes (e.g. [5], [7]). The last two sections showed that in nature faults are internally hetereogeneous, i.e. each of them contains a number of segments with distinct structural and hydraulic properties. For model calculations, segments are often represented as 1-dimensional tubes (representing channels) or as 2-dimensional areas (patches), depending on the genesis of the transmissive features.

Patches with elevated transmissivities (i.e. above the detection limit of the hydraulic tools) are called *channel patches*, while the remaining part of the fault area is allocated transmissivity values below the respective detection limits. Channel patches are the main conduits for flow, and their proportion within the whole feature area is called the *channeling fraction*. The channeling fraction at a given depth interval can be derived from the ratio of (hydraulically identified) inflow points to the total number of (structurally identified) faults in that interval. In the case of the Wellenberg site, the channeling fraction decreases with depth due to the decreasing trend of measured transmissivities (i.e. more faults have transmissivities below detection at greater depth when compared with shallow occurrences). Typical channeling fractions derived from integrated structural/hydraulic characterization of faults in fractured crystalline and sedimentary rocks in Switzerland ([5], [8]) are in the range 0.1 - 0.5. This means that 5 to 9 out of 10 faults penetrated by the boreholes have a transmissivity below detection in the vicinity of the boreholes (size of the tested volume characterized by the radius of visibility of the hydraulic tests), which does not preclude that they are more transmissive in regions farther away.

The simplest abstraction of the patchy structure of water-conducting features is a chessboard geometry with square patches (Figure 2). The spatial distribution of channel patches within the plane of the water-conducting feature can be chosen as random or, if geologic arguments are available, with some correlation length and anisotropy. Figure 2 (top) shows the trace and the 3-dimensional architecture of a fault mapped on the surface (taken from [5]). If indications exist that transmissivity is enhanced in fault steps that contain splay cracks, channel patches within the fault plane are arranged as shown in Figure 2, bottom left. For the stochastic model realizations, the elongated shape of higher-permeable sections within the fault plane is mathematically simulated by introducing a large correlation length of transmissivity in the direction parallel to the fault steps. If indications of enhanced transmissivity in faults steps or other internal features of faults are lacking, channel patches are distributed randomly in the fault plane, i.e. with a smaller and isotropic correlation length for transmissivity (Figure 2, bottom right).

3.   HETEROGENEITY WITHIN SINGLE WATER-CONDUCTING FEATURES: EXAMPLE 2 - LIMESTONE BEDS IN SHALES AND MARLS

Argillaceous rocks are known to have low permeabilities in many cases (e.g. [9]). Potential flowpaths in such rocks, apart from faults, are limestone beds intercalated with the shales or marls. In the Palfris-Formation at Wellenberg in the Central Swiss Alps, such limestone beds occur in clusters (10 - 20 beds in a sequence of rock 10 - 40 m thick). On a small scale, the limestone beds (each 10 - 100 cm thick) contain veins with conspicuous open druses, i.e. cavities of up to centimeters in aperture (Figure 3a). A part of these druses is interconnected and so constitutes a channel network in which advection may take place. The occurrence of these drusy veins, which result from brittle failure and dilation of the rock, is largely limited to the limestone beds (Figure 3b), whereas the embedding marls accomodate strain by ductile deformation processes. Outcrop evidence suggests that on a scale of meters, the limestone beds are not laterally continuous but dismembered by deformation processes such as boudinage and thrusting (conceptualized in Figure 3c). The space between individual limestone boudins is filled with argillic material, such that the boudins are structurally as well as hydraulically isolated features that may only play a role for water flow when they are cross cut by other water-conducting features, such as faults. On a scale of hundreds of meters, the rock sequences rich in limestone beds are folded and probably further dismembered (Figure 3d).

4.   UPSCALING: FROM SINGLE WATER-CONDUCTING FEATURES TO NETWORKS

Given the length scale of the total flowpath from a repository to the biosphere (typically hundreds of meters), the small-scale geometric and hydraulic characteristics of the host formations derived from borehole information need to be upscaled (see flowchart in Figure 4). The main purposes of the upscaling process include

-    the calculation of hydraulic parameters (such as K_{eff}, the large-scale hydraulic conductivity of the rock formation), and

-    the quantitative description of geological and hydraulic variability of the interconnected network of different water-conducting features on a scale comparable to the migration distance between repository and exfiltration areas (e.g. the relative proportions of different types of water-conducting features in which geosphere transport towards the surface occurs).

Other parameters derived from small-scale investigations are not scale-dependent and so do not require an upscaling process (such as mineralogy and diffusion-accessible wallrock porosity, see below and Figure 4).

4.1   EXAMPLE: WATER-CONDUCTING FEATURES IN FRACTURED ARGILLACEOUS ROCKS

Faults and fractured limestone beds as described above are typical water-conducting features in fractured argillaceous rocks. K_{eff} (large-scale hydraulic conductivity) is simulated by fracture network and flow modelling, which in turn requires information on the transmissivity, length, orientation, frequency and internal heterogeneity of both types water-conducting features.

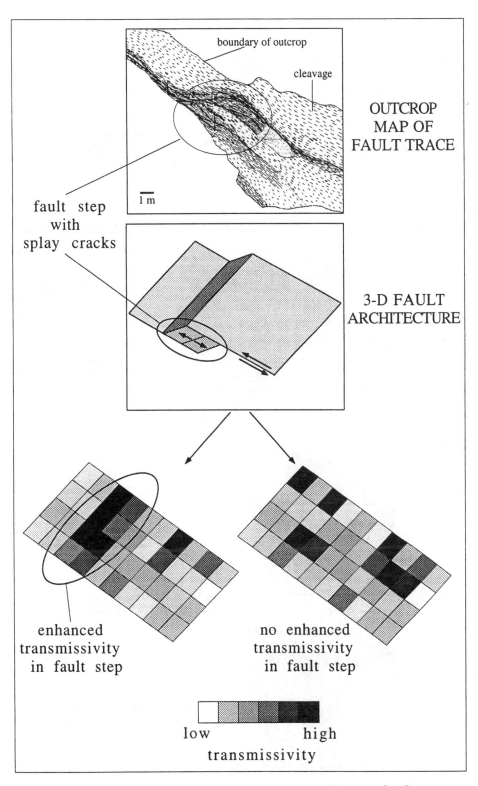

Figure 2. Conceptualization of a fault with complex internal architecture for fracture network flow modelling. **Top:** Fault trace in marls of the Palfris-Formation (Central Swiss Alps). **Centre:** Sketch of the 3-dimensional shape of the fault. **Bottom:** Conceptualization of the fault, accounting for heterogeneity by patches with variable transmissivity.

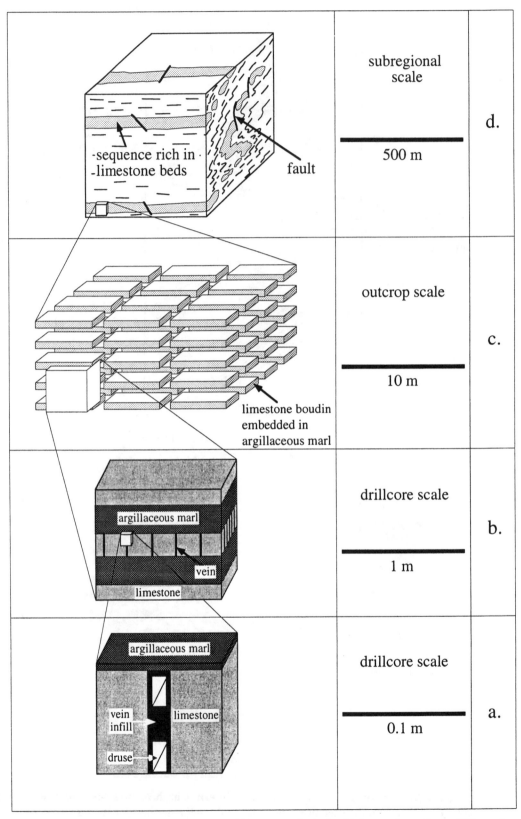

Figure 3. Conceptualization of spatial variability of limestone beds in shales and marls on different scales. Compiled from Nagra (1997).

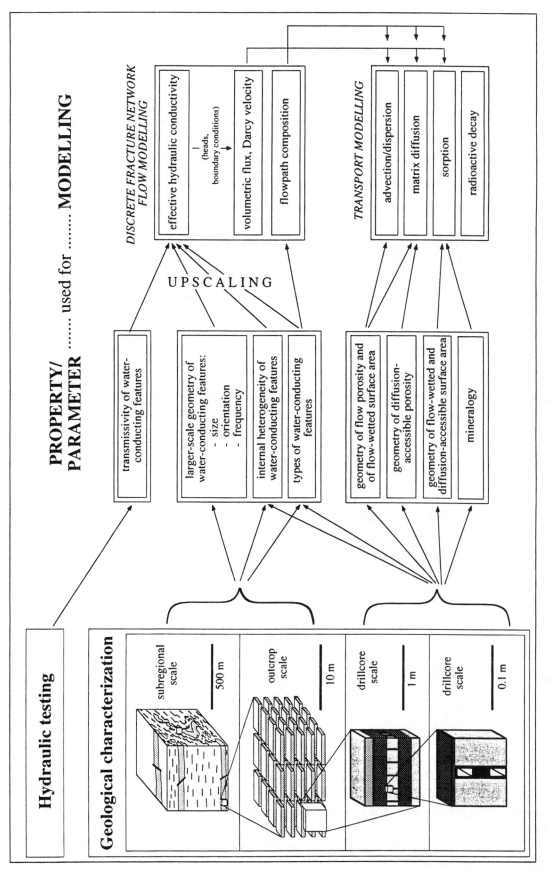

Figure 4. Overview of field-derived parameters used for contaminant transport modelling and flowchart of treatment of data derived on different scales, using limestone beds as an example.

Faults: The least adequately known parameter is the length distribution of the faults and of the individual channel-patch sizes within the faults. In the absence of representative surface outcrops that would allow direct observation, conceptual assumptions must be used, such as possible empirical relationships between fracture thickness (measurable in the drillcores) and fracture length ([10]). The uncertainty is then explored by means of sensitivity analyses, and results of the fracture network models are tested for their consistency with independent evidence, such as hydrochemistry (e.g. average groundwater residence times, flow directions). The conceptualization of the internal heterogeneity of faults for the sake of fracture network modelling is sketched in Figure 2. Transmissivity, frequency and orientation of water-conducting features are parameters that generally are well characterized on the basis of borehole measurements and drillcore logging.

Fractured limestone beds: Figure 3 shows that limestones embedded in argillic rocks are structurally heterogeneous on all scales between centimeters and hundreds of meters. Large-scale properties, i.e. the size, arrangement and clustering of the limestone boudins, are also conceptualized in Figures 3 and 4 on the basis of available information. The boudins, including the whole set of detailed geometric information, can either be incorporated explicitly into fracture network models, or effective parameters are calculated by averaging procedures prior to the fracture network calculations (e.g. [5]).

Together with measured or estimated hydraulic input parameters (transmissivity, boundary conditions of the modelled block, hydraulic heads), the relevant results of fracture network and flow modelling include Darcy velocity, volumetric flux and flowpath composition (i.e. the distribution of flow in a network consisting of different types of water-conducting features), all of which are then fed into contaminant transport models (Figure 4). Most critical for performance assessment is the estimation of the extent and connectivity of water-conducting features with small radionuclide retention potential, for example quartz- or calcite-lined veins, as these tend to dominate resultant calculated doses.

5.   SMALL-SCALE TRANSPORT PROPERTIES OF WATER-CONDUCTING FEATURES

On a small scale (mm - m), the following properties of the rocks are relevant to radionuclide transport and can be measured directly on a small scale, such as in boreholes or on core and groundwater samples:

-      flow-wetted channel-wall surface area per unit volume of flowing water

-      mineralogy and sorption characteristics of fracture infills and wallrocks

-      connected microporosity and matrix diffusion behaviour of fracture infills and wallrocks

-      groundwater compositions.

These properties are not scale-dependent and so do not require an upscaling procedure, and Figure 4 shows that their small-scale heterogeneity is fed directly into transport codes. Codes coupling advection/dispersion with matrix diffusion, sorption and radioactive decay are available and allow explicit representation of the geometric complexity shown in Figure 4 ([11]). Parameter values are reported either as ranges or as average values representative of the block size under consideration (e.g. scale of the total flowpath to the biosphere). The advection velocity of water flowing in the channels can be quantified on the basis of the Darcy velocity (calculated on the basis of fracture network modelling) and on estimations of flow porosity (best derived from hydraulic crosshole tests).

6. REPRESENTATION OF SPATIAL VARIABILITY IN PERFORMANCE ASSESSMENT TRANSPORT MODELS

Two different strategies are used to handle spatial variability in transport models:

- In the simpler approach, the transport properties of all types of water-conducting features are examined separately, and the feature type with the weakest retention performance is selected as the reference case for the entire flowpath. This approach is useful where the geological and hydrogeological database of a site is limited, even though the model results (e.g. rates of radionuclide release from the geosphere) may be over-conservative.

- A more realistic representation of spatial variability, considered in current Nagra projects, is to model the flowpath from a repository to the biosphere as a network of individual water-conducting features with variable properties. Depending on the reliability of data and on the availability of suitable computational tools, such a transport calculation handles either one selected (e.g. the fastest) flowpath, containing a number of water-conducting features in series between source and exfiltration point, or it handles the entire network comprising multiple flowpaths that branch and combine according to rules derived from fracture network models or direct observations and geological concepts. Heterogeneity within single water-conducting features is represented by attributing variable transmissivity values to different segments (patches) within each water-conducting feature. On an even smaller scale (mm - dm), the internal structure of the patches is represented explicitly in the model, including the heterogeneous distribution of different rocks domains (e.g. vug/channel, fracture coating, altered wallrock, unaltered wallrock).

The second approach requires a wealth of input parameters that can only be provided in the advanced stages of site characterization (detailed knowledge of the geologic structure and hydrogeology of the site, best combining information from boreholes, surface outcrops and surface-based geophysical investigation tools). However, this approach has the potential to reduce over-conservatism and represents a step towards reality when compared to simpler modelling concepts.

REFERENCES

[1]   Martel, S.J. & Pollard, D.D., Mechanics of Slip and Fracture Along Small Faults and Simple Strike-Slip Fault Zones in Granitic Rock, J. Geoph. Res., 1989, 94, 9417-9428

[2]   Peacock, D.C.P., Displacement and segment linkage in strike-slip fault zones, J. Struct. Geol. 1991, 13, 721-733

[3]   Peacock, D.C.P. & Sanderson, D.J., Displacement, segment linkage and relay ramps in normal fault zones, J. Struct. Geol. 1991, 13, 1025-1035

[4]   Mazurek, M., Bossart, P. & Eliasson, T., Classification and characterization of water-conducting features at Äspö: Results of investigations on the outcrop scale, SKB International Cooperation Report ICR 97-01, 1996, SKB, Stockholm, Sweden

[5]   Nagra, Geosynthese Wellenberg 1996 - Ergebnisse der Untersuchungsphasen I und II, Nagra Technical Report NTB 96-01, 1997, Nagra, Wettingen, Switzerland

[6]   Domenico, P.A. & Schwartz, F.W., Physical and chemical hydrogeology, John Wiley & Sons, 1990, 824 pp.

[7]   Dershowitz, W., Thomas, A. & Busse, R., Discrete fracture analysis in support of the Äspö Tracer Retention Understanding Experiment (TRUE-1), SKB International Cooperation Report ICR 96-05, 1996, SKB, Stockholm, Sweden

[8]   Thury, M., Gautschi, A., Mazurek, M., Müller, W.H., Naef, H., Pearson, F.J., Vomvoris, S. & Wilson, W.E., Geology and hydrogeology of the crystalline basement of Northern Switzerland - Synthesis of regional investigations 1981 - 1993 within the Nagra radioactive waste disposal programme, Nagra Technical Report NTB 93-01, 1994, Nagra, Wettingen, Switzerland

[9]   Neuzil, C.E., Low fluid pressure within the Pierre Shale: A transient response to erosion, Water Resources Research, 1993, 29, 2007-2020

[10]   Barnett, J.A.M., Mortimer, J., Rippon, J.H., Walsh, J.J. & Watterson, J., Displacement geometry in the volume containing a single normal fault, Amer. Assoc. Petrol. Geol. Bull. 1987, 71, 925-937

[11]   Jakob, A., Development of a new transport model (PICNIC) which accounts for the large- and small-scale heterogeneity, Proc. of "The Basis for Modelling the Effects of Spatial Variability on Radionuclide Migration", GEOTRAP workshop, Paris, France, 9-11 June 1997, OECD/NEA Project on the Transport of Radionuclides in Geologic, Heterogeneous Media (this volume)

The Geological Basis and the Representation of Spatial Variability in Sedimentary Heterogeneous Media

K.A. Cliffe, D.J. Franklin, P.I.R. Jones, E.J. Macleod and J.D. Porter

AEA Technology, United Kingdom

Abstract

This paper describes work that was performed in order to investigate the impact of different conceptual models of the heterogeneity of the Sherwood Sandstone Group (SSG) at Sellafield on calculations of flow and transport. The motivation for this work was the important role played by the SSG in calculations of the performance of a hypothetical repository at Sellafield. Detailed models of the heterogeneity of the Undifferentiated St Bees Sandstone (USBS) of the SSG were produced. The models took into account directly the geological structures at the facies level.

The software package STORM (STOchastic Reservoir Modelling), which was originally designed for modelling heterogeneous oil reservoirs, was used to construct the models. The USBS is a fluvial channel sandstone. The data required by the model are therefore those that characterise the geometry of the channel bodies and the properties of the various subfacies within the channels.

Using the detailed facies-scale models, upscaling calculations were performed in order to calculate the effective permeabilities in three orthogonal directions of larger-scale blocks of USBS. Variograms for the upscaled effective permeabilities were then calculated by averaging over the results from 100 realisations of the facies-scale models. The impact on the upscaled effective permeabilities of different distributions of the total variability in the small-scale permeabilities within the facies bodies was investigated. In one case, it was considered that all of the observed variability could occur within individual channels. In a second case, it was considered that the overall variability mainly arose from variability between individual channels. Intermediate cases were also considered.

It was found that for the case in which all of the variability was within channels, the larger scale permeabilities did not exhibit any significant correlation structure. This is consistent with the fact that the correlation lengths of the facies-scale permeabilities are less than, or comparable to, the block size used for upscaling. In the cases in which all of the variability was between channels, the correlation lengths of the facies-scale permeabilities are comparable with the corresponding channel dimensions. The upscaled effective permeabilities therefore also exhibited correlation lengths that were comparable with the channel dimensions.

Flow and transport calculations were also performed on 90 realisations of a detailed facies-scale three-dimensional representation of a larger block of the USBS. The calculations were performed for the case in which all of the variability was considered to be present within each of the channels. The dispersion associated with transport through the block was investigated using a particle tracking approach. It was found that the mean position in the direction of flow increased approximately linearly with time. The uncertainty in the mean position was small in the directions perpendicular to the flow but was greater in the direction of mean flow. These results are broadly consistent with the analytical results for transport through a random permeability field.

The results illustrate how detailed calculations based on a facies-scale model can provide valuable support for parameterisation of the larger-scale models that are used in a performance assessment. The results indicate the potentially significant role of the facies architecture and demonstrate how detailed calculations can be used to build understanding of the effects of heterogeneity.

1. INTRODUCTION

Nirex has been undertaking an assessment (Nirex 97) of the performance of a repository at Sellafield, updating the preliminary Nirex 95 assessment [1]. This paper describes work, supporting Nirex 97, that was performed in order to investigate the impact of different conceptual models of the heterogeneity of the Sherwood Sandstone Group (SSG) at Sellafield on calculations of flow and transport. At Sellafield the potential repository host rock, the Borrowdale Volcanic Group (BVG), is overlain by a sedimentary sequence that includes the SSG. In the Nirex 95 analysis of repository performance [1], most of the calculated travel time for groundwater moving from the repository to the biosphere was spent in the deep sandstones. The key heterogeneity in determining the uncertainty in the groundwater travel time was therefore that of the sandstones. It was also found that the dispersion that occurs during travel through the deep sandstones was an important determinant of the calculated risk. This dispersion arises from the heterogeneity of the sandstones. Thus, an understanding of the nature of the heterogeneity of the sandstones and its effects on flow and transport are key elements in a performance assessment. This paper describes calculations in which the heterogeneity that arises from the facies architecture of the sandstones was represented in considerable detail. Other potential sources of heterogeneity, such as discrete features, were identified but were not included in the numerical models developed for this study.

The SSG is similar in many ways to formations in which oil and gas are found, indeed the Wytch Farm oil and gas field in the south of England occurs in the SSG. It was therefore considered to be appropriate to make use of the accumulated experience of the hydrocarbon industry in generating models of heterogeneity in oil reservoirs, and apply similar techniques to the SSG at Sellafield. Models of the heterogeneity that take into account explicitly the geological structures at the facies level were used. The software package STORM (STOchastic Reservoir Modelling), which was originally designed for the modelling of heterogeneous oil reservoirs, was used to construct facies-scale models of the heterogeneity of the SSG. The specific aims of this work were:

- to investigate the effect of the distribution of variability at the level of the detailed facies model on the statistical structure of the upscaled effective permeability (the upscaled permeability is heterogeneous and can be characterised in terms of its correlation structure);

- to investigate the transport behaviour obtained from the detailed facies-scale models.

The rest of this paper is organised as follows. In Section 2, the relevant features of the geology of the SSG at Sellafield are briefly described. The conceptual model used to represent the geology is presented in Section 3. In Section 4, the numerical approach used in the study is outlined. In Section 5, the data used to build the stochastic models of the USBS are discussed. The results of the flow and transport calculations are presented in Sections 6 and 7 respectively. Finally, some conclusions are drawn from the study in Section 8.

2. GEOLOGICAL DESCRIPTION

The St Bees Sandstone is a fluvially dominated red bed succession. It has been provisionally subdivided into two members: the North Head Member and an unnamed upper member.

The two members have been differentiated mainly on the basis of geological log data. The St Bees Sandstone is composed of two main facies associations: sheetflood facies association and fluvial channel facies association. The sheetflood facies association only occurs in the lower North Head Member. The term Undifferentiated St Bees Sandstone (USBS) is used to describe the sandstones above the lower North Head Member which are composed of the channel facies association.

The channel fill sequences in the USBS are generally incomplete due to the erosion of the upper channel fill by subsequent channels. At the base of the channels, channel lags may be present. The lags typically have low permeabilities due to the presence of intraformational mud clasts (IMCs). Overlying the lags are the crossbedded sandstones which are the dominant facies identified in the USBS. The top of the channel may be composed of finer grained sand and mudstones which are likely to have lower permeabilities than the crossbedded facies. In this study the impact of this sedimentary structure is examined. The models involved are quite detailed and state-of-the-art for this type of modelling.

3.   CONCEPTUAL GEOLOGICAL MODEL

The first step in the development of a model of the flow and transport behaviour of a geological formation is to develop a conceptual geological model, which is a representation of the actual geological structure that captures all of the significant features of the system. The conceptual model of the USBS adopted for this study consists of a series of channels, cross-cutting and eroding each other, with the result that any given volume of the sandstone is entirely filled with channels. Systems broadly similar to the USBS have been previously modelled in this way [2].

Each channel has a given thickness and a width that can vary along the channel. The width and thickness can vary from channel to channel. The channels have a sinuosity (i.e. they meander) that is characterised by a wavelength and an amplitude. The channels are assumed to have rectangular cross-sections. This choice was made mainly for the convenience of the modelling, though it is thought to be a reasonable representation of reality given that the channels have a fairly large aspect ratio.

The channels themselves are not homogeneous, but have a definite structure that is characterised by a number of subfacies. The conceptual model for the subfacies structure consists of four layers. At the bottom of the channels there are lag deposits. These deposits do not necessarily cover all the base of the channel. They are represented by small cuboid bodies that occupy a specified proportion of the bottom of each channel. The next layer consists of cross-bedded sandstones, which make up the bulk of the channels. Above the cross-bedded sandstones there is a layer of ripple laminated sandstones, and finally plug deposits at the top of the channel. The erosion of older channels by younger ones is modelled by the fact that where two channels overlap, the higher one is assumed to have eroded the lower channel. This model of erosion allows for that fact that the ripple laminated sandstones and the plug deposits are frequently missing from the channels.

4.   NUMERICAL REPRESENTATION OF THE CONCEPTUAL MODEL

The software package STORM [3] was used to generate realisations of the facies structures and associated hydrogeological properties, that were consistent with the conceptual model described in the previous Section.

STORM (Stochastic Reservoir Modelling) is a software package primarily used in the petroleum industry. It consists of several different modules, that are applicable at different stages of the modelling process. Each module can be run independently of the others. The modules that were relevant for this study were the 'facies' module, which is used to generate stochastic realisations of facies distributions within a specified volume, the petrophysics module, which is used to generate stochastic realisations of petrophysical properties, and the upscaling module, which is used to upscale the petrophysical properties from the very fine grids on which they are initially generated to coarser grids practicable for use in flow and transport simulations of larger volumes.

The 'fluvial' program is the part of the facies module which was used for this study. It is designed to model both meandering and braided channel systems, consisting of either isolated channels or channel belts. The fluvial program requires input information on channel types, dimensions and orientations. It is envisaged as part of the conceptual model that the channels are further divided up into subfacies. As part of the work reported here, a new module to assign the four subfacies types discussed in the previous section to a channel body in a STORM model was written and incorporated into STORM.

The petrophysical modelling module is used to assign variable petrophysical properties such as porosity and permeability to each of the previously modelled facies. A multivariate Gaussian field model is defined for each facies. Depending on the structure that is considered appropriate, it is possible either to specify the petrophysical variables as independent for each facies body or to generate parameter values from the same Gaussian field for each facies body. The second approach greatly reduces the time needed for each simulation.

During the petrophysical modelling stage, a value for each of the modelled petrophysical parameters is assigned to each of the small-scale grid cells that are used to represent the facies bodies. It is generally impracticable to carry out flow and transport calculations for regions of the size of interest in assessment calculations using the small-scale mesh, because of the large number of grid cells involved. The upscaling module is used to obtain appropriate parameter values for larger-scale grid cells that can be used in models of flow in larger regions. The appropriate upscaling process depends on the parameter of interest. For parameters such as the porosity, when it is to be used to evaluate the total fluid volume, the upscaling procedure is not complex and may be performed by taking the arithmetic mean of the porosities assigned at the small scale. The upscaling process for permeabilities is more complicated because the requirement is then to reproduce the correct flow behaviour [4]. For this study the 'Perm Rate' method of upscaling implemented in STORM was selected.

5. MODEL CONSTRUCTION

Detailed models of the facies architecture were constructed using STORM. The grid cells were small enough to represent the details of the facies architecture. The size ranges of the facies bodies were based on outcrop and borehole data from the site and previous experience of modelling related systems. Data from outcrop studies, for example, suggested that appropriate mean values for the thickness and width of the channels were 4m and 30m, respectively. In order to resolve all of the features that the model was required to represent (e.g. the lag deposits) a resolution of 0.1m in the vertical (z) direction, 5m in the palaeoflow (x) direction and 2m in the remaining (y) direction, was necessary. An additional channel thickness of 4m at the top and at the base of the grid was required in order to ensure that edge effects associated with the facies program did not affect the simulation.

For transport calculations, the size of the region that is modelled must be large enough to contain several channels, in order to give a reasonable representation of the heterogeneity. The dimensions for the modelled volume were therefore as follows. The width was taken to be ten times the width of a single channel, i.e. 300m. the length was taken to be equal to the mean wavelength assigned to the sinuous channels, i.e. 600m. The depth was taken to be five times the thickness of a channel, i.e. 20m. When the additional channel thicknesses to account for edge effects is included, this dimension becomes 28m. The basic model size of 600m by 300m by 20m, is comparable to the size of a typical grid cell in a regional-scale groundwater flow model and so is an appropriate scale of interest for this study. The complete fine scale model therefore contained 120 x 150 x 280, i.e. 5.04×10^6 cells. This will be referred to as the full model. Even without the cells to take account of the edge effects the model contains 3.6×10^6 cells.

Calculations with the full model were very expensive, requiring significant amounts of CPU time, memory and disk storage. However, the full model was not really required for the flow calculations that were used to calculate the statistical structure of the upscaled permeability values. These calculations were therefore carried out using three smaller models, corresponding to blocks lying across the full model in the x, y and z directions, respectively. The dimensions of these models are as shown in Table 1. For convenience they will be referred to as the x, y and z direction upscaling models.

Table 1: Dimensions of the Upscaling Models

Direction	Mesh Size	Number of Cells	Total Number of Cells
x	600m x 30m x 12m	120 x 15 x 120	216,000
y	30m x 300m x 12m	6 x 150 x 120	108,000
z	30m x 30m x 28m	6 x 15 x 280	25,200

There was no information available on either the spatial correlation structure of the petrophysical parameters or on how much of the variability in the parameters is intra-channel variability and how much is inter-channel variability. At one extreme all the variability could be present in each of the channels (i.e. intra-channel variability). At the other extreme each of the channels could have more or less constant properties and all the observed variability could reflect variations between channels (i.e. inter-channel variability). In view of the lack of detailed information on these issues, seven scoping calculations were performed in order to assess the impact of different distributions on the structure of the upscaled porosities and permeabilities. The seven cases were as follows.

In Case 1, it was assumed that all of the variability in the petrophysical parameters occurs between channels. The variability was taken to be characterised by exponential variograms with ranges of 1200m, 60m and 4m in the x, y and z directions, respectively. The parameters for each facies body were simulated independently by STORM. The effect of the large variogram ranges is to make each the facies derived from a single channel body have nearly constant petrophysical values. The value assigned to each body is drawn from the appropriate Gaussian distribution.

In Cases 2, 3 and 4, it was assumed that all of the variability in the petrophysical parameters occurs within the channels. In Case 2, the variability was taken to be characterised by exponential variograms with ranges of 3m in each direction. In Case 3, the ranges were taken to be 10m, 10m and 3m in the x, y and z directions, respectively. In Case 4, the ranges were taken to be 30m, 10m and 3m in the x, y and z directions, respectively. The range parameters chosen for Case 4 were considered to be the 'best estimate' values for these parameters. In all of these cases, all bodies of the same facies type were simulated together in STORM. This effectively means that each channel samples all of the variability in the system.

In Cases 5, 6 and 7, it was assumed that the variability was a mixture of intra-channel and inter-channel variability. In all three cases, 30% of the variability was assumed to be associated with the inter-channel variability and 70% with the intra-channel variability. The variogram range parameters for Cases 5, 6 and 7 were the same as those for Cases 2, 3 and 4, respectively.

Figures 1 and 2 show for Cases 1 and 4, respectively, the distribution of the porosity on a slice through the x-direction upscaling model and illustrate how the different correlation structures and distributions of the variability affect the spatial variability of the porosity.

Figure 1: **Areal cross section through the x-direction model for Case 1 showing a realisation of the porosity distribution. The purpose of the figure is to show the correlation structure.**

The permeability assigned to the small-scale grid cells was taken to be anisotropic, which is consistent with the data from the site. The two horizontal components were taken to be equal. The vertical component was related to the horizontal by an anisotropy factor which was taken to be a Gaussian random field with the same correlation lengths as the horizontal permeability and a variance derived from site data on the anisotropy.

For each of the x, y and z direction upscaling models, the results obtained for the petrophysical properties in each realisation were upscaled to obtain values of the porosity and the three components of the permeability tensor on larger scale grid blocks. These upscaled parameters are still heterogeneous and there will also be variations in their values from one realisation to another. The x and y direction models were upscaled to coarse grid block sizes of 30m x 30m x 4m. The z direction model was upscaled to a coarse grid block size of 30m x 30m x 2m. This upscaling results in 10 coarse cells per realisation for the y and z direction models and 20 cells per realisation for the x direction models.

118

Figure 2:   **Areal cross section through the x-direction model for Case 4 showing a realisation of the porosity distribution. The purpose of the figure is to show the correlation structure.**

Cases 1 to 7 were run for each of the x, y and z direction upscaling models. Case 4 was run for the full model. Flow calculations were carried out for all the cases but transport calculations were only carried out using the full model for Case 4. For most of the calculations 100 realisations were generated. However, due to software difficulties, only 90 realisations were generated for the full model.

6. FLOW CALCULATIONS

The mean and standard deviations of the logarithm to base 10 of the upscaled porosity and the three principal components of the effective permeability tensor were calculated for the seven cases for each of the x, y and z direction upscaling models. In every case, variograms were also calculated for each of the upscaled parameters. It was not possible to determine the variograms by spatial averaging alone due to the small number of coarse mesh blocks in each of the upscaling models. The variograms were determined by averaging over the 100 realisations of each model. The variograms based on each block were then averaged over all of the blocks in the model to give an average variogram. It should be noted that the variograms are most reliable for smaller lags as these include more points within the averaging processes.

It was found that none of the cases showed much residual correlation structure in either the y or the z directions. This is to be expected. In Case 1, where all the variability is between the channels, this is because the channel dimensions in these directions is comparable to the size of the coarse mesh. In all the other cases the correlation length is smaller that the size of the coarse mesh cells. Thus, one would not expect to see much correlation from one coarse mesh cell to the next.

The only definite residual correlation structure seen in the results was that in the x-direction for Case 1, for which the correlation length is approximately 300m. This is illustrated by the results shown in Figure 3 for the variograms for the Kyy component of the permeability tensor for the various cases. Case 1 shows a clear correlation structure because the values at two points will be well correlated if they lie in the same channel. This is most likely to happen in the palaeocurrent direction, i.e. the x-direction. The correlation length of 300m is roughly what would be expected, given that the wavelength of the sinuosity of the channels is 600m. It can also be seen from Figure 3 that in all cases there is residual variability. The amount of variability depends on the correlation length, and, as would be expected, the greater the correlation length the greater is the residual variability.

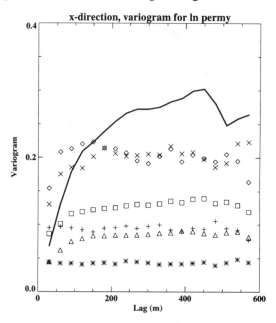

Figure 3: **Variogram for the logarithm of the Kyy component of the permeability calculated using the results from the x-direction STORM model. Solid line - case 1, ∗ -case 2, + - case 3, ◊ - case 4, Δ - case 5, ☐- case 6, × -case 7.**

Calculations of the upscaled permeabilities for a block the size of the full model (which were only performed for Case 4) showed very small residual variability, of the order of 10^{-3}. As these calculations only gave one value of the upscaled parameters for each realisation, it was not possible to examine the correlation structure of the upscaled parameters. However, any such structure would not be very significant given the very small values of the residual variability. This is an important result because it means that at the scale of the large model (600m by 300m by 28m), for the purposes of flow calculations, the heterogeneity due to the detailed facies architecture of the system can reasonably be treated using a homogeneous porous media model.

7. TRANSPORT CALCULATIONS

Transport calculations were performed for the representation of Case 4 on the full model using the SPVMG program [5]. In each of the coordinate directions in turn, a pressure gradient was applied to the two opposite faces of the model and no-flow boundary conditions were applied to the remaining four faces. SPVMG was used to solve the groundwater flow equations subject to these boundary conditions to give the pressure field at the scale of the fine grid. The pressure field was then used to calculate the pathline followed by a fluid particle starting from the centre of the upstream face and travelling through the grid due to advection alone. This calculation was performed once for each of the 90 realisations of the model to give the pathline coordinates for the particle as a function of time. The mean coordinates of the paths were then calculated over the realisations. The variance of the pathline coordinates as a function of time was also calculated.

In all the calculations presented here the number of realisations, or pathlines, was 90. Figures 4 and 5 illustrate the type of results obtained. All of the figures are for the case in which the imposed flow is in the y direction. The mean displacement in the direction of flow and the variance of the displacements in the x-direction (perpendicular to the flow) are shown in Figure 4. The variance of the displacements in the y-direction (parallel to the flow) and in the z-direction (perpendicular to the flow) are shown in Figure 5. The error bars on these figures were computed using the 90% Guttman confidence bounds due to Woo [6].

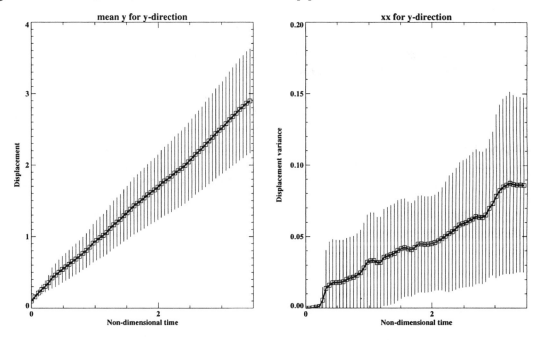

Figure 4:   **Mean particle displacement in the y direction (left Figure) and dispersion in the x direction (right Figure) when the mean flow is in the y direction.**

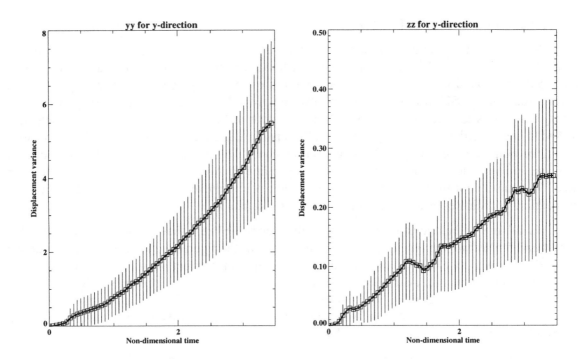

Figure 5: **Dispersion in the y direction (left Figure) and in the z direction (right Figure) when the mean flow is in the y direction.**

The results show that the mean position in the direction of flow increases linearly with time, as expected. They also show that the longitudinal spreading is much greater than the transverse spreading. It is not really possible to determine the rate of spreading from these figures. In particular it is not possible to determine whether or not the longitudinal dispersion is proportional to the time, which would indicate Fickian-like behaviour. However, the results appear broadly consistent with the analytical results of Dagan [4] and others for transport through a random permeability field that can be characterised by a simple multivariate Gaussian model. It is anticipated that different results would be obtained from a model in which the variability is predominantly between channels, i.e. in which the channel architecture is important.

8. CONCLUSIONS

Facies architecture modelling can be used to study flow and transport in sedimentary systems such as the St Bees Sandstone at Sellafield. The fact that these methods can be used to treat flow is not at all surprising, since they are now routinely used in the hydrocarbon industry to study such systems. However, their use to study transport is less wide spread.

This study has demonstrated that, depending on the distribution of the variability within a sedimentary rock, the geometrical arrangement of the various facies and the nature of the permeability variation within a particular facies could both play significant roles in determining the statistical structure of the effective permeability of large-scale blocks. For the parameter values used in this study it is clear that there would be very little residual variability in the effective permeability, at the scale of the grid blocks for a regional-sized groundwater flow model, arising from the detailed facies architecture of the system.

The results obtained illustrate how detailed calculations based on a facies-scale model can provide valuable support for parameterisation of the larger-scale models of flow and transport that are used in a performance assessment. The results indicate the potentially significant role of the facies architecture and demonstrate how detailed calculations can be used to build understanding of the effects of different distributions of the petrophysical properties.

ACKNOWLEDGMENT

This work was funded by United Kingdom Nirex Limited.

REFERENCES

[1]   Nirex 95: A Preliminary Analysis of the Groundwater Pathway for a Deep Repository at Sellafield. Volume 3 - Calculations of Risk. Nirex Science Report S/95/012, 1995.

[2]   Woods C.L., Franklin D.J., Todd J.R. and Gane B., Optimisation and Risk Assessment of a Water Shut-Off Treatment in a Wytch Farm Horizontal Well. Ninth European Symposium on Improved Oil Recovery, The Hague, 1996.

[3]   STORM Manual, Version 2.3, Geomatic, 1996.

[4]   Dagan G., Flow and Transport in Porous Formations, Springer Verlag, 1989.

[5]   Hartley L.J. and Cliffe K.A., SPVMG (Release 1.0) User Guide, AEA Technology Report AEA-ESD-0195, 1995.

[6]   Woo G., Confidence Bounds on Risk Assessments for Underground Nuclear Waste Repositories. Terra Nova **1**, 79-83, 1989.

Representation of Spatial Variability for Modelling of Flow and Transport Processes in the Culebra Dolomite at the WIPP Site

Lucy C. Meigs and Richard L. Beauheim

Sandia National Laboratories, USA

Abstract

The Waste Isolation Pilot Plant (WIPP) is a proposed repository for transuranic wastes constructed in bedded Permian-age halite deposits in southeastern New Mexico, USA. Site-characterization studies at the WIPP site identified groundwater flow in the Culebra Dolomite Member of the Rustler Formation as the most likely geologic pathway for radionuclide transport to the accessible environment in the event of a breach of the WIPP repository through inadvertent human intrusion. The Culebra is a 7-m-thick, variably fractured dolomite with massive and vuggy layers. Detailed studies at all scales demonstrated that the Culebra is a heterogeneous medium.

Heterogeneity in Culebra properties was incorporated into numerical simulations used for data interpretation and PA calculations in different ways, depending on the amount of data available, the certainty with which the effects of a given approach could be evaluated, and the purpose of the study. When abundant, spatially distributed data were available, the heterogeneity was explicitly included. For example, a stochastic approach was used to generate numerous, equally likely, heterogeneous transmissivity fields conditioned on head and transmissivity data. In other cases, constant parameter values were applied over the model domain. These constant values were selected and applied in two different ways. In simple cases where a conservative bounding value could be identified that would not lead to unrealistically conservative results, that value was used for all calculations. In more complex cases, parameter distributions were developed and single values of the parameters were sampled from the distributions and applied across the entire model domain for each of the PA Monte Carlo simulations. We are currently working to refine our understanding of the multiple rates of diffusion attributable to small-scale spatial variability. We hope to define a distribution of diffusion rates that can be used directly, or in simplified form, to represent the diffusion process more accurately at the PA scale.

Introduction

The Waste Isolation Pilot Plant (WIPP) is a proposed repository for transuranic wastes constructed in bedded Permian-age halite deposits in southeastern New Mexico, USA (Figure 1). Site-characterization studies at the WIPP site identified groundwater flow in the Culebra Dolomite Member of the Rustler Formation as the most likely geologic pathway for radionuclide transport to the accessible environment in the event of a breach of the WIPP repository through inadvertent human intrusion. The Rustler Formation represents the transition between the underlying thick evaporite beds of the Salado Formation (where the WIPP repository has been excavated) and the overlying clastic-dominated continental deposits of the Dewey Lake Redbeds. In the vicinity of the WIPP site, the Culebra is the most transmissive unit in the Rustler Formation.

The Culebra is a 7-m-thick, variably fractured dolomite with massive and vuggy layers. The Culebra is underlain by a mudstone unit and overlain by an anhydrite unit [1]. Over the last 20 years, many studies have been conducted on the Culebra geology and on flow and transport processes. These studies have included: geologic studies of core, shafts, and outcrops; measurements of core permeability, porosity, and formation factor; single- and multi-well hydraulic tests; single- and multi-well tracer tests of physical transport processes; laboratory diffusion tests; batch and core tests of chemical processes and properties; and both two- and three-dimensional regional groundwater flow modelling. The insights gathered from interpretation of data and numerical modelling have formed the basis for a detailed conceptual model of Culebra flow and transport processes. To apply data from laboratory and field observations to performance assessment (PA), the conceptual model must include an understanding of processes and their dependence on both temporal and spatial scales.

Background

Detailed studies at all scales demonstrated that the Culebra is a heterogeneous medium. Within the 41.4 km^2 area of the WIPP site, 44 wells and four shafts penetrate the Culebra (Figure 2), which is located about 230 m below land surface. Hydraulic tests have demonstrated that the transmissivity of the Culebra varies over at least six orders of magnitude in the vicinity of the WIPP site (Figure 2) [2]. Tracer tests at several sites suggest that transport properties also vary significantly. Detailed examination of core and shaft exposures suggests that multiple scales of porosity are present within the

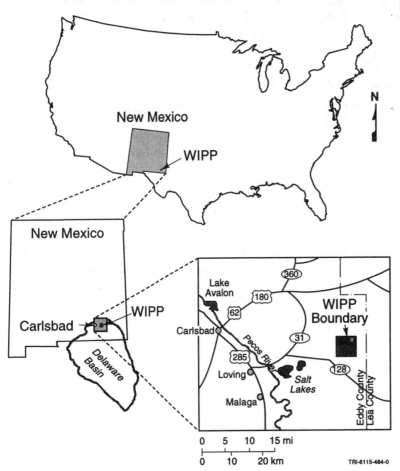

Figure 1. Location of the WIPP site

126

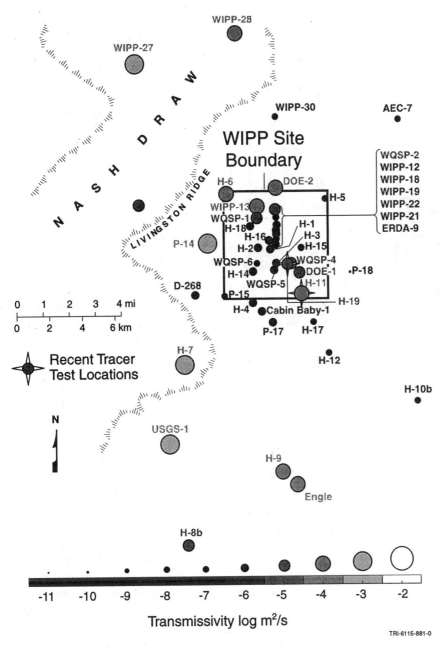

Recent Tracer
Test Locations

Transmissivity log m²/s

Figure 2. Variations in transmissivity values measured in Culebra wells near the WIPP site

TRI-6115-881-0

Culebra, including fractures ranging from <0.05 m to >1 m in length, vuggy zones, intercrystalline porosity, and interparticle porosity (Figure 3). Laboratory measurements of Culebra core plugs yield porosity values between 0.03 to 0.30 (median of 0.16), indicating the presence of significant porosity potentially available for advection and diffusion. Tracer-test results and geologic observations suggest flow can occur within fractures and, to some extent, within interparticle porosity and vugs where they are connected by fractures. Diffusion occurs within all connected porosity and may be the dominant transport mechanism in relatively low-permeability portions of the formation. The variation in peak arrival times in tracer-break-through curves between tests at different hydropads suggests that the types of porosity (fractures, vugs, interparticle) contributing to relatively rapid advective transport vary spatially.

Stratigraphic layering within the Culebra changes little across the WIPP area, apparently as a result of the large size of facies tracts within the Culebra depositional system [1, 3]. On the basis of shaft descriptions [4, 5, 6], core descriptions [1, 3], and RaaX (borehole) video logs, four distinct Culebra units (CU) can be identified (Figure 3) in the subsurface across the entire WIPP area [3]. On the WIPP site, CU 1 ranges from 2.5 to 3.2 m thick, and the lower three units range from 3.5 to 4.7 m in aggregate thickness [3]. Fractures and vugs are more common in the lower three Culebra units than in CU 1 and interparticle porosity is more common in the upper three Culebra units than in CU 4. Hydraulic and tracer tests indicate that the upper portion of the Culebra (CU 1; Figure 3) has a lower permeability than the lower Culebra (CU 2-4) and does not appear to provide a pathway for rapid transport.

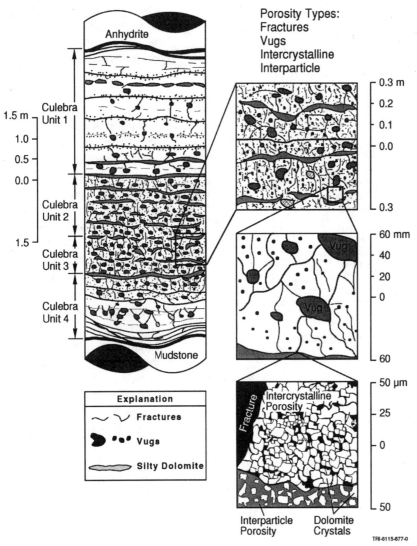

Figure 3. **Schematic of Culebra units and porosity types**

Modelling of Heterogeneity

Heterogeneity was incorporated into numerical simulations used for data interpretation and PA calculations in different ways, depending on the amount of data available, the certainty with which the effects of a given approach could be evaluated, and the purpose of the study. In some cases, the heterogeneity was explicitly included. For example, stochastic approaches were used to generate heterogeneous (conditioned or unconditioned) fields of transmissivity for flow and transport calculations at different scales. In other cases, a constant parameter value was selected in one of two ways and applied over the model domain. If a conservative bounding value could be identified that would not lead to unrealistically conservative results, that value was used for all calculations. If a bounding value could not be used without producing unrealistic results, a parameter distribution was developed and a single value of the parameter was sampled from the distribution and applied across the entire model domain for each of the PA Monte Carlo simulations. These approaches are discussed below in relation to several Culebra flow and transport numerical modelling activities.

The first example is a brief overview of the generation of heterogeneous transmissivity (T) fields for the WIPP Compliance Certification Application (CCA) [7]. To represent the heterogeneity in flow and transport simulations, 100 T-field realizations were generated and sampled on for the PA Monte Carlo simulations. The objective of this modelling activity was to take the available hydraulic data (including estimated measurement uncertainties) and generate numerous equally likely calibrated T fields, each with different spatial characteristics. The T fields were generated using an automated inverse code, GRASP-INV, that uses pilot points (synthetic measured-transmissivity locations) to improve the model fit to the data until acceptance criteria are met. This method is a refinement of the method used for the 1992 PA [8]. The initial information for the T fields came from the transmissivities and steady-state hydraulic heads measured at individual wells across the WIPP site area. The

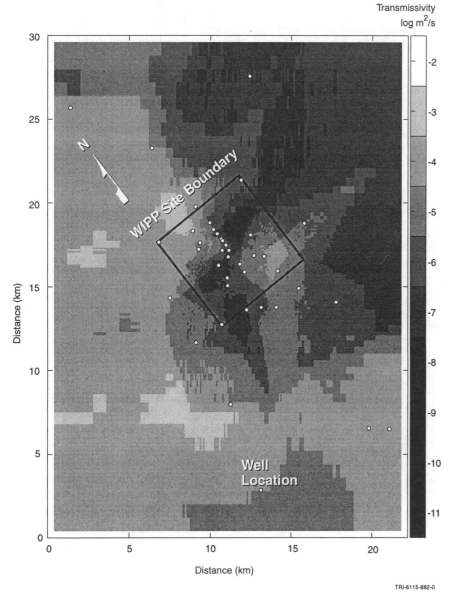

T fields were calibrated by comparing simulated responses to the observed transient pressure/water-level responses resulting from large-scale pumping tests of several months' duration and leakage into the WIPP shafts. Figure 4 shows the ensemble average of all 100 T fields. Figure 5 shows a more detailed view of one of the 100 T fields (no. 77).

Particle-tracking simulations were used to compare the 100 T fields on the basis of travel time. Particles were tracked from the location of a hypothetical intrusion borehole to the boundary of the WIPP site. The particle transport times varied by over two orders of magnitude because different T fields can have significantly different off-site transport pathways (Figure 6) and because similar pathways can have different transmissivities in different realizations.

Figure 4. The ensemble average of all 100 T fields

Heterogeneity caused by differences between the properties of different Culebra layers and by variations in the thickness of the Culebra was treated by excluding CU 1 from consideration and using a constant Culebra thickness of approximately 4 m, representing the average thickness of CU 2-4, in all transport calculations. This is a conservative approach because it reduces the advective porosity, thereby increasing the mean pore velocity, and reduces the diffusional porosity that acts to retard transport. The conservatism of this approach, however, is not thought to be grossly unrealistic such that it would obscure the importance of other parameters.

A third example of treating heterogeneity involves a set of numerical simulations of tracer tests conducted in the Culebra and the scaling of the parameters interpreted from these simulations for PA calculations. Data sets used for the numerical simulations consist of multi-well convergent-flow tracer tests conducted at the H-3 and H-11 hydropads in the 1980s and both single-well injection-withdrawal (SWIW) and multi-well convergent-flow tests conducted in 1995-96 at the H-11 and

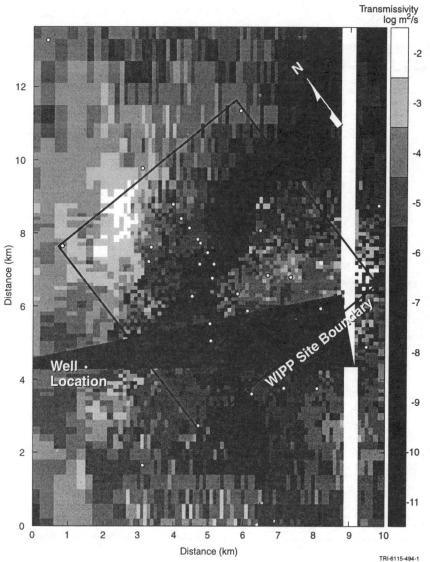

Figure 5. Example of a transient-calibrated transmissivity field (no. 77)

H-19 hydropads (Figure 2). The more recent tests benefited from numerous refinements, including repeated injections with different pumping rates at the central well and discrete injections into the upper and lower portions of the Culebra [9]. Objectives of the numerical simulations were to test the conceptual model of transport in the Culebra and interpret the tracer-test data to evaluate both the appropriate conceptual model and parameters for numerical simulations at both the tracer-test and PA scales.

Interpretations of the tracer-test data were conducted with both single- and double-porosity models. Spatial variability was incorporated into some of the simulations by generating heterogeneous, unconditioned, random fields of hydraulic conductivity at the tracer-test scale. The generated hydraulic-conductivity fields were intended to evaluate the effects the maximum plausible heterogeneity might have on the tailing in the breakthrough curves and were not intended to be an accurate representation of the actual spatial variability in hydraulic conductivity between wells. Interpretations of the SWIW tests indicate that the slow mass recovery observed cannot be explained by heterogeneity alone in a single-porosity ("fracture" only) conceptualization. The slow mass recovery would be expected, however, if diffusional mass-transfer between advective and non-advective porosity were controlling tracer recovery (Figure 7). Simulations of the multi-well data also suggest that the data cannot be modelled adequately without "matrix" diffusion [10].

Interpretations of the SWIW tests alone cannot be used to determine unique parameter fits because a recovery curve from a SWIW test is relatively insensitive to advective porosity (Figure 7), unlike a tracer-breakthrough curve from a multi-well test. The numerical simulations of the multi-well test data were used to bracket the appropriate values for both advective porosity and mass-transfer rate (matrix-block size was used as the fitting parameter and tortuosity was held constant). Some parameters (e.g., diffusive porosity, formation thickness, mass injected) were assumed

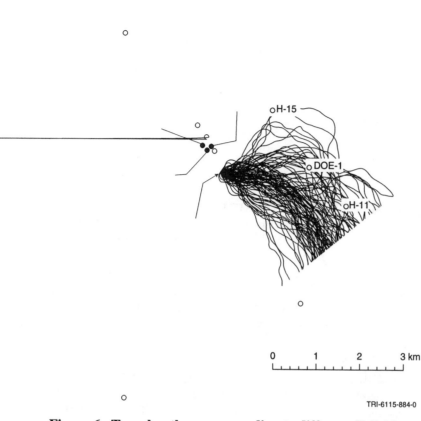

Figure 6. Travel paths corresponding to different T-field realizations

constant based on lab or field data or literature values after the sensitivity of the model to those assumed values was evaluated. The fitted-parameter ranges were found to be relatively insensitive to heterogeneity. The determined ranges were similar for simulations with both homogeneous and heterogeneous double-porosity models.

The tracer-test simulations were used as the basis for identifying distributions for some of the transport parameters used in the Monte Carlo simulations employed for the PA calculations. Whereas spatial variability in advective transport was represented directly by the T fields as described above, spatial variations in advective porosity were not directly represented. Instead, for each PA realization, a single advective-porosity value was sampled from a distribution and applied across the entire model domain. The selected distribution (log uniform between 10^{-2} and 10^{-4}) was intended to represent a conservatively low range of possible effective advective-porosity values across the entire off-site pathway consistent with the tracer-test results. Other transport parameters, such as matrix-block length and diffusive porosity, were also represented by parameter distributions from which single values of the parameters were sampled and applied across the entire model domain for each Monte Carlo simulation.

Both the PA calculations and the simulations of tracer-test data used to determine parameters employed a double-porosity model with a single rate of diffusion. Detailed modelling of the tracer-test data suggested that they are better described by a double-porosity model that incorporates multiple rates of diffusion [10]. Geologic descriptions and examination of Culebra core suggest that the spatial variability in porosity, variations in fracture spacing, and the tortuous nature of the pore space should result in significant variations in diffusion rates over relatively small volumes. Numerical simulations are being conducted with a double-porosity multirate model to evaluate the distributions of diffusion rates that best fit the tracer-test data and the significance of those distributions for the PA model. When properly implemented, the multirate model should provide a more direct method to transfer transport information between laboratory, field, and PA scales, allowing a more accurate representation of the diffusion process at the PA scale. This increased accuracy could provide a defensible basis for significant PA-model simplification, if desired.

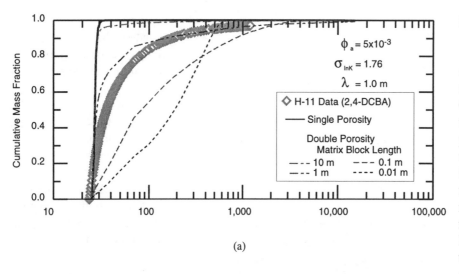

(a)

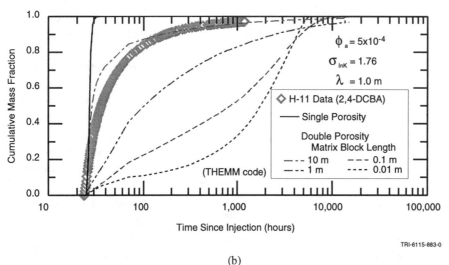

(THEMM code)

Time Since Injection (hours)

TRI-6115-883-0

(b)

Figure 7. Simulated mass recovery curves for H-11 SWIW test with single-porosity and double-porosity models compared to observed data. For the simulations, the advective porosity (ϕ_a) was 5×10^{-3} in (a) and 5×10^{-4} in (b). The heterogeneous field of hydraulic conductivity had a standard deviation (σ) in natural log space of 1.76 m/s and an exponential model with a correlation length (λ) of 1.0 m

Summary

We have been successful in incorporating heterogeneity in Culebra transmissivity, thickness, and transport properties in PA models. Heterogeneity in Culebra transmissivity was incorporated by generating numerous, equally likely, heterogeneous transmissivity fields and then using a different field for each PA Monte Carlo simulation. Heterogeneity arising from differences between the properties of different Culebra layers and from variations in Culebra thickness was addressed by eliminating the low-permeability upper Culebra (CU 1) from transport models and using a constant, reduced value for thickness, resulting in conservative estimates of radionuclide transport. The effects of heterogeneity in Culebra hydraulic conductivity were found to be inadequate in explaining slow tracer recoveries during tracer tests, leaving diffusion between advective and non-advective porosity as the most likely explanation. Tracer-test simulations were used to define ranges of different important transport parameters, and individual values were sampled from each range for each of the PA Monte Carlo simulations.

Interpretations of the recent tracer tests have resulted in a refined conceptual model of transport in the Culebra and the demonstration that transport is not limited to fractures. Our modelling to date has adequately defined parameters for use in the PA calculations. We recently initiated a series of laboratory diffusion experiments to examine the variability in diffusion rates due to porosity variations. These laboratory experiments, in combination with descriptive information from core samples and the tracer-test data, will be used to refine our understanding of the multiple rates of diffusion attributable to small-scale spatial variability. We hope to define a distribution of diffusion rates that can be used directly, or in simplified form, to represent the diffusion process more accurately at the PA scale.

Acknowledgments

Sandia is a multiprogram laboratory operated by Sandia Corporation, a Lockheed Martin Company, for the United States Department of Energy under contract DE-AC04-94AL85000. The methodology used to create the transmissivity fields was developed by Marsh LaVenue and Banda RamaRao of Duke Engineering & Services (DE&S, formerly INTERA, Inc). Yvonne Tsang of Lawrence Berkeley National Laboratory; Susan Altman, Jim McCord, and Sean McKenna of Sandia National Laboratories; Toya Jones and Joanna Ogintz of DE&S; and Roy Haggerty of Oregon State University assisted with numerical simulations of tracer-test data.

References

[1] Holt, R.M., and Powers, D.W., Facies Variability and Post-Depositional Alteration Within the Rustler Formation in the Vicinity of the Waste Isolation Pilot Plant, Southeastern New Mexico, 1988, DOE-WIPP 88-004, US DOE WIPP Project Office.

[2] Beauheim, R.L., and Ruskauff, G.J., Analysis of Hydraulic Tests of the Culebra and Magenta Dolomites and Dewey Lake Redbeds Conducted at the Waste Isolation Pilot Plant Site, 1998, SAND98-0049, Sandia National Laboratories.

[3] Holt, R.M., Conceptual Model for Transport Processes in the Culebra Dolomite Member, Rustler Formation, 1997, SAND97-0194, Sandia National Laboratories.

[4] Holt, R.M., and Powers, D.W., Geotechnical Activities in the Waste Handling Shaft, 1984, WTSD-TME-038, US DOE by TSC.

[5] Holt, R.M., and Powers, D.W., Geotechnical Activities in the Exhaust Shaft, 1986, DOE-WIPP 86-008, US DOE WIPP Project Office.

[6] Holt, R.M., and Powers, D.W., Geologic Mapping of the Air Intake Shaft at the Waste Isolation Pilot Plant, 1990, DOE-WIPP 90-051, US DOE WIPP Project Office.

[7] US Department of Energy, Title 40 CFR Part 191 Compliance Certification Application for the Waste Isolation Pilot Plant, Vol. XVIII, Appendix TFIELD, 1996, DOE/CAO-1996-2184, US DOE WIPP Carlsbad Area Office.

[8] Sandia WIPP Project, Preliminary Performance Assessment for the Waste Isolation Pilot Plant, December 1992, Volume 3: Model Parameters, 1992, SAND92-0700/3, Sandia National Laboratories.

[9] Beauheim, R.L., Meigs, L.C., and Davies, P.B., Rationale for the H-19 and H-11 Tracer Tests at the WIPP Site, in Field Tracer Experiments: Role in the Prediction of Radionuclide Migration, Synthesis and Proceeding of an NEA/EC GEOTRAP Workshop, Cologne, Germany, 28-30 August 1996, 1997, 107-118, OECD/NEA.

[10] Meigs, L.C., Beauheim, R.L., McCord, J.T., Tsang, Y.W., and Haggerty, R., Design, Modelling, and Current Interpretations of the H-19 and H-11 Tracer Tests at the WIPP Site, in Field Tracer Experiments: Role in the Prediction of Radionuclide Migration, Synthesis and Proceeding of an NEA/EC GEOTRAP Workshop, Cologne, Germany, 28-30 August 1996, 1997, 157-169, OECD/NEA.

The Discrete-Modelling Approach Adopted in SITE-94

Joel Geier
Clearwater Hardrock Consulting/Oregon State University, USA

Björn Dverstorp
Swedish Nuclear Power Inspectorate, Sweden

Abstract

A discrete-feature approach is applied to model 3-D, site-scale groundwater flow and to predict hydrological parameters for radionuclide migration in fractured crystalline rock, at the Äspö site in Sweden. The approach is based on networks of discrete, piecewise-planar elements, which represent transmissive features on scales ranging from individual fractures in the vicinity of waste canisters, to regional fracture zones that connect to the surface environment. At each scale within the model, the formulation is explicitly in terms of the geometry of the types of features that are indicated to control groundwater flow and solute transport. Thus the approach preserves the heterogeneous structure that is observed on scales ranging from fractures in core to regional lineaments, and avoids various smoothing errors that can result from continuum approximations. The approach integrates a variety of field data in a single model, and uses simulated hydrological testing, including hydraulic interference tests and tracer tests, to test the derived descriptions of site variability. Distributions of effective parameters from a suite of variational cases indicate that variability of hydrologic properties within site-scale fracture zones is a main source of uncertainty in geologic-barrier performance, given the type of site-characterization data that were available from Äspö. Types of data that are needed for reducing this uncertainty are discussed.

INTRODUCTION

A site-specific application of a discrete, 3-D modelling approach is used to predict hydrological parameters for radionuclide transport in the Swedish Nuclear Power Inspectorate's SITE-94 study [1]. SITE-94 is an exercise in the utilization of site-specific information to assess the performance of a hypothetical but site-specific repository design. The discrete-feature approach [2-4] has been used in SITE-94 to integrate data from multiple scales and geoscientific disciplines in a single hydrogeologic model for prediction.

The overall approach to performance assessment in SITE-94 emphasizes the use of multiple lines of analysis to evaluate the role of conceptual uncertainty. The discrete-feature approach, as described here, is one of several hydrogeological approaches that were developed in parallel to test the significance of alternative conceptual assumptions regarding groundwater flow. Other types of models used for quantitative predictions in SITE-94 include a site-scale stochastic continuum model [5] and a simple analysis based on 1-D Darcy's law and bounding estimates of hydrologic properties [6].

The main objective of the discrete-feature analysis for SITE-94 is to estimate effective hydrological parameters for subsequent analyses of radionuclide transport in the near-field and far-field environments. The effective parameters are derived by modelling advective-dispersive transport from specific, randomly chosen canister positions within a hypothetical repository layout. Spatial variability is evaluated as a function of canister (radionuclide source) location in multiple Monte Carlo simulations of the detailed-scale component. Model variations are used to evaluate the influence of specific sources of uncertainty.

OVERVIEW OF THE DISCRETE-FEATURE MODEL

A schematic illustration of the discrete-feature (DF) model for SITE-94 is shown in Figure 1. The DF approach is based on an explicit, discrete representation of hydraulically conductive features in the rock, on scales ranging from individual fractures on the canister scale to local and regional fracture zones on the site scale. At each scale within the model, the formulation is explicitly in terms of the geometry of the types of features that are indicated to control groundwater flow and solute transport. The aim is to preserve the heterogeneous structure that is observed on scales ranging from fractures in core to regional lineaments, and thus avoid the smoothing effects of continuum approximations.

The DF approach, as employed in SITE-94, is a mixed deterministic/stochastic representation of the heterogeneity in the rock. The deterministic component is explicitly based on geological and geophysical information regarding large-scale structural heterogeneity (i.e. fracture or fault zones). The use of large-scale structural information helps to reduce uncertainty in predictions and gives more precise estimates of spatial variability. However, deterministic interpretations of large-scale structures are associated with uncertainty that must be evaluated by means of structural variants.

As it is not possible to characterize all detailed-scale conductive features at a site (*e.g.* fractures), a stochastic component is included to allow evaluation of the uncharacterized part of the system. The stochastic component is based on statistical descriptions of the detailed-scale fractures and their hydraulic properties. Semistochastic realizations of the integrated model are generated by combining the deterministic component with successive, independent realizations of the stochastic component. Uncertainty regarding assumptions of statistical models was evaluated by testing alternative statistical models for the stochastic component.

Certain rock properties that are uncharacterized or very poorly characterized, such as pore structure and its influence on relationships among transport parameters in the large-scale fracture zones, are also evaluated by testing alternative variants. In contrast to equivalent-continuum representations of fractured rock, the discrete-feature approach contains an explicit representation of fractures. This makes it possible to express uncertainties in terms of alternative, explicit models of pore structure, fracture geometry, etc., and to directly evaluate the consequences of these alternative models for flow and transport predictions.

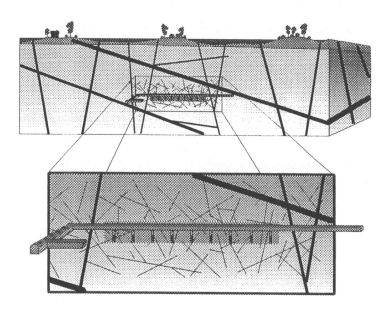

Figure 1. Schematic illustration of a discrete-feature model for performance assessment

A distinctive property of the DF approach is that features with a wide range of length scales are integrated in a single model. Continuity between the features on different scales is explicitly in terms of the intersections between individual features, rather than implicitly in terms of coupled boundary conditions between submodels on different scales. Thus flow and transport are modelled as a continuous process from detailed-scale fractures at the release point (the canister deposition hole) to large-scale structures at the discharge point (the biosphere). Due to practical (computational) constraints, it is not possible to represent all of the smaller-scale features throughout the model. Rather, the resolution is decreased at points distant from the release points; *i.e.* it is assumed that larger and more transmissive features dominate flow and transport on successively larger scales.

Boundary conditions are applied only on the exterior of the model, and along physical, internal boundaries such as boreholes, tunnels, or waste-deposition holes. Boundary conditions are specified along intersections between the discrete features and each external or internal boundary. Discrete representation of small boundaries such as waste-deposition holes makes it possible to evaluate transport for specific locations within the repository.

SITE DESCRIPTION

The SITE-94 DF model is based on site-specific data drawn from the preliminary investigations for the Swedish Hard Rock Laboratory (HRL) at Äspö [7]. Although subsequently data have been collected from underground tunnels at the HRL, the analysis is restricted to data that were obtained from surface-based (surficial or borehole-based) investigations The intent is that the basis for the model should be representative of data that would be available at an initial stage of site licensing, prior to excavation of a repository. Although the HRL is intended to be used as a research facility rather than as an actual repository, one of the stated objectives for the HRL was to demonstrate and validate the methods that are planned to be used in site characterization for an actual repository [8]. Hence the HRL dataset can be viewed as reasonably representative of the type of data that can be expected for an actual repository.

The site of the hypothetical repository is below the coastal island of Äspö, in southeastern Sweden (Figure 2a). The rock at the site is predominantly of granitic type, within which the dominant pathways for groundwater flow and radionuclide transport are via a complex network of fracture zones and individual fractures. The soil cover at Äspö is mainly thin (0 to 5 m) glacial till, and bedrock outcrops are common. Most of the site is less than 10 m above mean sea level.

The large-scale component is defined for a 5 km x 5 km x 1 km deep semiregional domain. The largest-scale features within this domain are constructed by projecting the traces of semiregional fracture zones to depth based on estimated dips as given by [9]. Within the 2 km x 2 km x 1 km deep site-scale domain, structures are specified as planar or undulating surfaces in 3-D (Figure 2b). The detailed-scale, stochastic DFN component of the DF model is defined within a 450 m x 320 m x 80 km thick block around the SITE-94 repository (Figure 2c). The design and layout of the hypothetical repository is based on the Swedish KBS-3 concept [1]. The waste form is assumed to be sealed in copper-steel canisters, which are individually emplaced in vertical deposition holes along a grid of nominally horizontal deposition tunnels.

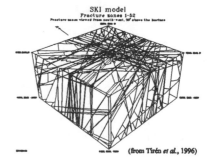

b) 3-D site-scale structural model.

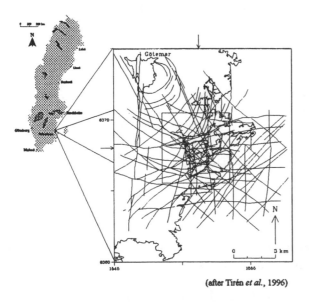

(after Tirén et al., 1996)

a) Location of Äspö site in Sweden, and regional
   structural model showing locations of semiregional
   and site-scale domains (squares in inset).

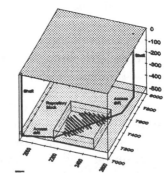

c) Repository layout and location of block within
   which detailed-scale component is simulated.

Figure 2. Site location and repository layout

Boundary conditions are imposed on the exterior of the semiregional domain and along boreholes (for model calibration) or deposition holes (for radionuclide transport predictions). The groundwater system on the semiregional scale is interpreted as being part of a much larger regional system [10], rather than being controlled by local hydrologic boundaries. Therefore the influence of different combinations of effective boundary conditions on the lateral and lower boundaries of the semiregional domain is examined in variational cases. The upper surface is modeled as a transmissive feature with undulations corresponding to the topography, and is divided into nine surface-boundary segments representing the Baltic, the freshwater lake Frisksjön, the mainland, and the six largest islands that lie within the semiregional domain. The Baltic and Frisksjön segments are modelled as fixed-head boundaries, while the subaerial boundary segments are modelled as fixed-infiltration boundaries (in the base-case DF model) or as topographic (hydrostatic) fixed-head boundaries (in variational cases). Internal boundaries include boreholes (for calibration runs) or waste-deposition holes (for runs to predict effective transport parameters under post-closure conditions). A total of 65 holes were drilled within the semiregional domain during the HRL preliminary investigations. These were subdivided by packers into a total of 127 distinct sections, each which is represented in the DF model by an internal boundary.

UTILIZATION OF SITE-SPECIFIC DATA ON HETEROGENEITY

Site-specific data from a variety of sources are used to deduce the structure of heterogeneity that controls flow and transport at the site. The following key types of information are used to develop a deterministic description of heterogeneity on the site (10^2 to 10^3 m) and semiregional (10^3 to 10^4 m) scales:

•   Structural interpretations of discrete faults and fracture zones in the Äspö vicinity [9] based on geological and geophysical data. This structural model in itself represents a synthesis of lineament interpretations (based on satellite imagery, topographic and bathymetric data, and above-ground geophysical surveys) with diverse geological and geophysical borehole logs.

•   Indications of transmissive features from 1291 packer tests in 8 boreholes and flowmeter logging in 14 boreholes.

•   Groundwater-level and pressure monitoring data from a total of 127 distinct sections in 65 observation boreholes, prior to and during a suite of nine short-term (2- or 3-day) and two long-term (3-month) pumping tests. These data are used in calibration to derive the hydraulic properties of the large-scale features.

•   A large-scale, radially-convergent multiple tracer test in fracture zones [11], which is used to test the large-scale transport characteristics of the integrated DF model .

The following types of data are used to develop a stochastic description of heterogeneity on the detailed scale (10^0 to 10^2 m):

•   Lithologic and fracture data in core, including data on fracture positions, orientations, and fracture minerals from 13 boreholes on Äspö and neighboring areas, and measurements of fracture trace length and orientation from outcrops and cleared trenches on the surface of Äspö. These data are used to derive the statistics that characterize the geometry (structure) of the discrete-fracture population for the stochastic DFN component, for fracture sets defined in terms of probability density functions (p.d.f.s) for fracture size and orientation.

•   Estimates of hydraulic conductivity from fixed-interval-length (FIL) injection tests in 8 of the core-drilled holes, from which are derived p.d.f.s for the effective transmissivity of single fractures, and stochastic models for location (clustering) of transmissive fractures.

- Transient hydraulic responses during the FIL injection tests in boreholes, yielding evidence of variable flow-path geometry, which are used to test the small-scale connectivity characteristics of the DFN component.

The use of hydraulic conductivity data to derive hydraulic properties of the detailed-scale, DFN component corresponds approximately to the way that these data are used to describe spatial variability in the stochastic continuum approach. The use of structural geological data and interpretations can be seen as providing additional constraints on the description of heterogeneity in the DF model.

DERIVATION OF PARAMETERS TO DESCRIBE HETEROGENEITY

The detailed-scale (stochastic) and large-scale (deterministic) components of the DF model are estimated/calibrated independently. The integrated model is tested by simulation of a large-scale, hydraulic-interference and radially-convergent tracer test.

Estimation of the detailed-scale stochastic component

The stochastic description of the detailed-scale component is derived by methods of discrete-fracture network (DFN) analysis, using a forward-modelling approach [12]. The general approach consists of simulating the characteristics that would be observed during site characterization (e.g. histograms of fracture trace length on outcrops) for an assumed statistical model of the fracture population, and optimizing the model parameters to obtain the best possible match of simulations to observations. For each lithological/structural domain of interest, fracture sets are defined by classifying the data according to fracture orientation. Stochastic processes for fracture location (clustering) are estimated by forward modelling of the histogram of fracture intensity in core; in the present analysis, fracture location is found to be best described by a fractal point process. Distributions of fracture size (disk radius) are estimated by forward modelling of trace lengths on outcrops; in general a power-law distribution is found to give the best match to the data. The probability that a given fracture is significantly transmissive and the p.d.f. of fracture transmissivity T_f (generally lognormal) are estimated by forward modelling with respect to the histogram of transmissivity values obtained from FIL injection tests. Details of the DFN analysis for SITE-94 are given by [2].

The connectivity of the detailed-scale component is tested by simulating the transient responses to FIL injection tests in Monte Carlo realizations of the DFN model, and statistically comparing with the responses observed in the actual tests on Äspö. The comparison is in terms of the bivariate distribution of transmissivity T_{GRF} and flow dimension D_{GRF}, as interpreted from the simulated and actual transient data according to a generalized-radial-flow model [13]. Results for the dominant lithologic type, Äspö granodiorite, show good agreement between simulations and observations.

Estimation of the large-scale deterministic component

The hydraulic properties of the deterministic, site-scale and semiregional structures (i.e., fracture zones) are estimated from hydraulic test data in a series of steps which include preliminary estimation based on hydraulic test data, and calibration of the large-scale model with respect to the undisturbed head distribution and steady-state drawdowns during pumping tests.

Preliminary estimates of structure transmissivity and hydraulic diffusivity are obtained from single-hole tests and by classical single-aquifer analysis of pressure-drawdown data from hydraulic interference tests. Site-scale structures that are clearly non-transmissive or hydrologically redundant are deleted prior to calibration, reducing the site-scale component to a total of 30 structures. The large-scale, deterministic component is formed by combining the site-scale and semiregional structures.

Calibration is accomplished by finite-element modelling of groundwater flow in the deterministic network, in response to both undisturbed conditions and perturbations due to the cross-hole testing. In the first stage of calibration, the infiltration rate per unit land surface area is optimized with respect to the undisturbed head distribution, by minimizing the sum of the absolute differences between observed and simulated heads at steady-state, undisturbed conditions, for the 127 monitoring points. In the second stage of calibration, the transmissivities T_i of selected site-scale and semiregional features are calibrated with respect to the initial equilibrium heads and the long-term (steady-state) drawdowns in two long-term pumping tests on Äspö.

Large-scale transport properties

In the Äspö preliminary investigations only a few (5) tracer tests were conducted that could serve to constrain the transport properties within large-scale structures; in all of these non-sorbing tracers were used. Therefore two different assumptions are evaluated regarding the local relationship between transmissivity T and effective transport aperture b_T (pore volume per unit area) within fracture zones:

PO   The relationship between T and b_T in fracture zones is the same as for single fractures.

P1   For all fracture zones, $b_T = 3$ cm uniformly.

The former amounts to an assumption that, at any point in the plane of a fracture zone, transport is localized in a single, dominant fracture; this gives a relatively low-porosity model. The latter assumption is based upon a high estimate from prior simple radial-flow interpretations of the LPT2 tracer tests [11], an d gives generally higher and more uniform porosity in the far-field.

Combined detailed-scale and large-scale model

To produce the final, base-case DF model for performance-assessment calculations, the calibrated, large-scale component is combined with the stochastic, detailed-scale component. Realizations of the integrated, semi-stochastic model are produced by combining the large-scale, deterministic structures with unconditional, Monte Carlo realizations of the DFN component. As a validation check prior to the predictive runs, the large-scale model is combined with the detailed-scale, DFN model and used to simulate transient drawdowns and tracer arrivals during the LPT2 experiment on Äspö. In aggregate the simulated responses in 121 monitoring sections compare well with observations, both in terms of the sequence and the relative magnitude of the responses in boreholes with multiple observation sections.

The radially convergent tracer test is simulated by advective-dispersive particle tracking from each of the six borehole sections that were used as sources in the experiment [11]. The model predicts tracer recovery from 2 to 4 of the six sources, depending upon which assumption (P1 or PO, respectively) is used for fracture-zone porosity. The actual tests produced definite arrivals from two of the sources, and ambiguous secondary peaks in the recovery data, which may represent two more of the sources, or alternatively secondary transport pathways. For both PO and P1 variants, two source locations are correctly predicted to produce no tracer arrival during the experiment. As a whole, the results demonstrate that the influence of uncertainty associated with the detailed flow geometry (*i.e.* distinct transport pathways) is substantial. The lower-porosity, PO variant gives the best results overall, but strongly overpredicts the advective velocity for one of the six sources. For other cases, even this variant overpredicts the advective velocity.

ESTIMATION OF EFFECTIVE PARAMETERS FOR PERFORMANCE ASSESSMENT

Effective hydrologic parameters estimated for near-field radionuclide transport codes include groundwater (Darcy) flux q_{nf}, mean fracture spacing S_{nf}, and effective transport aperture b_{nf}. Parameters estimated for far-field radionuclide transport codes include discharge location, groundwater (Darcy) flux q_{ff}, advective velocity u_{ff}, macrodispersion coefficients D_{Lff}, the specific wetted surface a_{rff} (wetted surface per unit volume of rock) and flow porosity Φ_{ff}. These parameters have been evaluated for a total of 42 calculation cases (including the base case), which comprise selected combinations of 25 distinct variations in the DF model (Table 1). The variations address major aspects of uncertainty in the integrated DF model, including: (1) alternative interpretations of the large-scale structures, (2) lithological variations, (3) effective semiregional boundary conditions, (4) geostatistical variation of hydraulic transmissivity within large-scale structures, and (5) variations in the relationships between hydraulic and transport properties.

For each calculation case, multiple semi-stochastic realizations are evaluated to gain an estimate of the uncertainty associated with the stochastic component of the model. The spatial variability within each realization is assessed by estimating near-field and far-field parameters at each of 40 randomly selected deposition holes. Near-field parameters are evaluated by sampling the flow field in the DFN network on three orthogonal panels placed at each deposition hole. Far-field parameters are evaluated by advective-dispersive particle tracking and by fitting analytical solutions of the 1-D advective-dispersion equation to the calculated residence-time distributions. Specific wetted surface and flow porosity values are estimated as residence-time weighted averages of the surface area and volume available to solute as it passes through the DF network.

RESULTS AND DISCUSSION

The strong spatial variability of the rock around the repository, as described by the stochastic DFN component, has important consequences for repository performance. The base-case DF model predicts that radionuclides would reach the biosphere from only a fraction of all failed canisters, even when retardation mechanisms such as sorption and matrix diffusion are neglected. Only 43% of the canister sites are *fractured* (intersected by transmissive fractures). Thus, in contrast to a continuum description in which all canister sites would be connective, the discrete description of near-field variability results in a prediction that most canisters would be hydraulically disconnected from the biosphere. Moreover, only 33% of the canister sites are *flowing* sites, in the sense that the predicted q_{nf} exceeds the numerical resolution of the model (about 10^{-6} m/yr), and of these flowing sites, only those with $q_{nf} > 10^{-4}$ m/yr tend to produce radionuclide arrivals at the surface within 5000 years of radionuclide release; these *active* canister sites account for only 8 to 12% of the total, in the base case, depending upon the model for fracture-zone porosity (P1 or P0, respectively).

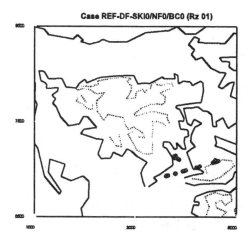

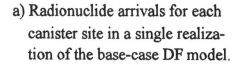

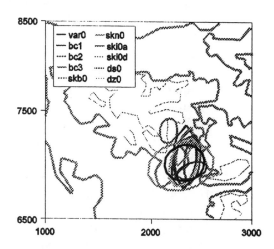

a) Radionuclide arrivals for each canister site in a single realization of the base-case DF model.

b) Fitted 90% ellipses to distribution of radionuclide discharge for selected variants of the DF model.

Figure 3. Predicted location of radionuclide discharge to the surface

Among near-field parameters, the highest spatial variability is seen for q_{nf}, which varies by more than two orders of magnitude. A weak negative correlation exists between q_{nf} and R, the distance from the canister to the nearest deterministic, large-scale structure; this confirms that improved performance will to some extent result from a practice of locating canisters far from major fracture zones. However, high values of q_{nf} do occur at canister sites with R > 40 m, and thus a large setback does not guarantee low flux through a canister site. Connectivity of canister sites is sensitive to the lithology in the vicinity of the repository (Variant NF1). The distribution of q_{nf} is found to be sensitive to uncertainty in the DFN transmissivity distribution (Variants NF0A and NF0B), in far-field boundary conditions (Variant BC2) and in the overall far-field hydrogeologic model (Variant SKB0). Each of these variants produces a significant, order-of-magnitude increase in median q_{nf} relative to the base case.

Variant	Type of uncertainty addressed
LO	Hydrologic: Low pore volume in site-scale structures based on single-fracture model.
HI	Hydrologic: High, uniform pore volume per unit area in all site-scale structures.
SKI0	Structural: Base-case SKI structural model for Äspö.
SKI1	Structural: Alternative interpretation of a major semiregional structure .
SKB0	Structural/hydrologic: Alternative geological and hydrological site interpretation.
SKN0	Structural/hydrologic: Alternative geological and hydrological site interpretation.
SKI0A	Hydrologic: Increase in transmissivity of gently vs. steeply inclined structures.
SKI0B	Hydrologic: Uniformly decreased transport porosity in site-scale structures.
SKI0C	Hydrologic: Decreased dispersion in site-scale structures.
SKI0D	Hydrologic: Effects of spatial variation of transmissivity within site-scale structures.
NF0	Lithologic: Base-case rock type (Äspö granodiorite) in the repository block.
NF1	Lithologic: Alternative rock type (Småland granite) in the repository block.
NF0A	Hydrologic (detailed): Increase in mean fracture transmissivity.
NF0B	Hydrologic (detailed): Increase in the variability of fracture transmissivity.
NF0C	Hydrologic (detailed): Increase in the effective transport porosity of all fractures.
BC0	Base-case boundary conditions.
BC1	Boundary conditions: Alternative conditions on upper and seaward boundaries.
BC2	Boundary conditions: Alternative conditions on upper, seaward, & lower boundaries.
BC3	Boundary conditions: Reduced rate of infiltration at the upper surface.
DZ0	Excavation effects: Effect of a disturbed rock zone (DRZ) around repository tunnels.
DZ1	Excavation effects: Decrease in DRZ porosity.
DZ2	Excavation effects: Increase in DRZ porosity.
DZ3	Excavation effects: Increase in DRZ conductivity.
DS0	Excavation effects: DRZ plus shaft sealing failure.
DS1	Excavation effects: DRZ plus shaft sealing failure, with increased DRZ conductivity.

Table 1. Variants analyzed with the discrete-feature model

Far-field discharge of radionuclides is focused mainly in the strait immediately southeast of Äspö, and on the southern tip of the island itself; thus the median net direction of transport from the repository is southeastward, and steeply upward; this is a consequence of strong, upward groundwater velocities in steeply dipping fracture zones along the eastern shore of Äspö. The controlling influence of individual structures is evident in Figure 3a, which shows arrival coordinates for particles for a single canister site and realization of the base-case model. Different groups of structures dominate release in different variants; however, the median centroids of release for the different variants (Figure 3b) are usually within about 50 m of that for the base case.

The predictions of far-field parameters are summarized in Figure 4a-k, in terms of the parameter ratio $F = a_{rff}L_{ff} / q_{ff}$ and the Peclet number $Pe = q_{rff}L_{ff} / \Phi_{ff}D_{Lff}$. Scoping calculations for SITE-94 indicated that the peak radiation dose to the biosphere (for a given radionuclide source term and fixed geochemical conditions) is essentially characterized by these two parameters, with F being by far the more important. High F implies a relatively low potential for retardation of radionuclides by sorption and matrix diffusion in the far field, and thus a higher dose to the biosphere in the event of a canister failure.

Most far-field parameters show a variability of about two orders of magnitude for the base case, P0 variant. In the P1 variant of the base case, the variability of far-field parameters in general is reduced, due to the more uniform model for fracture-zone porosity. Far-field parameters are in general weakly correlated or uncorrelated with respect to near-field parameters. A weak, negative correlation is found between F and q_{nf} (the two parameters that are most directly related to far-field and near-field performance [1]), which indicates a slight tendency for canisters that are exposed to high near-field flux to be associated with low-retardation pathways in the far-field.

Among the various aspects of uncertainty that have been evaluated, the most significant influences on far-field performance (capacity of the geological barrier to retard radionuclide release) relate to the spatial variability of porosity within large-scale structures, and its correlation to transmissivity. These issues are exemplified by the P0 and P1 variants, in the first of which pore volume is correlated to transmissivity, and in the second of which the pore volume is uniform in all large-scale structures. The effect is due to the relationship between advective velocities and the surface area available for sorption, in the different parts of the hydrologic system.

The generally increased variability in the P0 variant is largely due to the fact that radionuclides spend proportionally longer in the more variable, repository-scale fracture network, whereas in the P1 variant radionuclides spend proportionally longer in the less variable, large-scale structures. The contrast between these variants also reflects the fact that the relative porosity of the large-scale and detailed-scale structures affects the relative importance of retardation on the two scales. This uncertainty has practical implications, as the demands on a site-characterization program may depend upon which portion of the geosphere is viewed as the main locus of radionuclide retardation.

Significant effects are also produced by introducing stochastic variability of transmissivity within fracture zones (Variant SKI0D), and by variations in the relative transmissivity of gently vs. steeply dipping structures (Variant SKI0A), effective boundary conditions, particularly at the base of the model (Variant BC2), and the transport porosity of the stochastic fracture population (Variant NF0C).

These results suggest that uncertainty regarding the spatial variability of hydrologic properties within large-scale structures, and in particular regarding local correlations between hydraulic and transport properties, is of major significance for geologic-barrier performance. The types of data that were available from the preliminary investigations at Äspö are not adequate to constrain this uncertainty. In the stochastic continuum analysis for SITE-94 [5], it has been demonstrated that even if long-range correlation of hydraulic conductivity were present in the site-scale structures, the boreholes in which FIL packer tests were performed are too widely spaced to allow identification of such a correlation structure.

Given practical limits on the number and spatial density of boreholes at a future repository site, it seems unlikely that our ability to resolve large-scale spatial correlation of hydraulic properties will be markedly improved based on hydraulic testing alone. At best, the evaluated uncertainty might be reduced by developing improved calibration methods to restrict the locus of plausible realizations of a given stochastic model. In order to determine what relationships among hydraulic and transport properties are plausible within large-scale structures, multiple-tracer tests using both sorbing and nonsorbing tracers are essential, preferably under natural-gradient conditions so that the results reflect the characteristics of realistic flowpaths that would be active in the post-closure period of a repository. As such tests will most likely not yield unique estimates of relationships among flow and transport properties in large-scale structures, the emphasis should be on identifying the locus of plausible relationships by stochastic inverse techniques. Further constraints on these relationships can be gained by small-scale, multi-tracer experiments performed in well-characterized fracture zones. A number of such experiments, used as generic data, could greatly constrain the range of possibilities that need to be considered in analyzing large-scale tracer tests at a repository site.

Even if it is possible to reduce the large uncertainty in predictions that presently results from poor characterization of large-scale structures, true spatial variability may result in a wide range of predictions for sites in granitic rock, such as Äspö. In the present analysis, wide ranges of effective far-field parameters are predicted for virtually all of the model variants considered. For the parameters that can be related directly to far-field performance (F and Pe), the magnitude of this variability, from point to point in the repository, is comparable to or greater than the effects of most of the evaluated sources of uncertainty. The persistence of this spatial variability in a wide range of variants suggests that it is a true property of the site, rather than a consequence of uncertainty in the site characterization.

CONCLUSIONS

The discrete feature model has been demonstrated as a useful tool for the integration of site data, and for prediction of parameters relating directly to near-field and far-field performance assessment. Particular strengths of the approach include (1) the ability to integrate not only hydrologic data but also structural geological data and interpretations, (2) explicit integration of features with a wide range of length scales in a single model, thus avoiding many spatial averaging/upscaling issues, (3) the possibility to explicitly model the discrete nature of flow and transport in and around the repository, (4) the flexibility of the method for analyzing various aspects of conceptual and parametric uncertainty, and (5) the possibility to check for correlations between near-field and far-field hydrologic parameters, due to the multi-scale nature of the model.

Several current limitations of the approach are also identified. These include (1) inconsistent evaluation of overall site uncertainty due to the fact that not all variants are realizations are calibrated with respect to the full dataset, (2) possible sensitivity of results to choice of resolution in the detailed-scale DFN component, and (3) possible abstraction errors which occur in extracting effective 1-D transport parameters from a 3-D network model. Current research and development is underway to address these issues.

Application of the discrete-feature approach to the Äspö site indicates that spatial variation among different canister locations within the repository is a dominant component of the combined variability/uncertainty for this site. Consideration of results from all discrete-feature model variants indicates that, for this geological environment, heterogeneity leads to high spatial variability in the potential performance of the geologic barrier. This spatial variability appears to be a true property of the site, rather than a consequence of uncertainty in the site characterization. Among the evaluated aspects of uncertainty, the most important effects in terms of hydrologic parameter distributions are associated with assumptions concerning the relationships between transmissivity and pore geometry, and hence between advective velocity and sorption capacity, within the site-scale structures.

ACKNOWLEDGEMENTS

This work was supported as part of the SITE-94 project by the Swedish Nuclear Power Inspectorate (SKI). Data for SITE-94 were provided from the site investigations for the Äspö Hard Rock Laboratory by the Swedish Nuclear Fuel and Waste Management Co. (SKB).

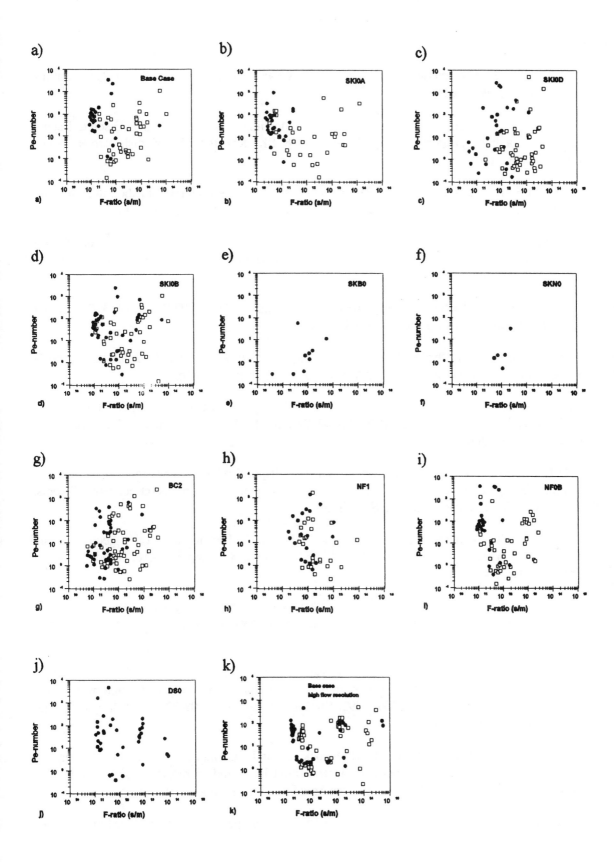

Figure 4. Scatter plots of F ratio vs. Peclet number for DF model variants

147

REFERENCES

[1]   SKI *SITE-94: Deep repository performance assessment project, 1996, SKI Report 96:36, Swedish Nuclear Power Inspectorate, Stockholm.*

[2]   Geier, J.E., and Thomas, A.L., Discrete-feature modelling of the Äspö site: 1. Discrete-fracture network models for the repository scale (SITE-94), 1996, SKI Report 96:5, Swedish Nuclear Power Inspectorate, Stockholm.

[3]   Geier, J.E., Discrete-feature modelling of the Äspö site: 2. Development of the integrated site-scale model (SITE-94), 1996, SKI Report 96:6, Swedish Nuclear Power Inspectorate, Stockholm.

[4]   Geier, J.E., Discrete-feature modelling of the Äspö site: 3. Predictions of hydrogeological parameters for performance assessment (SITE-94), 1996, SKI Report 96:7, Swedish Nuclear Power Inspectorate, Stockholm.

[5]   Tsang, Y.W., Stochastic continuum hydrological model of Äspö (SITE-94), 1996, SKI Report 96:9, Swedish Nuclear Power Inspectorate, Stockholm.

[6]   *Dverstorp, B., Geier, J., and Voss, C., Simple evaluation of groundwater flux and radionuclide transport at Äspö (SITE-94), 1996, SKI Report 96:14, Swedish Nuclear Power Inspectorate, Stockholm.*

[7]   Stanfors, R., M. Erlström, and I. Markström, 1991. Äspö Hard Rock Laboratory: Overview of the investigations 1986-1990, 1991, SKB Technical Report 91-20, Swedish Nuclear Fuel and Waste Management Co., Stockholm.

[8]   SKB RD&D Programme 95: Treatment and final disposal of nuclear waste -- Programme for encapsulation, deep geological disposal, and research, development and demonstration, 1995, Swedish Nuclear Fuel and Waste Management Co., Stockholm.

[9]   Tirén, S., Beckholmen, M., Voss, C., and Askling, P., Development of a geological and structural model of Äspö, southeastern Sweden (SITE-94), 1996, SKI Report 96:16, Swedish Nuclear Power Inspectorate, Stockholm.

[10]   Provost, A., Voss, C., and Neuzil, C., Glaciation and regional ground-water flow in the Fennoscandian shield (SITE-94), 1996, SKI Report 96:11, Swedish Nuclear Power Inspectorate, Stockholm.

[11]   Rhén, I. (ed.), U. Svensson (ed.), J-E Andersson, P. Andersson, C-O Eriksson, E. Gustafsson, T. Ittner, and R. Nordqvist, Äspö Hard Rock Laboratory: Evaluation of the combined longterm pumping and tracer test (LPT2) in borehole KAS06, 1992, SKB Technical Report 92-32, Swedish Nuclear Fuel and Waste Management Co., Stockholm.

[12]   Dershowitz, W., Lee, G., Geier, J., Hitchcock, S., and Lapointe, P., FracMan Version 2.4: Interactive discrete-feature data analysis, geometric modeling, and exploration simulation, User documentation, 1994, GAI report 923-1089, Golder Associates Inc., Redmond, Washington.

[13]   Barker, J.A., A generalized radial flow model for pumping tests in fractured rock, *Water Resources Research,* 1988, 24, 1796-1804.

Approaches Dealing with Fractured Media within the SKB Äspö Hard Rock Laboratory Project

Gunnar Gustafson
Chalmers University, Sweden
Ingvar Rhén
VBB Viak AB, Sweden
Anders Ström
SKB, Sweden

Abstract

The Äspö Hard Rock Laboratory is a research establishment run by the Swedish Nuclear Fuel and Waste Management Co (SKB). The laboratory is situated on the SE coast of Sweden in a heterogeneous granitic bedrock. Important questions have been to identify relevant key issues for characterisation and modelling in appropriate scales. The approach to handling spatial variability in groundwater flow modelling has been to model large scale features deterministically and small scale fracturing as a stochastic continuum. In the characterisation work the reliability of the existence, geometry and properties of different features have been found to be of great importance, and a formal procedure for handling these questions was worked out. Before the excavation of the laboratory the effects of the laboratory on flow to the tunnel and on groundwater pressure were predicted. The modelling results were then compared to the outcome during excavation. A good agreement was in general found but also some systematic deviations, that could be explained by featured tha were not found in the characterisation work. Some of the modelling results were not possible to verify because of characterisation difficulties. Parallel studies of a large scale flow and tracer test as well as the laboratory draw-down were made within the framework of the Äspö Task Force on Modelling of Groundwater Flow and Transport of Solutes. These studies were made by different modelling groups and with different approaches. It was found that in the site scale at Äspö, results were similar even if some approaches could not reproduce certain features of the results from the experiments.

Introduction

The Äspö Hard Rock Laboratory[1] is built in a granitic rock under an island close to the coast of SE Sweden. The Äspö HRL was built to provide an opportunity for research, development and demonstration of nuclear waste storage in a realistic and undisturbed rock environment down to the depth planned for a future deep repository for spent nuclear fuel. Throughout the project modelling of groundwater flow and transport has been an important item, since one of the main objectives of the project is to test groundwater models at different scale.

Key Issues and Scales

An important task throughout the Äspö project has been to characterise features pertinent to performance and safety assessment of a future repository. Thus a number of key issues[2] were identified as in order to structure the characterisation process. The identified key issues and their rationale are:

- *Groundwater flow* – since the only escapeway for dissolved radio-nuclides is by groundwater flow from the repository to the surface.
- *Groundwater chemistry* – since it governs the solubility of the nuclides and has influence on the corrosion rate of the fuel canisters.
- *Transport of solutes* – since the solubility of the radioactive species and their interaction with the geologic media will govern the magnitude of the release of radio-nuclides to the biosphere.
- *Mechanical stability* – since without it there is a risk for mechanical damage to the fuel canisters both during construction an for geologic time scales.
- *Geology* – since it is the framework for all pertinent processes and without a thorough understanding and knowledge of the geology of a site no reliable predictions of its performance can be made.

In a heterogeneous geological environment such as Äspö no complete description of all features can be achieved. This means that the properties of the rock have to be generalised with respect to the scale of the studied issue. In order to systemise characterisation and modelling the studies were standardised to four geometrical scales, identified by the size of a representative block: of rock:

- *Regional scale -*   > 1000 m
- *Site scale -*   100 – 1000 m
- *Block scale -*   10 – 100 m
- *Detailed scale-*   0 – 10 m

Groundwater modelling of Äspö and parts of it has been made on all scales depending on the problem studied. This paper will mainly deal with the Site scale modelling of the site, since heterogeneities of different kinds and scales had to be addressed simultaneously.

The Geology of the Äspö Area

Äspö, which is situated on the East coast of Southern Sweden, lies in a bedrock province of the Scandinavian shield called the Trans-Scandinavian Igneous Belt. It consists mainly of plutonic rocks with an age of about 1.7 Ga. The rocks of the Äspö area[3] are heterogeneous granitic rocks containing both xenolites of older metavulcanites and veins of a brittle fine-grained granite within the locally dominating dioritic variety, Äspö Diorite, of the Småland granite series.

Discontinuities occur in all scales from joints in the rock mass to regional fractured zones with an extension of several tens of kilometres. A common feature of the major discontinuities is that they are initiated early after the forming of the rock and have been reactivated at several tectonic events since then. This gives them heterogeneous properties and different styles at different locations. In Äspö and its neighbourhood the trend of the major fractured zones is E-W to NE, ad they are generally dipping steeply to the North. These major fractured zones are clearly indicated both topographically and geophysically. They are also major hydraulic conductors. There are no evidences of subhorizontal major fractured zones in the Äspö area.

Throughout Äspö there are a number of minor fractured zones trending NW to N-S and steeply dipping. Their geological and geophysical signatures are weak, but they are in many cases good hydraulic conductors.

The small scale fracturing is dominated by a steeply dipping fracture set trending NW. Other major sets are oriented in N-S and NE and there is also a sub-horizontal set. The fracture system distribution mapped at surface has shown to be persistent also at depth.

Site Characterisation and Conceptual Modelling

In the site characterisation of Äspö an integrated approach was taken where geological, hydrogeological and hydrochemical data were co-evaluated. The work was conducted in stages and after each stage a model, that was succesively updated, was presented. This meant that the models became considerably more comprehensive during the project, both regarding the concepts and the parameter-base involved.

During the work it became evident that it was necessary to keep better track of the conceptual models, the data on which the parameters of the models were derived and the approach used for the numerical modelling. This need was even more accentuated since different approaches were used in different scales and for different aspects of every key issue. Thus a condensed description scheme[4] for the used models was adopted and applied to all modelling, see Table 1.

Table 1 format for a condensed description of models in the Äspö HRL project[4]

<table>
<tr><td colspan="2" align="center">**MODEL NAME**</td></tr>
<tr><td colspan="2" align="center">**Model scope or purpose**
Specification of the intended us of the model</td></tr>
<tr><td colspan="2" align="center">**Process description**
Specification of the process accounted for in the model, definition of constitutive eqations</td></tr>
<tr><td>CONCEPTS</td><td>DATA</td></tr>
<tr><td colspan="2" align="center">**Geometrical framework and parameters**</td></tr>
<tr><td>Dimensionality and/or symmetry of the model. Specification of what the geometrical (structural) units of the model are and the associated geometrical parameters (the ones fixed implicitly in the model and the variable parameters).</td><td>Specification of the size of the modelled volume. Specification of the source of data for geometrical parameters (or geometrical structure). Specification of the size of the geometrical unit and resolution.</td></tr>
<tr><td colspan="2" align="center">**Material properties**</td></tr>
<tr><td>Specification of the material properties contained in the model (it should be possible to derive them from the process and geometrical units). Specification of the source of data for material</td><td>parameters (could often be the output from some other model). Specification of the value of material parameters.</td></tr>
<tr><td colspan="2" align="center">**Spatial assignment method**</td></tr>
<tr><td>Specification of the principles for the way in which material (and if applicable geometrical) parameters are assigned throughout the modelled volume. Specification of the source of data for model,</td><td>material and geometrical parameters, as well as stochastic parameters. Specification of the result of the spatial assignment.</td></tr>
<tr><td colspan="2" align="center">**Boundary conditions**</td></tr>
<tr><td>Specification of (the type of) boundary conditions for the modelled volume. Specifications of the source of data on boundary and</td><td>initial conditions. Specification of the initial and boundary conditions.</td></tr>
<tr><td colspan="2" align="center">**Numerical or mathematical tool**
Computer code used</td></tr>
<tr><td colspan="2" align="center">**Output parameters**
Specification of the computed parameters and possibly derived parameters of interest.</td></tr>
</table>

The table has proven to be very useful, and has given us a good opportunity to keep track of how things were done and also a possibility to compare the differences between different modelling approaches.

The Groundwater Flow Model of Äspö

Based on the conceptual and structural models i site scale numerical groundwater flow models were set up. Calculations were made for certain flow cases that could be characterised in the next stage of the site characterisation. The models were then compared to the measured outcome and on this basis scrutinized in order to evaluate their validity.

Since the rocks at Äspö make up a fractured heterogeneous medium, the modelling approaches taken had to cope with the difficulties involved with that from the very start. Another demand on the models was that they had to be operative and give guidelines for further characterisation as well as the construction of the site. The latter implied that predictions of constructability, i.e. required reinforcements of the tunnel as well as possible water inflows and necessary sealing of these by grouting. The construction problems, as well as the highest groundwater fluxes in the rock, are related to the major discontinuities. This lead us to an approach where the major hydraulic conductors were treated in a different way than the averagely jointed rock.

The major hydraulic conductors were treated deterministically, and were modelled as approximately planar features with a defined transmissivity. The identification of the hydraulic conductor domains (HCD) was made by an integrated analysis of geological, geohydrological and hydrochemical data. Depending on the abundance of identified features of the HCD a classification of reliability of existence, ranging from possible to certain, was adopted, see table 2.

Table 2 Classification of the level of reliability of structural features

Reliability class	Possible	Probable	Certain
Surface features	Single topographic identifications	Several topographic identifications or geophysical identification	Several topographic identifications and geophysical identification
Borehole identification	Fracturing Hydraulic anomaly	Fracturing and/or Hydraulic anomalies in several boreholes	Fracturing and/or hydraulic anomalies in several boreholes and radar anomaly and/or hydraulic contact between boreholes
Underground identification	Fractured area in tunnel Groundwater leakage	Fractured area in tunnel and/or groundwater leakage in more than one part of the tunnel	Fractured area in tunnel and/or groundwater leakage and/or geophysical and hydraulic connections
Combinations		Several combined possible indicators	Several combined probable indicators

From this followed that targeted characterisation of the HCDs of importance for the tunneling was carried through until the required degree of certainty was achieved. The transmissivity of important HCDs was assessed by crosshole pumping tests, where also the continuity and extension of them could be determined. For others transient packer tests or single hole pumping tests with a flowmeter survey was used. For a few HCDs outside the target volume a professional judgement based on the style of the conductor and statistics of the properties of those where data were at hand were used. On Äspö thus 16 HCDs were deterministically identified, see Figure 1

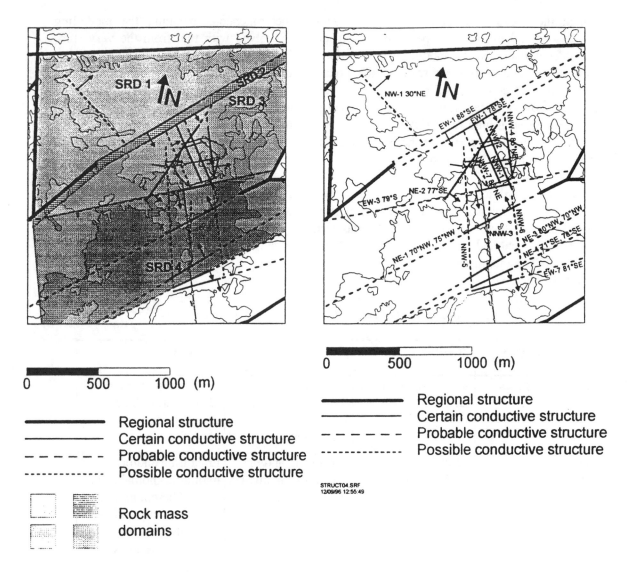

Figure 1   The major hydraulic conductor domains (HCD) and site scale hydraulic rock mass domains (SRD) at Äspö.

The averagely jointed rock between the HCDs was analysed by data mainly from core boreholes. For this purpose the "transport drilling" between the targeted HCDs was used. Core mapping, geophysical logging and hydraulic packer tests in different scales were used. Based on these data, the hydraulic properties of different subvolumes and lithological units of the target area were assessed. Basically the conductivity distributions on different scales for the different site scale hydraulic rock mass domains (SRD) were analysed. The correlation structure of the hydraulic conductivity was also analysed, but the correlation length was found to be short and data had a large nugget effect. For each SRD thus a distribution of the hydraulic conductivity, normally a lognormal distribution, and a scale dependency was defined, see Figure 2.

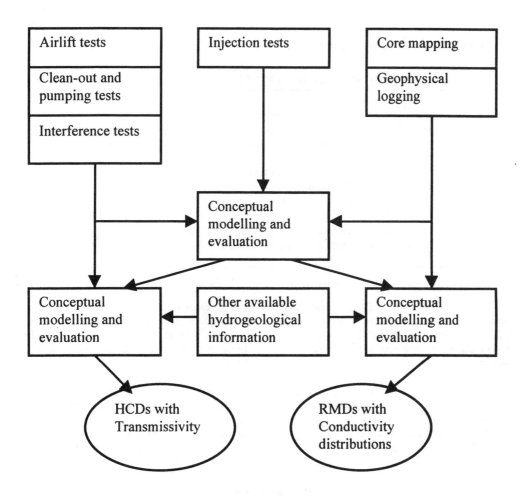

Figure 2 Flow chart for the assessment of model parameters

Numerical Modelling

The site scale numerical modelling within the Äspö project was mainly carried out with a stochastic continuum approach using the finite volume code PHOENICS[5]. The code is of a general solver type giving maximum flexibility in introducing new concepts into the studies. The modelling capabilities have thus grown considerably during the project and a number of concepts have been introduced such as: Density driven flow, 2D HCDs in a regular cell mesh, different transport algorithms and an efficient groundwater recharge algorithm. Also visualisation capabilities have been improved considerably during the project. The modelled volume has thus been treated as a stochastic continuum in between a network of deterministically determined hydraulic conductors. A condensed description of the site scale groundwater model of Äspö is given in table 3.

The hydraulic conductivity of the cells of the continuum have been assessed randomly according to the property distributions of the RMDs. In order to control the validity of these distributions a study of the effective conductivity of 50 m blocks was carried through with a discrete fracture network approach[6] (Fracman-Mafic). This study gave, however, the result that flow through the blocks could not always be expected with the assumed fracture network.

Table 3 A condensed description of the site scale groundwater flow models of Äspö

SITE SCALE GROUNDWATER FLOW MODEL OF ÄSPÖ

Model scope or purpose
Modelling of groundwater flows and pressures at Äspö under natural and perturbed conditions.
Stochastic continuum model.

Process description
The model is based on:
- Continuity equation (mass balance equation)
- Equation of motion (Darcy's law including density driven flow)
- Equation of state (Salinity – density relationships)
- Time evolution by consecutive pseudo-steady states

CONCEPTS	DATA
Geometrical framework and parameters	
2D hydraulic conductor domains (HCD) and defined 3D rock mass units (RMD) The modelled volume is a rectangular block covering Äspö and its surroundings.	HCDs determined by geological inference and hydraulic tests in site characterisation. RMDs are discretised in normally cubic blocks with a side of 20 m.
Material properties	
Transmissivities, deterministically determined for HCDs.	Transmissivities from analysis of data from hydraulic tests.
Stochastic distribution of hydraulic conductivity for RMDs.	Hydraulic conductivities from analysis of data from hydraulic tests in different scales. Empirical adjustment for scale.
Spatial assignment method	
Transmissivities deterministically allocated to defined HCDs.	Transmissivities from analysis of data derived from site characterisation.
Hydraulic conductivities of cells in RMDs randomly generated.	Hydraulic conductivity distributions derived from site characterisation.
Boundary conditions	
Upper boundary: Fixed recharge rate on Äspö. Constant head at sea and peat areas. Lower boundary: No flow. Side boundaries: Prescribed pressure. Tunnel boundaries: Prescribed pressure with flow resistance or defined sinks.	undisturbed water table. No flow lower boundary based on general assuimption of decreasing K with depth. Hydrostatic pressure with assumed salinity distribution on side boundaries. Skinfactor and sink strengths from professional judgement.

Recharge rate based on calibration using

Numerical or mathematical tool
PHOENICS

Output parameters
- Groundwater pressures
- Groundwater fluxes
- Salinity distribution
- Flow trajectories

The results from the modelling was evaluated[2] and can be summarised as follows:

- The outcome of the cumulative flow to the tunnel was 84 – 93 % of the prediction.It shall however be borne in mind that a large portion of this was estimated because of the necessary grouting and the working conditions in the tunnel.
- Major inflows to the tunnel is as expected connected related to the HCDs. Their positions were predicted within reasonable range.
- Draw-downs and groundwater pressures were well predicted until the elevator shafts were opened. A minor but very conductive fracture zone was penetrated by the shaft which locally gave large draw-downs. Introduction of this feature into the model and recalculation with actual sink strengths gave reasonable accuracy.
- Verification of the flux distribution has not been possible because of measurement biases in boreholes and tunnel.
- The original and resulting salinity fields were reasonably well reproduced.
- Flow trajectories have not been possible to verify. Tracer test made for assessment of connectivity of the system do not contradict calculated trajectories.

Parallel International Studies

In order to test different approaches to the problem of modelling groundwater flow and transport in a heterogeneous rock volume such as at Äspö the work in the Äspö Task Force on Modelling of Groundwater Flow an Transport was Initiated. The Task Force itself is a peer group where modelling problems related to the Äspö project are selected and initiated as modelling tasks. These tasks are then carried through in parallel by modelling groups tied to the different participating organisations. So far two tasks have been carried through[7,8] where groundwater flow and transport were modelled for the whole target volume, i.e. the large scale pumping test LPT2, which also included a tracer test, and the draw-down caused by the construction of the laboratory access tunnel. As a reference case the modelling of the same experiments by the project modelling team was used. Thus different approaches, see Table 4, like equivalent continuum, stochastic continuum using FEM, discrete fracture network models and discrete channel network models were used.

In total some ten parallel studies for each task were carried through. The results, similarities and differences were later evaluated in the framework of predefined performance measures. The final conclusions of these studies were that the results of different approaches were more similar than what was expected, and that the models used and on the scale in question give similar, if not equivalent, results.

Table 4 Approaches used for parallel modelling studies of the LPT2 experiment

Study	Continuum/ Discontinuum	Stochastic/ deterministic	Treatment of heterogeneity	Density effects	Transport approach	Numerical method
CRIEPI	Continuum	Deterministic	By domains	No	Direct coupl.	FEM
VTT I	Continuum	Deterministic	By domains	No	-	FEM
VTT II	Discontinuum	Deterministic	No	No	Direct coupl.	Analytical
BRGM I	Continuum	Deterministic	By domains	Yes	Part. track.	FDM
BRGM II	Continuum	Deterministic	By domains	No	Direct coupl.	FEM
CFE	Continuum	Stoch/determ.	Domains/stoch	Yes	Part.track.	FVM
Hazama	Continuum	Stochastic	Stochastic K	No	Part. track	FEM
AEA	Cont./disc.	Stochastic	Stochastic K	No	Part. track.	FEM
Golder	Discontinuum	Stochastic	Frac. Zones	No	Part. track.	FEM
Itasca	Discontinuum	Deterministic	Frac. Zones	No	Part. track	FEM
KTH	Discontinuum	Stochastic	Flow channels.	No	Part. follow.	FDM

References

[1] SKB, Aspö Hard Rock Laboratory, 1997, SKB TR 97-08

[2] Rhén I, Bäckblom G, Gustafson G, Stanfors R and Wikberg P, ÄSPÖ HRL – Geoscientific evaluation 1997/2, Results from pre-investigations and detailed site characterisation, Summary report, 1997, SKB TR 97-03

[3] Rhén I, Gustafson G, Stanfors R and Wikberg P, ÄSPÖ HRL – Geoscientific evaluation 1997/5, Models based on site characterisation 1986 – 1995, 1997, SKB TR 97-06

[4] Olsson O, Bäckblom G, Gustafson G, Rhén I, Stanfors R and Wikberg P, The structure of conceptual models with application to the Äspö HRL project, SKB TR 94-08

[5] Spalding D B, A general purpose computer program for multi-dimensional one- and two-phare flow, 1981, Math. Comp. Sim., 8, 267-276. See also http://www.cham.co.uk

[6] Axelsson C, Jonsson E-K, Geier J and Dershowtz W, Discrete fracture modelling, 1990, SKB PR 25-89-21

[7] Gustafson G and Ström A, The Äspö Task Force on Modelling of Groundwater Flow and Transport of Solutes, Evaluation report on Task No 1, the LPT2 large scale field experiments, 1995, SKB, ICR 95-05.

[8] Gustafson G, Ström A and Vira J, The Äspö Task Force on Modelling of Groundwater Flow and Transport of Solutes, Evaluation report on Task No 3, the Äspö tunnel drawdown experiment, 1997, SKB, ICR 97-06

Spatial Variability in the Geosphere Models used for AECL's Long Term Performance Assessment of the Disposal of Canada's Nuclear Fuel Wastes

C.C. Davison, T. Chan, N. Scheier, T. Melnyk
A. Brown, M. Gascoyne and T.T. Vandergraaf
AECL, Whiteshell Laboratories, Canada

Introduction

In 1978 Atomic Energy of Canada Limited (AECL) was given the responsibility by the governments of Canada and Ontario to conduct research and development on the concept of disposing of Canada's nuclear fuel waste in a deep underground repository in intrusive igneous rock of the Canadian Shield (Joint Statement, 1978). In 1981 the governments of Canada and Ontario further announced that no disposal site selection would be undertaken until after the disposal concept was accepted (Joint Statement, 1981). Since then AECL has developed a multibarrier disposal concept that involves isolating the waste in long-lived, corrosion-resistant containers emplaced in a sealed vault at a depth of 500 to 1000 m in plutonic rock of the Canadian Shield. The disposal vault would be a network of horizontal tunnels and disposal rooms excavated in volumes of relatively low permeability plutonic rock between fracture zones, with vertical shafts extending from the surface to the tunnels. The technical feasibility and social aspects of the concept, and an assessment of the impact of the concept on the long-term future environment and human health, have been documented in an Environmental Impact Statement (EIS) (AECL 1994) and in a set of nine primary references. During 1994 to 1997 AECL's EIS has been reviewed by a federal Environmental Assessment Panel that was established to judge the safety and acceptability of the disposal concept. As part of their review, the Panel organized a series of public hearings to solicit input from a broad crossection of Canadian society including the scientific and technical community. The public hearings concluded in 1997 March and the Panel is currently developing its recommendations to assist the governments of Canada and Ontario in reaching decisions on the safety and acceptability of AECL's disposal concept and the next steps that must be taken to ensure the safe long term management of nuclear fuel wastes in Canada.

Canada's Atomic Energy Control Board requires that quantitative estimates will need to be made of the radiological risk associated with a nuclear fuel waste disposal vault up to 10 000 years following vault closure (AECB 1987). Radiological risk is defined as the probability that an individual of the "critical group" will incur a fatal cancer or serious genetic defect due to exposure to radiation. The critical group is a hypothetical group of people assumed to live at a time and place and in such a way that they are at greatest risk to exposure of any future radiation releases from the disposal vault. The AECB has specified that the radiological dose to individual members of the critical group should be less than 0.05 milliseiverts per year (about 1% of exposures received from natural background radiation) to ensure a radiological risk of less than 10^{-6} per year.

Part of the evidence that AECL presented to the Panel for the postclosure safety of the disposal concept involved constructing two case studies of contaminant releases from possible disposal systems (consisting of vault, geosphere and biosphere components), evaluating the long-term postclosure radiological exposure that would be received by members of a hypothetical critical group, and comparing the results to the AECB risk/dose criteria. In order to achieve the objectives of these case studies AECL believed it was important that realistic information needed to be used and that important relationships be reflected in the models and data. For instance, the structural, hydrogeological, chemical and physical conditions of a disposal site are interrelated and AECL believed their interrelationships and interactions with the disposal vault could not be defined in non-site-specific terms. Therefore to perform these case studies of possible postclosure effects, AECL defined site specific geosphere characteristics and positioned the disposal vaults within these site-specific geospheres. The site-specific geosphere characteristics were based on field and laboratory data from the plutonic rock research area AECL had investigated most thoroughly, the Whiteshell Research Area in southeastern Manitoba, Canada. The hypothetical disposal vaults were assumed to be situated at a depth of 500 m within this geosphere at a location that coincided with the location of the site of AECL's Underground Research Laboratory where a substantial amount of borehole and underground information was available to define the spatial variability of the geosphere characteristics.

Spatial Variability

The case studies AECL used for its long term safety assessments of nuclear fuel waste disposal incorporated geosphere models that were based on known or inferred conditions at the site of the Underground Research Laboratory. Thorough subsurface investigations had revealed the plutonic rock at the URL site was intersected by an interconnected network of large scale fracture zones that controlled groundwater flow and hydrogeochemical patterns in the rock. The known geometric arrangement of these fracture zones was explicitly represented in 2-D and 3-D groundwater flow and solute transport models of the site using planar and volume elements that were available in AECL's MOTIF finite element flow and transport code (Davison et al. 1994; Chan et al. 1996). Where information was lacking and the extent or interconnection of some of these fracture zones was unknown, the geometric model of the rock's structure was constructed by assuming the fracture zones extended as projected and they interconnected with other fracture zones. The degree of fracturing elsewhere in the rock at the URL site was also spatially variable which caused both a trend of decreasing permeability with depth as well as a pronounced anisotropy in the permeability of the upper 300 m of the rock (i.e. enhanced vertical permeability). The observed trend of decreasing permeability with depth and the permeability anisotropy were incorporated into the groundwater flow model by representing the rest of the rock at the URL site (ie. other than the major fracture zones) by layers of finite elements that were assigned spatially-variable hydraulic characteristics. Subsequent sensitivity analysis with the groundwater flow model was used to investigate the feasibility of some of these assumptions and the geometric model was modified where appropriate (Scheier et al. 1992). A final check of the flow model of the URL site involved comparing simulated groundwater ages with the ages determined by isotopic analyses of groundwater samples from the URL site.

Once the geometric framework was established for the groundwater flow and transport model of the URL site, the location was chosen for the hypothetical disposal vault and the layout of the waste emplacement areas of the vault and the positions of the access shafts and tunnels were defined. These were also incorporated into the flow model using different finite elements to represent the rooms and shafts themselves, excavation damage zones surrounding the rooms and buffers, backfills and other engineered seals within the waste emplacement rooms. The disposal vault was purposefully located near a major, low dipping fracture zone in the site model so the effects of this geologic structure on contaminant transport could be examined.

The final feature that was built into the groundwater flow model was a representation of various hypothetical groundwater supply wells that could extract water from the groundwater flow paths contaminants would follow in moving from the vault to discharge locations in the surface biosphere. This representation allowed the effects of wells at different depths and with different pumping rates to be simulated. The wells could penetrate the rock to depths up to 200 m and were located in the groundwater discharge area of the low-dipping fracture zone that passed near the vault.

Particle tracking was used on the MOTIF flow model results to trace the pathways contaminants would follow through the geosphere along groundwater flow paths from the vault to the groundwater wells and to surface discharge locations in the biosphere (Nakka and Chan 1995). Particle tracking was also used to determine how the geometry of the groundwater flow field would change in response to the wells pumping at different depths and rates as well as how the flow field would change over time due to the heating effects of the heat-generating wastes in the disposal vault. The results of these particle tracking analyses of the advective groundwater pathways as well as work to identify important diffusive pathways in the groundwater were used to construct a simplified network of one-dimensional paths that could describe the three-dimensional geometry of all of the important transport paths from the vault to the biosphere during a future timeframe of 10,000 years. This assembly of one-dimensional geosphere pathways comprised a network that was in turn used to construct a model of radionuclide and contaminant transport from the disposal vault to the biosphere for an overall assessment of the long term performance of the complete disposal system (vault + geosphere + biosphere) (Goodwin et al. 1994). This model of the transport of vault contaminants through the geosphere from the vault to the biosphere is the geosphere model component of AECL's integrated system model for performance assessment. The geosphere model is referred to as GEONET (Davison et al. 1994).

After the geometric elements of all the possible transport paths from the vault through the geosphere were established, the spatial variability of the geochemical characteristics in the geosphere was also mapped onto the GEONET network to be able to simulate chemical effects on the transport of vault contaminants. This involved incorporating field and laboratory information on the spatial variation of redox conditions, the groundwater chemistry and the chemistry of the minerals occurring along the flow and diffusion pathways that could affect the transport of some of the vault contaminants (Davison et al. 1994; Vandergraaf and Ticknor 1993). In some cases the one-dimensional network of geosphere transport pathways used in GEONET was divided into extra segments to reflect the spatial variability that was known to exist in these geochemical characteristics from field studies at the URL site.

Discussion of Issues

The geosphere models AECL used for long term postclosure assessment of the disposal concept were constructed to describe and account for the effects that spatial variability in the physical and chemical characteristics of the geosphere would have on the transport of vault contaminants through the geosphere for particular, yet hypothetical site. They were based on models and analyses that used actual field and laboratory data of the spatial variability known to exist in the rocks at the URL site, rather than a generic understanding of the spatial variability that could exist at any possible site in the Canadian Shield. Scientific and technical reviewers of AECL's EIS raised many issues on the geosphere modelling aspect during the recent CEAA Panel hearings (AECL 1996). Many reviewers were concerned about the ability of a network of one-dimensional geosphere pathways (GEONET) to properly describe the three-dimensional transport of vault contaminants through the geosphere. Other concerns related to how to account for random variations in the spatial variability of flow and transport properties in the geosphere within the GEONET model AECL used for performance assessment calculations. Some reviewers believed that the only way to address some of these issues was to develop fully coupled (thermal, fluid, mechanical, chemical), transient, three-dimensional flow and solute transport models of the geosphere in the final models being used for integrated total system performance assessments of nuclear fuel waste disposal. Another issue was raised on how best to incorporate spatial uncertainty into the geosphere modelling approach. Although the geosphere models of the URL site used conservative or pessimistic assumptions when there was uncertainty in the spatial aspects of the groundwater flow characteristics, many reviewers believed wider ranges of hydrogeological parameter values should be incorporated into the models of integrated performance assessment. AECL's second case study was developed in response to some of these issues and it demonstrated the robustness of the multibarrier disposal concept to provide a wide margin of safety even if the spatial characteristics of the geosphere are less favorable to waste containment.

References

AECB (Atomic Energy Control Board). 1987. Regulatory policy statement. Regulatory objectives, requirements and guidelines for the disposal of radioactive wastes – long-term aspects. Atomic Energy Control Board Regulatory Document R-104, Ottawa, 1987 June 5.

AECL. 1994. Environmental impact statement on the concept for disposal of Canada's nuclear fuel waste. Atomic Energy of Canada Limited Report, AECL-10711, COG-93-1.

AECL. 1996. Response to request for information. Atomic Energy of Canada Limited Report, AECL-11602-VI, COG-96-237-VI.

Chan, T., V. Guvanasen, B.W. Nakka, J.A.K. Reid, N.W. Scheier and F.W. Stanchell. 1996. Verification of the MOTIF code Version 3.0. Atomic Energy of Canada Limited Report, AECL-11496, COG-95-561-I.

Davison, C.C., T. Chan, A. Brown, M. Gascoyne, D.C. Kamineni, G.S. Lodha, T.W. Melnyk, B.W. Nakka, P.A. O'Connor, D.U. Ophori, N.W. Scheier, N.M. Soonawala, F.W. Stanchell, D.R. Stevenson, G.A. Thorne, T.T. Vandergraaf and P. Vilks. 1994. The disposal of Canada's nuclear fuel waste: The geosphere model for postclosure assessment. Atomic Energy of Canada Limited Report, AECL-10719, COG-93-9.

Goodwin, B.W., D.B. McConnell, T.H. Andres, W.C. Hajas, D.M. LeNeveu, T.W. Melnyk, G.R. Sherman, M.E. Stephens, J.G. Szekely, P.C. Bera, C.M. Cosgrove, K.D. Dougan, S.B. Keeling, C.I. Kitson, B.C. Kummen, S.E. Oliver, K. Witzke, L. Wojciechowski and A.G. Wikjord. 1994. The disposal of Canada's nuclear fuel waste: Postclosure Assessment of a reference system. Atomic Energy of Canada Limited Report, AECL-10717, COG-93-7.

Joint Statement. 1978. Canada/Ontario radioactive waste management program. Joint statement by the Minister of Energy, Mines and Resources Canada and the Ontario Energy Minister, 1978 June 5. In Unpublished Documents Cited in the EIS and Primary References. Atomic Energy of Canada Limited Technical Record, TR-567, COG-92-27.

Joint Statement. 1981. Canada/Ontario joint statement on the nuclear fuel waste management program. Joint statement by the Minister of Energy, Mines and Resources Canada and the Ontario Energy Minister, 1981 August 4. In Unpublished Documents Cited in the EIS and Primary References. Atomic Energy of Canada Limited Technical Record, TR-567, COG-92-27.

Nakka, B.W. and T. Chan. 1995. A particle tracking code (TRACK 3D) for convective solute transport modelling in the geosphere: Description and users manual. Atomic Energy of Canada Limited Report, AECL-10881, COG-93-216.

Scheier, N.W., T. Chan and F.W. Stanchell. 1992. Sensitivity analysis using two-dimensional models of the Whiteshell geosphere. Atomic Energy of Canada Limited Technical Record, TR-572.

Vandergraaf, T.T. and K.V. Ticknor. 1993. A compilation and evaluation of radionuclide sorption coefficients used in the geosphere model of SYVAC for the Whiteshell Research Area. Atomic Energy of Canada Limited Report, AECL-1056, COG-92-59.

An Integrated UZ Flow and Transport Model
for Yucca Mountain, Nevada, U.S.A.

Gilles Y. Bussod, Bruce A. Robinson, Andrew V. Wolfsberg,
Carl W. Gable, and Hari S. Viswanathan
Los Alamos National Laboratory, USA

Abstract

The unsaturated zone (UZ) is one of the primary barriers to the migration of radionuclides to the accessible environment from the potential repository at Yucca Mountain, Nevada . This paper describes the overall methodology and highlights the results from the first integrated flow and transport model for this geologic repository. The model incorporates the results of mineralogic, structural, hydrologic, and geochemical data from field and laboratory measurements. The purpose of the integrated modeling studies is to predict the migration paths and times of radionuclides from the potential repository to the water table in a heterogeneous geologic setting.

Introduction

A versatile, three-dimensional, multi-phase flow and transport code, FEHMN [1, 2, 3] is coupled with an efficient and flexible computational grid generation method [4, 5, 6], to incorporate the structural, mineralogic, hydrologic and geochemical heterogeneities which characterize the potential geologic repository at Yucca Mountain, Nevada. The computations honor databases and are populated with flow and transport parameters in three-dimensional space so that an integrated process model can be built and abstracted for Performance Assessment calculations.

Stratigraphic and Structural Characteristics

The potential geologic repository site at Yucca Mountain, Nevada is being designed as the first United States repository for commercial, high-level radioactive waste. It is located in the northern Mojave desert, 150km northwest of Las Vegas, Nevada (cf., location map, Figure 1). The unsaturated zone varies between 500m and 750m in thickness and is comprised from the surface to the water table of hydrogeologic units which include alluvium (0-50m), the Tiva Canyon welded tuff (TCw), the Paintbrush nonwelded tuff (Ptn), and theTopopah Spring welded tuff (TSw) in which the potential repository horizon is located. Below the potential repository horizon lies the Calico Hills formation (CHn) which consists of interbedded welded (mostly zeolitic) and non-welded (mostly vitric) tuffs of varying zeolite mineral abundances. Because of its mineralogy and hydrologic properties, this formation represents the principal barrier to radionuclide migration from the repository to the water table and accessible environment. The welded units are generally associated with fracture flow and characterized by low matrix permeabilities and high fracture densities, whereas the non-welded units are associated generally with matrix dominated flow, characterized by high matrix permeabilities and low fracture densities. Major high angle normal fault systems in the area trend approximately North-South.

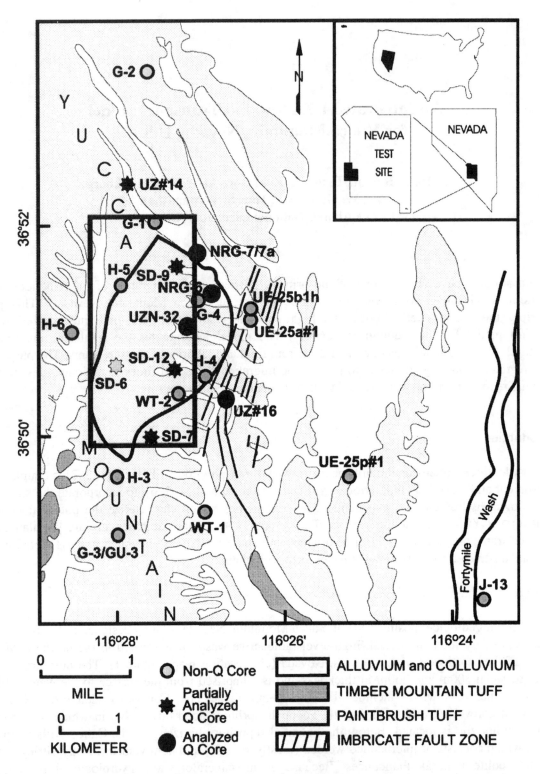

Figure 1. Location map and model area. A three-dimensional model is built for the entire map area. Area outlined in rectangular box is where a 2000m x 4000m x 300m block with 40m grid spacing geostatistical model is built from the potential repository horizon to the water table. Boreholes G-2, UZ#14, NRG-7/7a, SD-9, SD-12, and SD-7 are where perched water has been observed. Also shown is the location of 21 drill holes where 3817 measurements of zeolitic abundance are used for geostatistical modeling.

Mineralogic Characteristics

The mineralogy of the unsaturated zone at Yucca Mountain plays an important role in the prediction of hydrologic and transport processes. The distribution of secondary minerals such as zeolites and clays present in the UZ, can strongly influence the flow patterns of percolating fluids through the UZ. In addition, the presence of these minerals is linked to the sorption of many radionuclides, which is strongly controlled by the mineral distribution in the flow pathways between the repository and the water table. For these reasons, we have developed an automated method for incorporating mineralogic data and interpretations directly into our site-scale model.

In this work, the distribution of zeolites in the unsaturated zone is modeled using this information either through the assignment of a threshold percentage for each stratigraphic layer above which the rock is given the properties of zeolitic tuff, or by populating the model with a geostatistical model of zeolite abundance that is used as an indicator in assigning flow and transport properties.

Hydrologic properties

The hydrologic characteristics of the UZ system are investigated by integrating laboratory and field hydrologic measurements with environmental isotope data and perched and pore water chemistry data to understand the past and present hydrologic behavior of the mountain. The hydrologic parameters used in the model were obtained from the recent report on the unsaturated zone flow system [7]. These parameters include unsaturated hydrologic property values for all units defined in the geologic framework model, as well as the hydrologic properties of zeolitic tuffs. Two- and three-dimensional flow simulations, using the spatially varying infiltration map [8], and variations of this map, are performed to provide either steady-state or transient flow fields that are used to simulate the transport of major-ions or environmental isotopes, and to predict radionuclide migration. The major-ion chemistry and environmental isotope modeling studies provide important constraints on the hydrologic and transport system; therefore, we devote considerable effort toward developing models that are consistent with these data sets.

Steady-state, dual permeability fluid flow fields are calculated for different hydrologic property sets, after which particle tracking transport simulations of ^{99}Tc and ^{237}Np from the repository to the water table are computed. The transport metrics used to identify which property sets yield unique or distinctively different results are (i) the first arrival time of radionuclide at the water table, (ii) the relative amount of radionuclide arriving early, and (iii) the degree of sorptive delay of ^{237}Np to zeolitic tuffs. In addition, the relative flow split between the fractures and matrix in the zeolitic tuffs was used as a hydrologic metric for judging the property sets. Because transport of radionuclides escaping the repository is the subject of the study, the emphasis is on predicted flow and transport below the repository horizon.

In order to understand the overall effect of heterogeneities on the transport characteristics of the unsaturated zone, environmental isotope data and numerical modeling are coupled to identify flow and transport processes as well as constraining the property sets used to model them [9, 10, 11, 12]. The most significant finding from this work is that minor changes in hydrologic parameters lead to very different flow distributions in matrix and fractures. We find that property sets that do not lead to significant fracture flow in the Ptn, except in fault zones, provide a better match with the substantial set of analyses of ^{36}Cl/Cl ratios measured for the Exploratory Studies Facility (ESF) samples [9, 10, 11, 13, 14]. This finding supports the assumption that episodic infiltration events in the PTn are damped except in fault zones.

In this study, two- and three-dimensional models are used to assess the effects of spatially distributed infiltration and dipping strata of varying thickness and extent in a faulted system. One-dimensional models are then used to examine, in detail, the effects of episodic transients, fault zone properties, and the relationship between material properties, infiltration rate, and travel time from the surface to the potential repository horizon.

In order to understand the past characteristics of the hydrologic system(s) at Yucca Mountain, perched- and pore-water chemistry databases are coupled with hydrologic parameters to evaluate different conceptual models of the hydrologic behavior of the mountain during past wet and dry climatic conditions (i.e., for high and low infiltration rates). The premise for this exercise is that chemical data, when combined with hydrologic data, provide the most relevant information for ìvalidatingî radionuclide transport models, as the movement of naturally occurring solutes is closely related to the potential migration of radionuclides. The existence of perched water bodies at Yucca Mountain (i.e., boreholes , G-2, UZ-14, NRG-7/7a, SD-9, SD-12, and SD-7; Figure 1) indicate that barriers to flow exist in the unsaturated zone. These barriers constitute heterogeneities related to the presence of both fast pathways and lateral flow. The perched waters are characterized by low chloride concentrations and relatively older apparent ^{14}C ages compared to pore waters located in the overlying TSw and underlying CHn units. The apparent disequilibrium between pore water and perched water chemistries is consistent with a model of transient flow, and we demonstrate that the stable isotope data and apparent ^{14}C and ^{36}Cl/Cl ages can be interpreted as resulting from a mixture of late Pleistocene/early Holocene water with modern waters [15]. The existence of these mixed ancient/modern perched water bodies therefore implies that an additional retardation mechanism exists above the Calico Hills formation. The ubiquitous young ^{14}C ages in the Calico Hills formation pore waters beneath the perched water bodies and in other areas also imply that a lateral component of flow occurs in the vitric Calico Hills.

Geochemical Properties

The radionuclides examined in this work include the actinides ^{237}Np, ^{233}U, ^{242}Pu, and ^{99}Tc. The nonsorbing ^{99}Tc can be thought of as a surrogate for other conservative radionuclides such as ^{36}Cl, ^{14}C, and ^{129}I, so the others are not modeled explicitly though they are potentially important to total system performance. The geochemical parameters used for transport simulations are derived from laboratory based studies of radionuclide migration [16, 17, 18, 19, 20, 21]. Sorption coefficients for these radionuclides have been measured on a variety of tuff types at different chemical conditions.

The radionuclides ^{99}Tc, ^{14}C, and ^{129}I are all assumed to be conservative, nonsorbing radionuclides in the unsaturated zone, based on measurements and their inferred oxidation states in the Yucca Mountain fluids. ^{14}C is perhaps better described by a multi-phase model that includes partitioning between the liquid and gas phases, and precipitation as calcite on the rock.

The actinides exhibit small to moderate amounts of sorption to the Yucca Mountain tuffs, especially to zeolites such as clinoptilolite and mordenite, which are present beneath the potential repository. ^{237}Np and ^{233}U sorption appears to be minimal on vitric and devitrified tuffs that have not been altered by secondary mineralization. The sorption coefficient of ^{237}Np onto zeolitic tuff is slight but significant and an appropriate value for this radionuclide is about 2 cc/g on zeolitic tuff; we conservatively assume a value of 0 elsewhere, although measurable sorption has been observed on other tuff samples associated with other trace mineral abundances (e.g., clays). Uranium exhibits somewhat higher sorption coefficients on zeolites (3-20 cc/g, depending on concentration), but similarly small values on unaltered tuff samples. Plutonium sorbs to all types of tuff at Yucca Mountain; for this radionuclide we use a nominal value of 50 cc/g on all tuffs. Ordinarily, this degree of sorption would imply very slow migration although colloid-facilitated transport of plutonium is also considered.

The matrix diffusion model for fracture transport requires the input of a diffusion coefficient for each radionuclide into the tuff. Values in the range of 10^{-10} to 10^{-11} m^2 s^{-1} have been measured under saturated conditions [18, 19, 20, 21]. In addition, fracture coatings do not appear to inhibit the diffusion into the rock matrix. Given the rather small sensitivity to diffusion coefficient, we choose a fixed diffusion coefficient of 10^{-11} m^2 s^{-1} for all simulations in this section. Dispersivity also had a relatively small impact on the predictive results. Therefore, to simplify the interpretation, we conservatively assume no hydrodynamic dispersion. When this assumption is made, all spreading of radionuclide is caused by large-scale heterogeneities captured explicitly in the model, such as flow partitioning between the fractures and matrix and variations in vitric and zeolite thicknesses with position beneath the repository.

UZ Field Transport Test: Scaling and Validation

The effects of scaling and the remaining uncertainties in the behavior of the UZ barrier requires the characterization of its response to different hydrologic conditions so that we may take credit for its mitigating role in radionuclide migration under different future climates. In spite of outcrop and borehole studies, there is a paucity of information on both the physical and chemical properties of the Calico Hills lithologic package and the structural and lithologic discontinuities it exhibits. In addition, little is known of the applicability of laboratory transport data to the field environment. Finally there remain critical uncertainties in key parameters used in Performance Assessment for License Application.

An unsaturated zone field transport test now underway at Busted Butte, located 9 km southwest of the potential repository site at Yucca Mountain, aims to directly address these data needs [22]. The underground field test is designed to scale/validate/calibrate our transport models for the UZ barrier with the following goals:

(1) confirm the validity of our dual permeability unsaturated zone transport process models,
(2) validate our laboratory databases on sorption and matrix diffusion,
(3) confirm the validity of our minimum Kd approach at the field scale,
(4) assess the role of heterogeneities such as fractures in the Calico Hills formation (e.g., fracture matrix interaction within the Chn),
(5) investigate colloid mobility in fractured welded and nonwelded rocks,
(6) determine fracture flow and transport mechanisms in unsaturated rocks,
(7) investigate the effect of permeability contrasts at welded/nonwelded contacts and,
(8) develop our testing capabilities for possible future experiments beneath the repository horizon (i.e., east-west (EW) drift extension).

The results are incorporated directly into updated process flow and transport models using the methodology described above and abstracted into performance-based calculations.

Incorporation of Heterogeneities into a 3-D Computational Mesh

Structural Heterogeneities

The distribution of major and minor structural discontinuities (i.e., faults and fractures) has been organized into a numerical model using the Lynx [23] and STRATAMODEL software. Structural heterogeneities such as fractures and faults are incorporated into the numerical grid schemes by using both discrete curviplanar representations (e.g., faults) as well as the dual permeability conceptual model (e.g., fractures and fault zones).It has been shown previously [17] that dual permeability representations of fracture/matrix interactions provide a more robust hydrologic setting on which to predict transport of solutes through the unsaturated zone.

Computational mesh generation involves translating the Integrated Site Model, ISM [23,24], of geologic and hydrologic structure and stratigraphy to a finite element mesh. This process maintains the geometry defined by the ISM model while meeting additional geometric constrains on element size and geometry for stable and accurate flow and transport solutions (Figure 2).

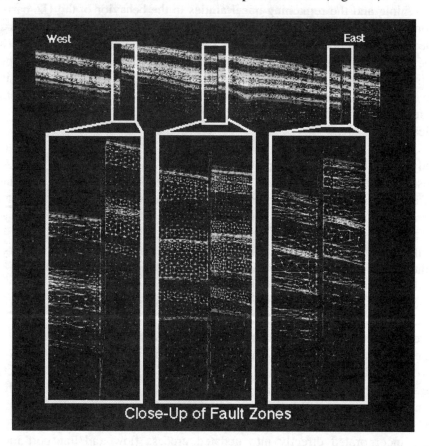

Figure 2. Two-dimensional mesh generated from YMP model ISM showing faults as part of the computational grid. The faults from west (left) to east (right) are Solitario Fault, Ghost Dance Fault and Bow Ridge Fault. The cross section is bounded on the west and east by unnamed faults in ISM. Colors represent different hydrogeologic units. Close-up views of the grid are shown at each of the faults.

The GEOMESH/X3D grid generation toolkit [4, 5, 6, 25], has been developed to provide an integrated software package for all grid generation steps, from initial ISM model import to quality checking of input data, mesh optimization, mesh post processing and quality checking and interfacing with the FEHMN flow code [1, 2, 3].

Coupled Mineralogic, Geochemical and Hydrologic Heterogeneities

The distribution of minerals has been organized into a numerical model using the STRATAMODEL software. An interface between the mineralogic model [26, 27], and the flow and transport simulation models developed in the present study has been developed to make the investigation of mineralogic controls on flow and transport as seamless as possible. The distribution of zeolites in the unsaturated zone is modeled using this information either through the assignment of a cut-off percentage above which the rock is given the properties of zeolitic tuff, by populating the model with a zeolitic abundance or by using a stochastic formulation for zeolite abundances that are then used in assigning flow and transport properties.

Sorptive zeolitic minerals in the Calico Hills, located between the potential repository and the water table strongly affect travel times of radionuclides such as [237]Np through the unsaturated zone. Prior to 1997, the site-scale geologic model for the potential repository at Yucca Mountain was represented by a binary model comprised of layers of either zeolitic or non-zeolitic material in order to model the sorptive behavior of the zeolites. This division drew upon knowledge of whether a layer was sparsely or abundantly populated with zeolitic minerals. For purposes of transport calculations, those layers containing abundant zeolites were assigned zeolitic properties (i.e, sorptive properties) while the others were given vitric properties (i.e, non-sorptive properties). This assumption was found to be problematic because zeolites are formed through secondary alteration, and their distribution does not always follow the primary geologic stratigraphy.

The greatest weakness of such a binary framework models lies in its effect on species retardation: if a layer is not as completely zeolitic as portrayed, retardation of contaminant species may not be accurately captured. Specifically, zeolitic tuff is less permeable than vitric tuff, and a radionuclide such as [237]Np which sorbs to zeolites may therefore bypass portions of heterogeneous zeolitic layers leading to less retardation of [237]Np. Therefore, the homogenous zeolitic layers of the site-scale model used in previous studies [16, 17], may have overpredicted the sorption of the zeolites (Figure 3a).

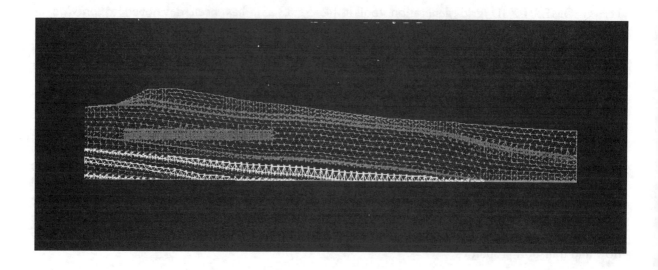

Figure 3.   Antler Ridge cross section
(a) Continuous zeolitic layers (light-shaded area) of the Site-Scale Model [17]

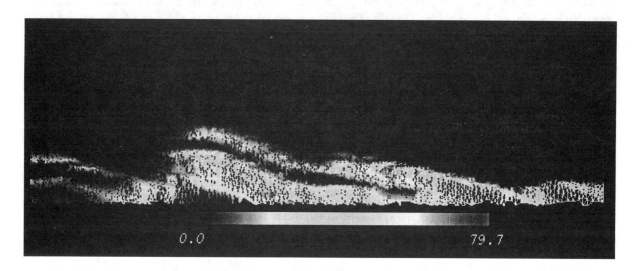

0.0   *79.7*

(b) Zeolitic distributions are mapped from the 3-D mineralogical model [26] to the FEHMN finite element mesh. The profiles look different because (b) accounts for fault offsets.

Recently the mineralogic framework was revised by integrating mineralogic abundances obtained from drill core data [26]. Incorporation of these data has provided a more accurate representation of the 3-D mineralogic distribution, and therefore zeolite abundance distribution, at Yucca Mountain. Previously simplistic homogeneous zeolitic or vitric layers have now been replaced by spatially distributed mineralogic abundances. Specifically, zeolitic properties extracted from the mineralogical model can now be mapped onto the site-scale model providing a more accurate representation of the zeolites (Figure 3b) . After the mapping process, the percent alteration (% zeolite) is prescribed at each node of the site-scale model and the appropriate flow and transport properties are assigned. In earlier simulations, we chose a simple threshold of 10% to flag nodes as either zeolitic or unaltered. This threshold was used because, at that time, there were no correlations between the degree of zeolitization and the hydrologic and chemical properties. Hydrologic characterizations for purely altered (zeolitic) and unaltered (vitric) matrix materials have been developed [28]. Therefore, the threshold provides a method for assigning either vitric or zeolitic properties to each node (i.e., nodes having less than 10% zeolite are assigned vitric properties and nodes having greater than 10% zeolite are assigned zeolitic properties). In this work we replace the zeolitic threshold method with a more rigorous geostatistical model to determine the applicability of using a threshold cutoff for zeolitic abundance.

Quantitative mineralogy, as obtained by XRD methods, is used as the mineralogic input into the 3-D mineralogic model. Data from a particular drill hole are included in the mineralogic model if a reasonable sample distribution exists for the drill hole or if the drill hole is located in an important location such as at the edge of the modeling area (Figure 1).

Six classes of geologic materials are incorporated in the 3-D mineralogic model. These minerals, mineral groups, or glasses used in the model are:

1. Smectite + Illite
2. Zeolites (clinoptilolite, heulandite, mordenite, chabazite, erionite, stellerite)
3. Tridymite
4. Cristobalite + Opal-CT
5. Glass
6. Analcime

The commercially available computer code STRATAMODEL is used to interpolate mineral abundances onto volumes where no measurements are available. A common feature of software available for property interpolation (STRATAMODEL, RC2, GSLIB) is that they operate on a logically rectangular grid. However, the ISM model [23, 24] is not represented on a logically rectangular grid. Therefore, for property interpolation and distribution, but not during construction of the computational grids, the ISM model is interpolated onto regular grids. The geologic model is generated by importing files of surfaces that contain x (easting), y (northing) and z (elevation in meters) coordinates. STRATAMODEL then assigns a regular grid spacing to these irregular grids and generates continuous surfaces. Using knowledge of the structural relationships between the surfaces specific to this site, a stratigraphic framework model is constructed which captures any pinchouts, onlap, offlap or truncations that are known to exist.

The model volume is then populated with attributes by importing the mineralogic data [26, 27] that occur at specific x, y, and depth locations within the model. STRATAMODEL uses a $1/r^2$ interpolation scheme to distribute the properties within each layer of the three dimensional volume. That is, in a model cell for which there is no mineralogic information, STRATAMODEL assigns mineralogic values by searching within a stratigraphic unit and a certain radius from that cell and giving weight to each data point found. The weight given is based on the distance (r) of the data point from the cell being filled. Data points farther from that cell are given less weight. STRATAMODEL is guided in this step by a user-specified search radius which is based on knowledge of the site, such as drill core spacing. The $1/r^2$ interpolation scheme acts to smooth the mineralogic distributions.

The 3-D model has a grid spacing of 243.84 m, which covers a ground area of approximately 166×10^6 m^2, and contains information from 17 boreholes [26]. Figure 1 shows the location of the wells in the mineralogical model. The model volume is divided into 335 layers to adequately capture the vertical density of mineralogic data. The result is a model with 1,034,470 cells. Each cell of the model contains eleven values, including mineral abundance for the six mineral classes, cell location, stratigraphic layer, elevation in meters, cell volume and cell thickness. Any cell within the model

173

volume can be queried to determine any of these values. Figure 4 compares the Chipera et al. model [26] with the updated mineralogical model [27] at the Antler ridge cross section which is used in the two dimensional flow and transport calculations. This figure shows that the updated model resolves more distinct zeolite layers, but in general, the two models are similar for this section.

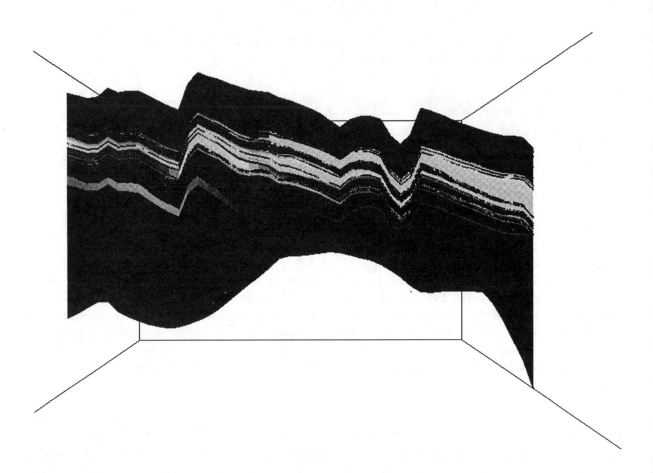

Figure 4.   Comparison of the (a) Chipera et al. model [26], and

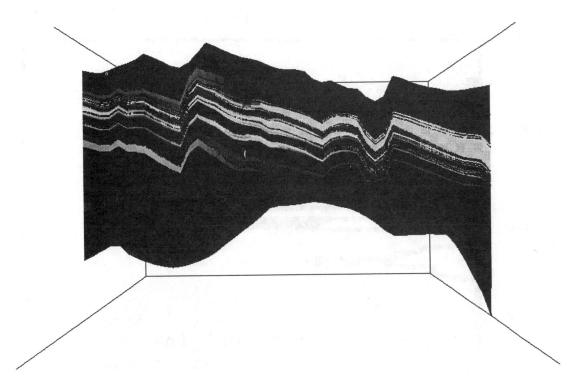

(b) the updated mineralogical model [27] for the Antler Ridge cross section.

Figure 5 compares zeolite and smectite abundance for the Chipera et al. model [26] and updated mineralogical model [27] at locations close to several wells. As expected, the profiles for zeolitic abundance at a location near G1 are nearly identical because G1 is present in both models (Figure 5a). The profiles for zeolitic abundance are only slightly different at a location near UZ-14 even though UZ-14 data was not present in Chipera et al. model [26], (Figure 5b). The Chipera et al. model [26] captures the correct zeolitic profile near UZ-14 for two reasons: G1 is close to UZ-14, and zeolites have large correlation lengths in the horizontal direction. Therefore, the data from UZ-14 and G1 are consistent and act to decrease the overall uncertainty in the model. The additional well data used in the updated well model buttresses the zeolitic abundance data from [26]. In general, the maps for zeolitic abundance are very similar between the models. However, the additional well data affects the distributions of smectite which is not as highly correlated and Figure 5c shows that the two models show significant differences in smectite abundance at a location close to NRG-6.

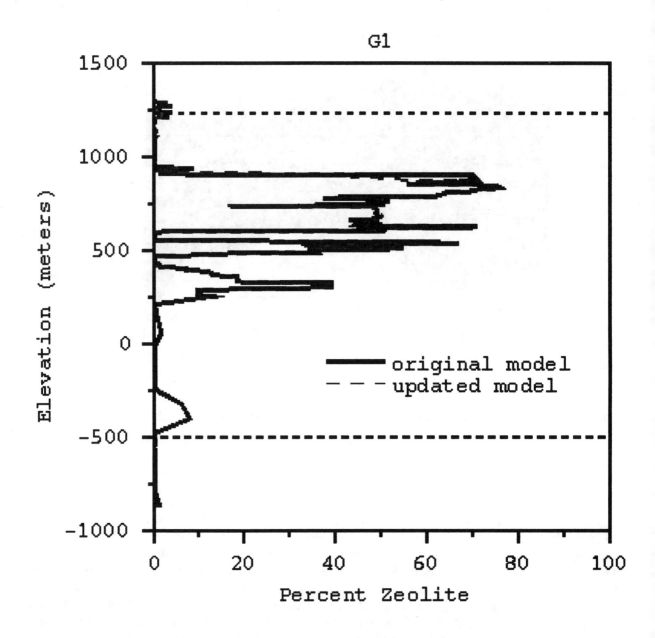

Figure 5. Comparison of Chipera et al. [26] and updated mineralogical model [27] at locations close to wells (a) zeolite abundance at G1 (G1 existed in both models)

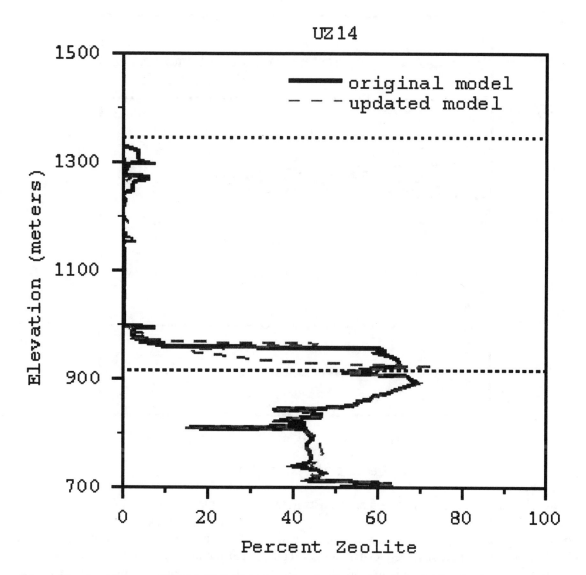

(b) zeolite abundance at UZ-14 (UZ-14 only exists in the updated model but is close to G1)

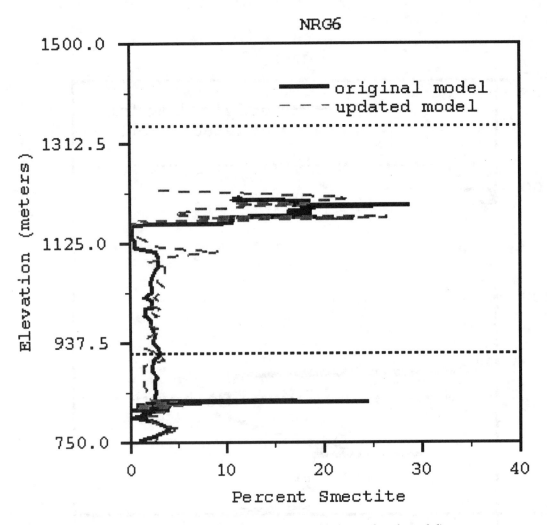

(c) smectite abundance at NRG6 (NRG-6 only exists in the updated model).

Capturing spatial variability is necessary to assess the site performance as it relates to uncertainty in attribute (permeability, porosity, K_d) distributions. This is particularly important for the zeolitic minerals between the basal vitrophere and the water table due to their role in either retarding migrating radionuclides or enhancing fracture flow. The geologic model of the UZ is based only on primary formation processes and, hence, does not capture the structure of the secondary alteration of vitric minerals to zeolites. Thus, capturing both the geologic structure and the mineralogic structure requires automatic integration of multiple sources of information. An approach and companion tools have been developed to couple the relevant models and generate appropriate attribute distributions to perform the analyses of the impact of zeolite spatial variability on repository performance. This same approach and tool set can then be applied to studies of the impact of uncertainty in attribute distribution.

The zeolitic threshold method is replaced with more rigorous geostatistical modeling. With this more sophisticated modeling approach, we examine whether using a threshold cutoff for zeolitic abundance is an appropriate upscaling method for characterizing small scale heterogeneity.

The methodology for integrating geostatistical techniques into the mineralogical model has been developed elsewhere [15]. These geostatistical techniques are used to develop variograms, kriging maps, and conditional simulations of zeolitic abundance. The kriging maps are used to estimate the uncertainty in the mineralogical model. The conditional simulations result in equally probable maps of zeolitic abundance. Eventually, simulating flow and transport for many conditional simulations, we will be able to estimate the uncertainty in ^{237}Np travel times using Monte Carlo techniques.

Using grid-to-grid interpolation alogarithms, a conditional simulation of zeolite abundance was mapped onto the computational mesh in three dimensions. We investigate the relationship between percent alteration, permeability changes due to alteration, sorption due to alteration, and their overall effect on radionuclide transport.

Relationships exist between zeolitic abundance and hydraulic conductivity (empirical); porosity and hydraulic conductivity (empirical); permeability and the distribution coefficient (theoretical); and α and hydraulic conductivity (empirical). As experimentalists gather more data, knowledge of these relationships may improve. With these correlations, the relevant properties can be distributed using a random field generator. The ultimate goal of the simulation process is to generate from a map of zeolitic abundance realizations of multiple properties (permeability, relative permeability, porosity, and Kd) which honor the univariate statistics of each property, the covariances between properties and the spatial statistics of each property as well as the spatial cross-covariances between properties. Figure 6 shows the permeability distribution derived from the conditional simulation of zeolitic abundance . The geologic formations are given permeabilities which correspond to zeolitic abundance using the correlation shown in Figure 6.

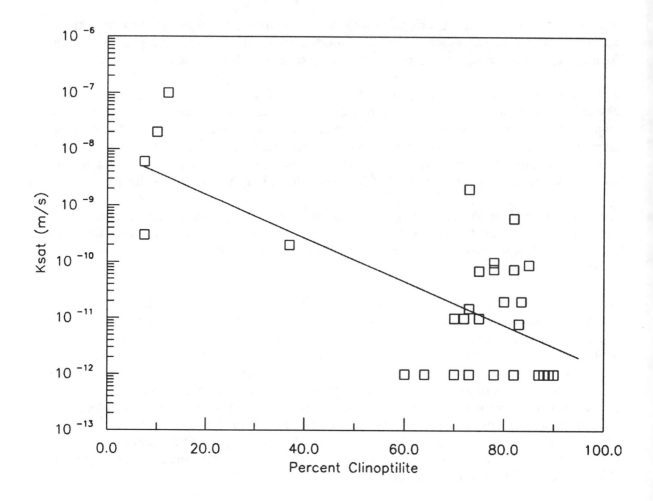

Figure 6. Data relating per cent clinoptilolite abundance to K_{sat}, (c.f., [28]).

Similar relationships exist between α and K_{sat} and are used to assign values for a in the Calico Hills. No exact relationship currently exists which relates zeolitic abundance to K_d, although experiments by Triay et al. [18-21] clearly have shown that K_d increases with zeolitic abundance. Results from the UZ transport field test [22] are expected to provide this relationship in the future. At present we assume that as zeolitic abundance increases, the surface area available for sorption increases resulting in a higher K_d. We used a simple linear relationship to relate Kd to zeolitic abundance. At 0% zeolite, we assume that no sorption of ^{237}Np will occur. We then choose different slopes to relate the zeolitic abundance to Kd. By running simulations at several different slopes we bracket the variability in K_d. Figure 7 shows the matrix and fracture saturations for the conditional simulation. In general the saturation profiles look similar to the saturation profiles which resulted from using a 10% threshold cutoff.

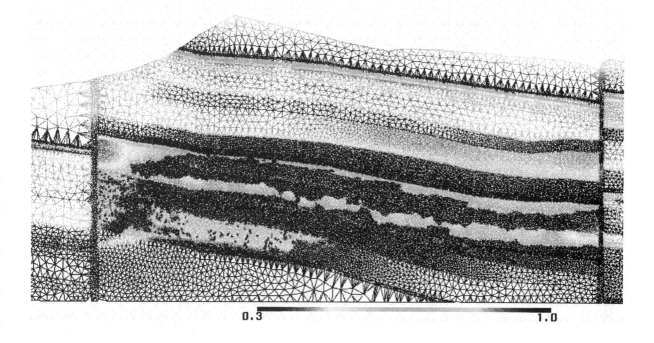

0.3   1.0

Figure 7.   Saturation profile of the conditional simulation. (a) matrix saturation

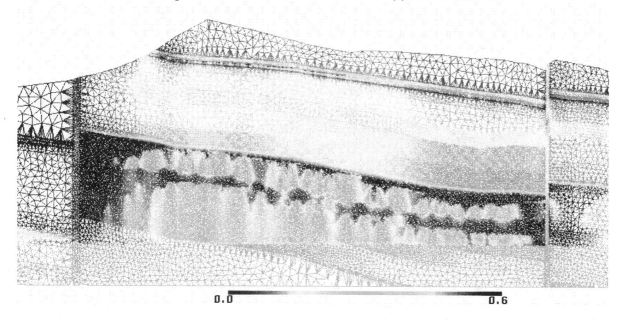

0.0   0.6

(b) fracture saturation.

When we compare the 10% zeolite cutoff results to the conditional simulation results for a conservative tracer, the breakthrough curves look similar (Figure 8). This result is not surprising since the saturation profiles for the two cases are similar. The differences in the simulations only appear as the retardation of ^{237}Np is included. The correlation between permeability, a and K_d then play an important role.

181

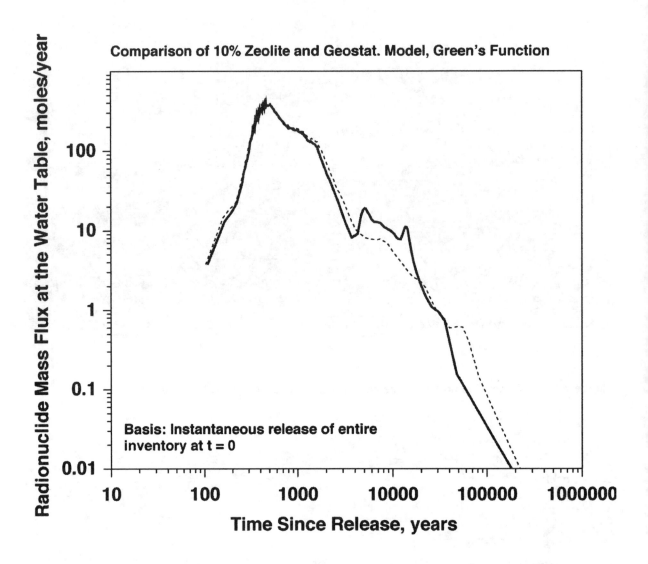

Figure 8.   Comparison of the 10% zeolite cutoff runs with a conditional simulation for a conservative tracer.

A key conclusion to these studies is that the retardation due to sorption predicted by a conditional simulation is much larger than the retardation predicted by the zeolite threshold method. The reason for larger retardation of [237]Np for the conditional simulation is a small but significant K_d at locations with zeolite abundance less than 10%. At these locations, the K_d is low (less the 1 cc/g) but permeability is large enough for the flow to be matrix dominated. By contrast, the increased retardation is not due to high K_d values which occur at high zeolite abundance because very little flow travels through these low permeability regions. Modifying the 10% threshold case to include some sorption at locations with zeolite abundances less than 10% closely matches the more rigorous conditional simulation. For the purposes of abstracting these results, setting nonzero K_d's in the vitric tuffs may be an effective way to mimic the more rigorous conditional simulations.

Rocks with low but non-zero zeolitic abundance exhibit measurable sorption of radionuclides like [237]Np and Uranium. Rocks with high zeolitic abundance may not be as effective in retarding these radionuclides since these rocks also in general have low permeability and contaminants can only enter these regions through molecular diffusion. In addition, any fractures in these highly zeolitized rocks will provide a mechanism for fluid and radionuclides to bypass these regions. However, small but significant sorption of rocks with low zeolite content may have a very large impact on the travel times for ^{237}Np and other radionuclides through the unsaturated zone. The UZ transport test at Busted Butte should provide results for testing this hypothesis.

Two- and Three-Dimensional Radionuclide Transport

Although a two-dimensional approach to transport calculations is commonly used, here we also illustrate the full three-dimensional transport model, with its implications for lateral flow, spatial variability of flux, and underlying strata thicknesses. Compared to the two-dimensional models, greater spreading in the travel time distribution curve is simulated due to the additional flow paths sampling a greater range of property variability that occur in three dimensions. This variability leads to cumulative breakthrough curves at the water table spanning the range from 10 years to 10,000 years. The three-dimensional model predicts some sorption onto the zeolites, similar to the two-dimensional model. However, pathways bypassing the highly zeolitized layers still exist and may lead to peaks in the mass flux at the water table which are not affected by retardation in the zeolites.

Prior to the 3-D modeling, representative two-dimensional transport calculations were performed on an East-West cross section model. Percolation flux is one of the key uncertainty parameters that will influence the mass flux values and arrival time distributions at the water table. Strictly speaking, to simulate the flow field assuming a different percolation flux, a new set of hydrologic parameters should first be developed. However, given that matrix saturations are relatively insensitive to percolation flux over the range of uncertainty at Yucca Mountain (about a factor of three higher or lower than the base case map), we assume that the hydrologic parameter sets developed using the base-case map can be applied directly for fluxes a factor of three higher or lower than the base case. This concept is demonstrated in Figure 9, a plot of the difference in predicted fluid saturations in the matrix (top figure) and fractures (bottom figure) for an increase in percolation flux of a factor of three at all locations on the base infiltration map. The matrix saturations, the only ones for which data are available, exhibit changes of less than 0.05 at almost all locations, with differences of 0.1 only in a few locations. At the higher percolation rate, the fracture saturations increase more dramatically, a characteristic of increased fracture flow. Nevertheless, we conclude from this comparison that as a first approximation, the use of the parameter values determined with the base map is warranted for this range of percolation flux values.

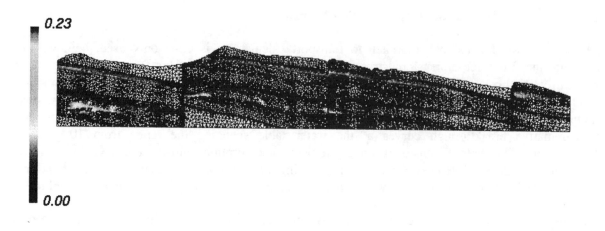

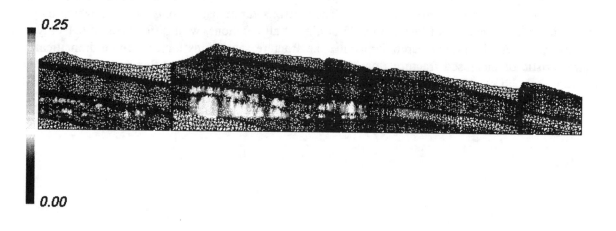

Figure 9. Difference in predicted fluid saturations from an increase in the base-case percolation flux map by a factor of three at all locations. Top figure: matrix saturation difference; bottom figure: fracture saturation difference.

In a similar way, many of the uncertainties of the flow and transport model were investigated in 2-D, including different hydrologic parameter sets, infiltration rates, and transport parameters. A great number of processes and parameters can be examined quickly in two dimensions without compromising the validity of the result through dimensionality reduction. However, there are some issues that can only be examined thoroughly in three dimensions. These include any heterogeneities that vary in the direction not modeled in the two-dimensional simulation (the North-South direction in the present study), variations in hydrologic layer thicknesses or mineralogic thicknesses, and spatial variability in infiltration. In addition, lateral flow effects can only really be studied thoroughly in three dimensions.

We use the three-dimensional model to investigate these and other issues, and present results of variations in transport properties not examined explicitly in the two-dimensional results, such as lateral flow, and diffusion and dispersion parameters, using the process for incorporating the three-dimensional geologic framework model into a numerical grid described earlier (Figure 10). As with the two-dimensional simulations, zeolite abundance from the min-pet model was mapped onto the three-dimensional numerical grid using a cut-off value approach . Locations with zeolite abundance greater than 10% are given zeolite properties. At other locations in the model, the unaltered properties of the geologic units are used to populate the model with property values that are homogeneous within a layer.

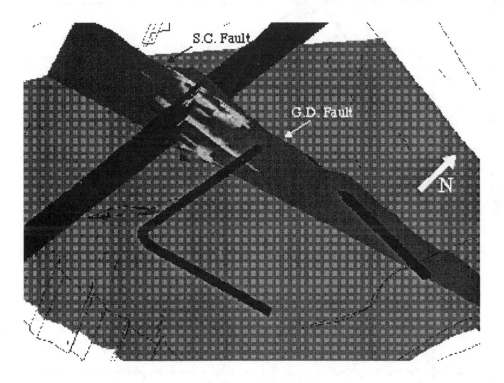

Figure 10.   Three-dimensional simulation of transport for particles injected at the surface above the potential repositorey. Image represents a cut-away fence diagram from the 3-D grid. Represented are the Solitario Canyon (S.C.) and Ghost Dance (G.D.) faults and the ESF. Particles undergoing matrix flow are represented with shaded areas corresponding to a high particle concentration. Significant lateral diversion is predicted in the vitric CHn matrix (beneath the TSw basal vitrophyre) and to the Ghost Dance fault.

185

As an example of a set of simulations, we compute the radionuclide transport functions for base case hydrologic properties, with the base-case infiltration map from Flint et al. [8]. For sorption, we examine the behavior assuming sorption only in zeolitic rocks, where the K_d value is varied from 0 (no sorption anywhere in the model) to 20 cc/g. This series of simulations covers the nonsorbing radionuclides (^{99}Tc, [129]I, 14C, and [36]Cl) and sorbing radionuclides [237]Np and the various isotopes of uranium. The Green's function, or response of the system to a pulse of radionuclide released instantaneously at time 0, is used to characterize transport through the three-dimensional model to the water table. Figure 11 shows the results of the simulations for this case. With the current hydrologic property set, there is a tendency for some fraction of the inventory to travel rapidly through fractures to the water table. This fracture transport is through the zeolitic rocks where, for the current property sets, the matrix permeability is insufficient to transmit the flow. There is a hint of a bimodal distribution of transit times in the $K_d = 0$ curve, but, in general, the overall response exhibits a wide range of residence times without much tendency of clustering into distinct arrival times. This behavior is due to the wide range of hydrogeologic conditions present across the repository. The three-dimensional model samples the entire range of this variability, rather than biasing the results through the selection of a particular cross section. The spatial variability issue is addressed further in a moment.

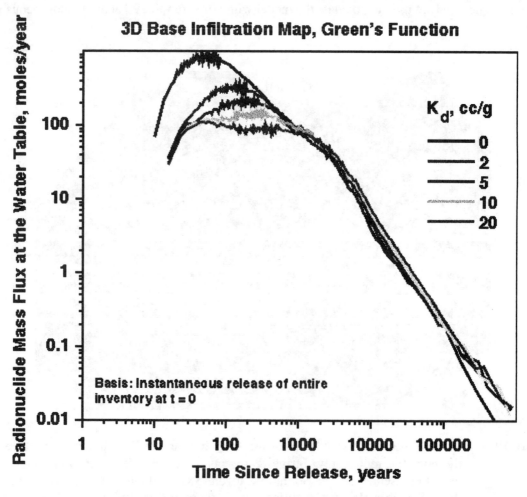

Figure 11. Green's function for radionuclide transport in the three-dimensional site scale model. Property set 6541, base infiltration map (cf., [15]).

As sorption coefficients in the zeolitic tuffs are increased, retardation reduces the amount of early arriving radionuclide through the matrix diffusion/sorption process. In addition, radionuclide percolating through the zeolite matrix is delayed. The result is a reduction in peak concentration in the Green's function, and a shift to longer residence times of a larger portion of the inventory. This effect is seen more readily in the integrated cumulative breakthrough curves in Figure 12. The portion of the inventory with transit times less than 100 years is small, and the distribution of times for different fractions of the radionuclide inventory to reach the water table spans a range of 10 years to greater than 10 ky. This variability is greater than for the two-dimensional simulations presented previously, and is an essential feature that must be captured in any total system performance assessment. Sorption onto zeolites shifts the curves to longer residence times, which means that for a given time such as a hypothetical 10 ky compliance period, less of the radionuclide is predicted to reach the water table. Thus the benefit of sorption onto zeolites is present in the three-dimensional model to perhaps a greater degree than for the two-dimensional simulations.

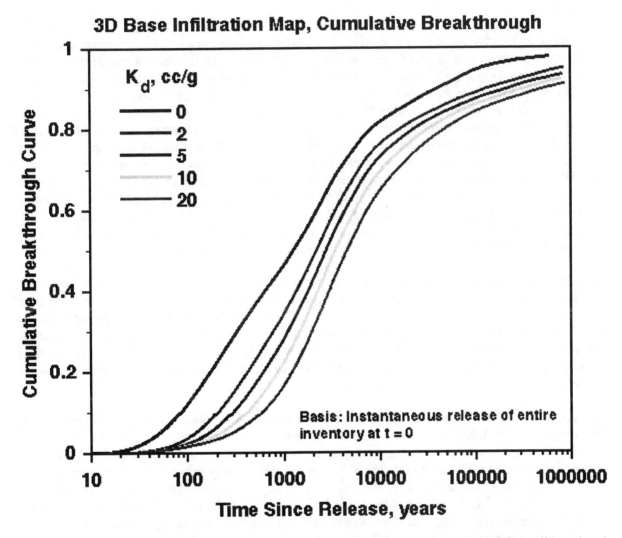

Figure 12. Cumulative breakthrough curves for radionuclide transport in the three-dimensional site scale model. Property set 6541, base infiltration map (cf., [15]).

The three-dimensional transport model is used to examine the difference in travel time distribution as a function of the release position in the potential repository. Radionuclides released from the northern end of the repository are more likely to experience rapid transport to the water table than radionuclides released in the southern portion of the potential repository. The behavior is attributed to less uniform distribution of zeolites in the southern portion of the domain. This results in slower matrix flow in the nonzeolitic units in the south compared to fracture dominated flow in the zeolitic units in the north. As with the two-dimensional simulations, this finding highlights the critical need to develop greater certainty on the material properties assigned to the zeolitic units. A slightly higher matrix permeability in the zeolites leads to substantially greater retardation of radionuclides.

The diffusion coefficient influences the breakthrough curves simulated with the three-dimensional model, especially for the portion of the inventory with the shortest travel times. Diffusion into the rock matrix and sorption onto zeolites retard the transport of ^{237}Np traveling in fractures. The dispersion coefficient is shown to have very little impact on the UZ transport.

Conclusions

The site-scale unsaturated-zone transport model for Yucca Mountainintegrates information from the field, the laboratory, and other Project modeling efforts into an assessment of the geologic barrier from the potential repository to the water table. The integration begins with the Project's geologic and mineralogic models describing the structural features and the zeolite distribution in the unsaturated zone. Our software integration system brings these two different models together into one structural and geochemical representation of the study system. The automatic grid generation capabilities play a large role in integrating these data sets as well as generating reliable finite-element grids on which accurate flow and transport calculations can be performed. These grids capture essential structural controls at Yucca Mountain represented in the Integrated Site Model of geology. Surfaces which are output from that model feed directly into the GEOMESH grid generation package. Thus, the dipping, faulted beds of spatially varying thickness and extent are captured with as high a resolution as the geologic model puts out. Figures throughout this document show that even the most complex features such as layer pinchouts and fault offsets are captured with high resolution.

The transport simulations presented in this study are performed with the FEHMN equivalent continuum, dual permeabilty and particle tracking modules. Particle tracking is used for unsaturated transport simulations because the limitations of standard finite-element of finite-difference methods (small time steps and spatial resolution) are mostly eliminated with this cell-based approach. Accurate simulation of dual-permeability systems in which there is a vast disparity in the travel times depending on whether the transport is in the fractures or the matrix are achieved efficiently with this method.

We apply the FEHMN particle tracking method to our study of environmental isotopes in the hydrologic system. Like the property assessment study, this study serves to evaluate the multiple different parameter sets generated recently by the site-scale flow model to describe the material properties. Whereas the property sets study focused on the pathway between the potential repository and the water table, this analysis focuses on the pathway to the potential repository and the ESF, where sampling for ^{36}Cl/Cl ratios has led to a substantial database. The flow and transport simulations are performed in one-, two-, and three-dimensions and serve to describe a complex system where travel times through the PTn depend of infiltration flux, PTn thickness, and the parameters representing the Ptn properties. All of these can vary in space thus leading to a system where responses such as rapid transport in fractures may occur in one portion of the site but not in another. The general conclusions from this study are that transient responses at the potential repository are more likely to occur in the southern portion of the site, but the infiltration may be low enough that they never occur. Moving north, the PTn thickens and the capacity to imbibe even large pulses of short duration increase. In the simulations, the proportion of flow that stays in the fractures of the Ptn is sensitive to the model of fracture-matrix coupling. Until now, we have used geochemical databases to constrain and evaluate the validity of our flow and transport models in order to provide a strong scientific basis for the abstractions of these products used by Performance Assessment. We have also focused on the water chemistry databases for pore waters and perched waters, in order to identify and discuss alternate models for flow and transport in Yucca Mountain.

Alternative conceptual flow and transport models for Yucca Mountain are being entertained based on hydrologic and hydrochemical databases. Explanations for the water chemistry variations in perched and pore waters at depth in the vicinity of the potential repository site are needed so as to further evaluate the hydrologic system. We have integrated the chemical and hydrologic databases to evaluate different model conceptualizations of the hydrologic behavior of the mountain at low and high infiltration rates. Chemical data, when combined with hydrologic data, provide the most relevant information for ìvalidatingî radionuclide transport models, as the movement of naturally occurring solutes is closely related to the potential migration of radionuclides.

All of the data above are consistent with a model of transient flow, and furthermore, the stable isotope data and apparent ^{14}C and $^{36}Cl/Cl$ ages can be interpreted as resulting from a mixture of late Pleistocene/early Holocene water with modern waters. The existence of these mixed ancient/modern perched water bodies therefore implies that an additional retardation mechanism exists above the Calico Hills formation. The ubiquitous young ^{14}C ages in the Calico Hills formation pore waters beneath the perched water bodies and in other areas, also imply that a lateral component of flow occurs in the vitric Calico Hills and that this formation is not bypassed.

For radionuclide transport predictions, a variety of two- and three-dimensional simulations are presented. These studies demonstrate the overriding importance of proper characterization of the zeolitic units for performance assessment. Presently, there is significant uncertainty in the characteristics of these units due to data limitations. Enhanced characterization efforts, including the results of the UZ field transport test at Busted Butte, Nevada, should reduce the uncertainty in these calculations and lead to a better understanding of the true nature of the hydrologic and transport behavior of the Chn.

Comparing two-dimensional and three-dimensional model simulations, we found that greater spreading in the travel time distribution curve is simulated in three dimensions due to the additional flow paths sampling a greater range of property variability that occur in three dimensions. This variability leads to cumulative breakthrough curves at the water table spanning the range from 10 years to 10,000 years. Part of this spreading is caused by a propensity for radionuclides released from the northern end of the repository to experience rapid transport to the water table compared to radionuclides released in the southern portion of the potential repository.

In general, this type of variability has a greater influence on unsaturated zone performance than diffusion or dispersion mechanisms. The diffusion coefficient influences the breakthrough curves simulated with the three-dimensional model, especially for the portion of the inventory with the shortest travel times. Diffusion into the rock matrix and sorption onto zeolites retard the transport of ^{237}Np traveling in fractures. The dispersion coefficient is shown to have very little impact on the unsaturated zone transport.

The potential for small scale variability of chemical and hydrologic properties on radionuclide migration, due to the correlation between mineral distribution and hydrologic and chemical properties, and the distribution of minerals such as zeolites, to affect the flow patterns of percolating fluids and the sorption of many radionuclides is high.

Large-scale models typically do not capture these processes in detail. We used geostatistical techniques to develop variograms, kriging maps, and conditional simulations of zeolitic abundance. This conditional simulation was mapped onto the geologic model. The investigation of the relationship between percent alteration, permeability changes due to alteration, sorption due to alteration, and their overall effect on radionuclide transport supports the conclusion that the retardation due to sorption predicted by a conditional simulation is much larger than the retardation predicted by the zeolite threshold method. The reason for larger retardation of ^{237}Np for the conditional simulation is a small but significant Kd at locations with zeolite abundance less than 10%. At these locations, the K_d is low (less the 1 cc/g) but permeability is large enough for the flow to be matrix dominated. This effect has a large beneficial impact on performance, especially because of the tendency in other simulations for zeolites to be bypassed. For the purposes of abstracting these results, setting nonzero Kd's in the vitric tuffs may be an effective way to mimic the more rigorous conditional simulations.

Acknowledgments

This work was supported by the Yucca Mountain Site Characterization Office as part of the Civilian Radioactive Waste Management Program. This project is managed by the U.S. Department of Energy, Yucca Mountain Site Characterization Project. The authors would like to acknowledge Dr. G. Zyvoloski for his review of the manuscript.

N.B. Coloured versions of all the figures included in this paper are available upon request from Gilles Bussod at LANL (see List of Participants).

References

1   Zyvoloski, G. A., Z. V. Dash, and S. Kelkar, FEHMN 1.0: Finite element heat and mass transfer code, Technical Report LA-12062-MS, Rev. 1, Los Alamos National Laboratory, 1992.

2   Zyvoloski, G. A., B. A. Robinson, Z. V. Dash, and L. L. Trease, Users manual for the FEHMN application, Technical Report LA-UR-94-3788,Rev. 1, Los Alamos National Laboratory, 1995a.

3   Zyvoloski, G. A., B. A. Robinson, Z. V. Dash, and L. L. Trease, Models and methods summary for the FEHMN application, Technical Report LA-UR-94-3787, Rev. 1, Los Alamos National Laboratory, 1995b.

4   Gable, C.W, T. A. Cherry, H.E. Trease, and G. A. Zyvoloski, GEOMESH grid generation, Technical Report, LA-UR-95-4143, Los Alamos National Laboratory, 1995.

5   Gable, C.W, H.E. Trease, and T. A. Cherry, Automated grid generation from models of complex geologic structure and stratigraphy, Santa Barbara: National Center for Geographic Information and Analysis, proceedings paper, Technical Report, LA-UR-96-1083, Los Alamos National Laboratory, 1996a.

6   Gable, C. W., H. Trease and T. Cherry, Geological applications of automatic grid generation tools for finite elements applied to porous flow modeling, Numerical Grid Generation in Computational Fluid Dynamics and Related Fields, ed. B. K. Soni, J. F. Thompson, H. Hausser and P. R. Eiseman, Engineering Research Center, Mississippi State Univ. Press, 1996b.

7   Bodvarsson G.S., T.M. Barndurraga, and Y.S. Wu. The site-scale unsaturated zone model of Yucca Mountain, Nevada, for the Viability Assessment, Technical Report LBNL-40378, Lawrence Berkeley National Laboratory, 1997.

8   Flint, A.L, J. Hevesi, and L.E. Flint, Conceptual and numerical model of infiltration for the Yucca Mountain Area, Nevada, USGS WRIR MOL 19970409.0087, GS960908312211.003DOE Milestone, 3GUI623M., U.S. Geological Survey, (in preparation), 1996a.

9   Fabryka-Martin, J. T., P. R. Dixon, S. Levy, B. Liu, H. J. Turin, and A. V. Wolfsberg, Summary report of chlorine-36 studies: systematic sampling for chlorine-36 in the Exploratory Studies Facility, Los Alamos National Laboratory YMP Letter Milestone number 3783 AD, 1996.

10 Fabryka-Martin, J.T., A.L. Flint, D.S. Sweetkind, A.V. Wolfsberg, S.S. Levy, G.J.C. Roemer, J.L. Roach, L.E. Wolfsberg, and M.C. Duff. Evaluation of flow and transport models of Yucca Mountain, based on chlorine-36 studies for FY97, Los Alamos National Laboratory, Yucca Mountain Project Milestone Report SP2224M3, 1997.

11 Fabryka-Martin Fabryka-Martin, J.T., A.V. Wolfsberg, P.R. Dixon, S.S. Levy, J. Musgrave, and H.J. Turin. Summary report of chlorine-36 studies: systematic sampling for chlorine-36 in the Exploratory Studies Facility, Los Alamos National Laboratory, Yucca Mountain Project Milestone Report 3783M, 1996, LANL 1998 Technical Report, in press.

12 Wolfsberg, A.V., J.T. Fabryka-Martin, and S.S. Levy, Use of chlorine-36 and other geochemical data to test a groundwater flow model for Yucca Mountain, Nevada, in OECD Proceedings :Use of Hydrogeochemical Information in Testing Groundwater Flow Modelsî, Borgholm (Sweden), 1-3 September, 1997, OECD/NEA, 1998.

13 Sweetkind, D.S., J. Fabryka-Martin, A. Flint, C. Potter, and S. Levy, 1997, Evaluation of the structural significance of bomb-pulse 36Cl at sample locations in the Exploratory Studies Facility, Yucca Mountain, Nevada, U.S.Geological Survey memorandum transmitting Level 4 milestone SPG33M4 to R. Craig, Yucca Mountain Project Branch, 29 August, 1997.

14 Levy, S. S., D.S. Sweetkind J.T. Fabryka-Martin, P.R. Dixon, J.L. Roach, L.E. Wolfsberg, D. Elmore, and P. Sharma. Investigations of structural controls and mineralogic associations of chlorine-36 fast pathways in the ESF. Los Alamos National Laboratory YMP milestone SP2301M4, 1997.

15 Robinson, B.A., A. V. Wolfsberg, H. S. Viswanathan, G.Y. Bussod, C.W. Gable and A. Meijer, The site-scale unsaturated zone transport model of Yucca Mountain, Los Alamos National Laboratory YMP Milestone SP25BMC, 1997.

16 Robinson, B.A., A.V. Wolfsberg, G.A. Zyvoloski, and C.W. Gable, An unsaturated zone flow and transport model of Yucca Mountain, Los Alamos National Laboratory YMP Milestone 3468, 1995.

17 Robinson, B.A., A.V. Wolfsberg, H.S. Viswanathan, C.W. Gable, G.A. Zyvoloski, and H.J Turin,. Modeling of flow, radionuclide migration, and environmental isotope distributions at Yucca Mountain, Los Alamos National Laboratory YMP Milestone 3672, 1996.

18 Triay, I. R., A. Meijer, J. L. Conca, K. S. Kung, R. S. Rundberg, and E. A. Streitelmeier, Summary and synthesis report on radionuclide retardation for the Yucca Mountain Site Characterization Project, Los Alamos National Laboratory YMP Milestone 3784, 1996a.

19 Triay, I.R., C.R. Cotter, S.M. Kraus, M.H. Huddleston, S.J Chipera, D.L. Bish, Radionuclide sorption in Yucca Mountain tuffs with J-13 well water: neptunium, Uranium, and Plutonium, Los Alamos National Laboratory Yucca Mountain Site Characterization Project Milestone 3338, Technical Report LA-12956-MS, 1996b.

20   Triay, I.R., C.R. Cotter, M.H. Huddleston, D. E. Leonard, S. C. Weaver, S. J. Chipera, D.L. Bish, A. Meijer, and J. A. Canepa, Batch sorption results for neptunium transport through Yucca Mountain tuffs, Technical Report LA-12961-MS, Los Alamos National Laboratory , 1996c.

21   Triay, I. R., A. C. Furlano, S. C. Weaver, S. J. Chipera, and D. L. Bish, Comparison of neptunium sorption results using batch and column techniques, Technical Report LA-12958-MS, Los Alamos National Laboratory, 1996d.

22   Bussod, G.Y., B.A. Robinson, D.T. Vaniman, D.E. Broxton and H.S. Viswanathan. UZ Transport Test Plan, Rev. 1, Demonstration of the Applicability of Laboratory Data to Repository Transport Calculations, Los Alamos National Laboratory YMP Milestone SP341SM4, 1997.

23   Clayton, R. W., W. P. Zelinski and C. A. Rautman, (CRWMS), ISM2.0: A 3-D geological framework and integrated site model of Yucca Mountain: Version ISM1.0, Doc ID B00000000-01717-5700-00004 Rev 0, MOL.19970122.0053, Civilian Radioactive Waste Management System Management and Operating Contractor, February 1997.

24   Zelinski, W.P., and R.W. Clayton, A 3D geologic framework and integrated site model if Yucca Mountain, version ISIM 1.0: Civilian Radioactive Waste Management System Management and Operating Contractor Document B00000000-01717-5700-00002, TRW Environmental Safety Systems Inc.,Las Vegas, Nev., 1996.

25   Trease, H., D. George, C. W. Gable, J. Fowler, A. Kuprat and A. Khamyaseh, The X3D grid generation system, Numerical Grid Generation in Computational Fluid Dynamics and Related Fields, ed. B. K. Soni, J. F. Thompson, H. Hausser and P. R. Eiseman, Engineering Research Center, Mississippi State Univ. Press, 1996.

26   Chipera, S.J., K. Carter-Krogh, D.T. Vaniman, D.L. Bish, and J.W. Carey, Preliminary three-dimensional mineralogical model of Yucca Mountain, Nevada, Los Alamos National Laboratory YMP Milestone SP321AM4, 1997a.

27   Chipera, S.J., D.T. Vaniman, D.L. Bish, and J.W. Carey, Mineralogic variation in drill holes USW NRG-6, NRG-7/7a, SD-7, SD-9, SD-12, and UZ#14: New data from 1996-1997 analyses, Los Alamos National Laboratory YMP Milestone SP321BM4, 1997b.

28   Flint, L.E., Matrix properties of hydrogeologic units at Yucca Mountain Nevada, U.S. Geological Survey Open File Report, MOL 19970324.0046, GS950308312231.002, (in preparation), 1996b.

SESSION III

New Areas for Investigation: Two Personal Views

Chairmen: G. Gustafson (Chalmers Univ., Sweden) and E. Fein, (GRS, Germany)

Impact of Uncertainties in Chemical and other Entities on Radionuclide Migration from a Repository for Spent Nuclear Fuel

Ivars Neretnieks
Royal Institute of Technology, Sweden

Abstract

The migration of radionuclides in the geosphere is influenced by many factors. The primary factors that influence their mobility in the geosphere are: The water flowrate, the sorption properties in the rock matrix, the matrix diffusion properties and the flow wetted surface along the flowpath. There are also some secondary factors such as hydrodynamic dispersion, flowpath geometry, fracture filling materials etc.

The water flowrate and flowpaths are assessed by hydrodynamic modelling using data from primarily borehole measurements. Some inherent uncertainties emanating from both conceptual as well as measurement uncertainties will be briefly discussed.

Sorption data are uncertain due to several reasons. The water chemistry, especially the salinity of the water, its pH and redox conditions as well as encountered minerals can vary along the flowpath. In addition there is a variability i.e. uncertainty in the measured data. Small amounts of fracture filling and alteration materials will not much influence the migration over long times whereas larger amounts of strongly sorbing alteration minerals can have a strong impact on the short lived nuclides.

Diffusion of the nuclides into the matrix of the porous crystalline rock and sorption on the internal surfaces is a very strong retardation mechanism. Diffusion properties in the same rock matrix can vary considerably over short distances and there is also a considerable uncertainty in data.

The at present largest uncertainty is due to the largely unknown flow wetted surface encountered by the mobile water. This entity has been much less studied than sorption and diffusion properties. It is also the entity that is most difficult and expensive to assess because it must be done in the field and on a large scale.

The paper discusses the variabilities and uncertainties of the primary factors and exemplifies and illustrates the impact on the mobility of some of the nuclides of most concern by some sample calculations.

Introduction and background

To assess the impact of uncertainties in chemical and other transport parameters on the migration of radionuclides it is necessary to understand how these mechanisms influence transport. The main mechanisms are: Sorption on fracture surfaces and in possible gouge material, diffusion into the rock matrix and sorption on the inner surfaces. The contact surface between the flowing water and the rock is thus a key entity. The water flux also is an important entity but lies somewhat outside the scope of this paper. Its impact will nevertheless be shown and compared to that of the other entities.

For illustration purposes the simplest of models is presented where the main mechanisms are accounted for. It starts with a very simple geometry to highlight the mechanisms and then is generalised to handle variable properties along the flowpath. Water flows in a fracture with an aperture 2b with a flowrate Q. The solute can sorb onto the fracture surface with a surface sorption coefficient K_a. There is some fracture filling material that can sorb the nuclides. The solute can also diffuse into the porous rock matrix which has a porosity ε. The solute can in addition sorb on the inner surfaces of the rock matrix. The distribution coefficient is $K_v = K_d \rho$. The surface sorption can also be thought of as resulting from very fast sorption in a thin layer with thickness d of some material that is readily accessed by the nuclide. K_a is the simply the product of the thickness of this material and its volume sorption coefficient. $K_a = d\, K_{da} \rho$.

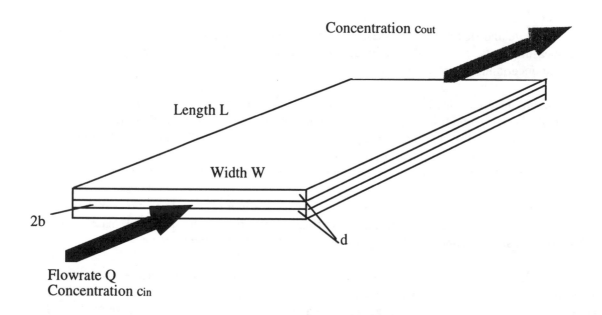

Figure 1. A fracture with aperture 2b and fracture alteration material of total thickness 2d has a water flow of Q, a length L and width W.

The rock on each side of the fracture has an altered region with a total thickness of 2d. The flowrate of water in the fracture is Q m^3/s, the width is W m and the length is L m. The mass sorption coefficients of the alteration layer is K_{da}, that of the good rock is K_d and that of the filling material is K_{df}.

The concentration at the inlet of the fracture is suddenly raised to c_{in} and kept so except for the decay. The inlet concentration thus will decrease over time as $c = c_{in}e^{-\lambda t}$. If we assume that there is no diffusional resistance in the alteration layer and that this and the gouge material is instantaneously equilibrated with the water in each location, the outlet concentration from the fracture is obtained as [1]

$$c_{out} = c_{in}e^{-\lambda t}Erfc\left[\frac{LW\sqrt{D_eK_v}}{Q\sqrt{t-t_o}}\right] \tag{1}$$

$K_v = K_d\rho$ is the volumetric sorption coefficient of the rock matrix, $K_{vf} = K_{df}\rho_f$ and $K_{va} = K_{da}\rho_a$. D_e is the effective diffusivity in the good rock and

$$t_o = \frac{LW}{Q}2(b + bK_{vf} + dK_{va}) \tag{2}$$

$$t_o = t_w\left(1 + K_{vf} + \frac{d}{b}K_{va}\right) = t_wR_a \tag{3}$$

t_o is the tracer residence time if it were only affected by the water residence time and the retardation by fracture filling (gouge) and alteration materials.

The retardation factor caused by instantaneous reversible sorption on the fracture filling and alteration materials R_a is

$$R_a = 1 + K_{vf} + \frac{d}{b}K_{va}$$

Equation (3) warrants some comments. If the two volumetric sorption coefficients for the gouge and the alteration material are zero, the expression reduces to be the water residence time in the fracture. If the nuclide is sorbed on the gouge material, the second term in the parenthesis is not zero and the time t_o is the nuclide residence time. Because the volumetric sorption coefficient K_{vf} is much larger than 1, t_o will be very much larger than the water residence time and in fact the water residence time will no longer influence the residence time of the sorbing nuclide. The same applies for the sorption in the alteration layer.

We note that the entity that has a dominating influence on the nuclide transport is the ratio of the flow wetted surface, FWS, to the water flowrate wetting this surface 2LW/Q. Uncertainties in the matrix diffusivity and the matrix sorption coefficient D_e and K_v do not have as strong an influence as the flow wetted surface and the local flowrate because of the square root effect. This means that an uncertainty of a factor of ten in the FWS or the local flowrate will influence the group $\dfrac{LW}{Q}\sqrt{\dfrac{D_eK_v}{(t-t_o)}}$

by a factor of ten when an uncertainty in De and K will influence it by a factor of the square root of ten, which is slightly more than three.

Accounting for parameter variabilities along a flowpath

The very simple model presented above does not account for the variability of the entities along the flowpath. The width of the flowpath W can vary considerably, as can the matrix diffusivity, the

197

sorption coefficients of the rock, the amounts of alteration material and gouge. Also the fracture aperture and the water velocity can vary.

Fortunately, it is possible to account for these variations in a simple manner. We follow a packet of water from inlet to outlet. The packet of water will encounter sections that are narrow and sections that are wide, it will be exposed to locally a large FWS as the flowpath widens and sections that have high and low diffusivities, sorption coefficients and different amounts of fracture filling materials etc. For such a complex flowpath the solution to the mass balance equations is [2]

$$c_{out} = c_{in}e^{-\lambda t}Erfc\left[\frac{\psi(z)}{2\sqrt{t - \phi(z)}}\right]$$ (5)

Where

$$\phi(z) = \int_0^z \frac{R_a(z)}{u(z)}dz$$ (6)

and

$$\psi(z) = 2\int_0^z \frac{W(z)\sqrt{D_e K_v}}{Q}dz$$ (7)

Equation (6) actually gives the total nuclide travel time in a pathway where there is no matrix diffusion and the retardation is caused by sorption on the filling and alteration materials. It is equivalent to Equation (2). For a sorbing nuclide the influence of water residence is negligible. For short lived nuclides such as ^{137}Cs and ^{Sr}90 the retardation due to sorption on the filling and alteration materials can be quite effective if their thicknesses are a mm or more. For long lived nuclides the matrix diffusion effects are often more important than the surface sorption.

The above treatment illustrates that variabilities along the flowpath in a variable aperture and width fracture can be accounted for. Equations (6) and (7) show that it is possible to account for variable rock properties by integration along the flow path. It is difficult to see how we will ever be able to obtain such detailed information. However, the equations show that if there is information available on the frequency distributions of K_v, D_e and W and if these distributions are not correlated, then it is correct to use the averages of W, $\sqrt{K_v}$ and $\sqrt{D_e}$ in Equation (7).

The above treatment can be extended to apply to the conditions in a stream tube with variable cross section A_o. Consider a streamtube, which is an imaginary tube within which the water flows. There is no transport of water or solutes through the wall of the tube and thus the water that enters one end will be the same water all along the stream tube. The stream tube can vary very much in cross section along the path. In the low conductivity "good"rock it will have a very large cross section and the water velocity will be low. As the water in the stream tube enters a high flow region its cross section will become small and the velocity will be large. Along the flowpath the water will move through the fractured rock and encounter varying amounts of local FWS. For a given location in the rock it has a specific FWS, a_r m^2 surface per m^3 rock volume. This entity may also vary along the flowpath. We deem it to be a rock property and assume that it can be measured.

Equation (7) can be rewritten to include a_r. It becomes

$$\psi(z) = \int_0^z \frac{a_r(z)\sqrt{D_e K_v}}{u_o(z)} dz = \int_0^z \frac{A_o(z)a_r(z)\sqrt{D_e K_v}}{Q} dz \qquad (8)$$

$u_o(z)$ is the local flux, i.e. flowrate per cross sectional area. In the Hydrology literature it is often called Darcy velocity. This entity is assessed by hydrological modelling in the form of fluxes along streamlines or in stream tubes. In this paper it is assumed to be known. Also the flowpath along which to integrate is assumed to be obtainable from the hydrological modeling.

These very simple equations give results that are very similar to those obtained by a very much more complex model, the Channel Network Model, CNM,[3,4]. In this model a multitude, hundreds of thousands, of independent channels form a three dimensional network. Each channel can have its own stochastically generated properties such as channel conductance, aperture, width etc. The more channels there are along the flowpath the more the solution resembles that of the above equations. The above equations will thus well describe situations where a large number of channels have been encountered along the flowpath.

In this paper the influence of the variabilities and uncertainties in the sorption and diffusion properties and in the specific FWS a_r will be studied using the above equations.

Data, variabilities and uncertainties

By variability in data is meant the stochastic nature of the data due the differences between samples or measurement points and due to random measurement differences.

By uncertainties is meant here that there may be doubts if the method used to evaluate an entity from a measurement e.g. is correct and meaningful. An example is the hydraulic measurements using packer tests in boreholes to obtain some measure of a hydraulic conductivity. The methods may be reasonable of perhaps even correct for a homogeneous porous medium but it is questionable if they give meaningful results when applied to a fractured rock system.

The use of the notion of "Flow Wetted Surface" and the data that are available is another area that is questionable and uncertain. Another cause of uncertainty is if the samples or the experiment have used a representative piece of rock or location. This uncertainty arises in many field situations where the heterogeneity is very large and only a few measurements can be made.

In this paper it will be attempted to separate the variability from the uncertainty. As will be seen this is not easy and sometimes not possible.

Altered zone and fracture filling material

There is very little systematic information on the amounts and properties of fracture filling and alteration materials. Some recent observations at Äspö indicate that it is not uncommon at this site to find alteration materials with a thicknesses of 1 mm and porosity of 5 %. Also thicknesses of 5 mm and sometimes several tens of mm with porosities of 1-2 % have been observed by Tullborg [5]. Scoping calculations indicate that at least for the short lived nuclides
^{90}Sr and ^{137}Cs this can have strong retarding effect [6].

Matrix diffusion data

Recently Ohlsson and Neretnieks [7] compiled available data on matrix diffusion in crystalline rocks. The variability between different measurements for the same nuclides is very large. This variability is to a large extent due to the variability between different rock samples. This was noted already by Skagius and Neretnieks [8,9] in laboratory measurements and by Birgersson and Neretnieks [10] in field measurements . Porosities varied from less than 0.1 % up to 1 % with many data around 0.3-0.4%. No formal statistical analysis of the data has been made but he variability in effective diffusivities was typically 1 to one and a half order of magnitude. It is deemed that most of the differences are due to stochastic variations and only a lesser part due to uncertainty.

Very seldom have measurements using different sorbing nuclides in the same rock sample been performed. This is because the experiments take a very long time. <for the sorbing nuclides a larger part is due to uncertainty because there is less data and the sorption mechanisms are not fully understood.

The effective diffusivity, D_e, which is used to determine the steady state flux of a diffusing species through a piece of the rock depends on the geometrical properties of pore network and on the diffusing species. It also depends on the diffusing species.

For a specie that does not interact with the mineral surfaces it is commonly assumed that a rock property called the formation factor F_f is an invariant. this entity multiplied by the diffusivity of the specie in bulk water will then give the effective diffusivity of the specie in the rock. F_f can be determined by diffusion experiments using e.g. tritiated water or methane or some other small non-charged molecule. This approach seems to be basically sound. Larger molecules will probably be influenced by steric hindrances as well, but for most nuclides of interest this is not a major concern.

There are two major additional effects that must be considered. One is the influence of the electrical double layer that can influence the flux of charged species. A high salinity will diminish the impact of the double layer whereas a low salinity will cause negatively charged (an-)ions to be repulsed from the naturally negatively charged mineral surfaces. This leads to what is called ion exclusion that prohibits the whole pore volume to transport anions. The opposite effect, a concentration of cat-ions at the surface leads to a higher flux in the pore because the concentration and thus the gradient will be larger than for an uncharged species with the same bulk concentration outside the piece of rock. The mobility in this more concentrated layer is called surface diffusion.

Figure 2 below illustrates the mean concentration profile of a non-charged, specie, an an-ion and a cat-ion outside and inside a rock sample.

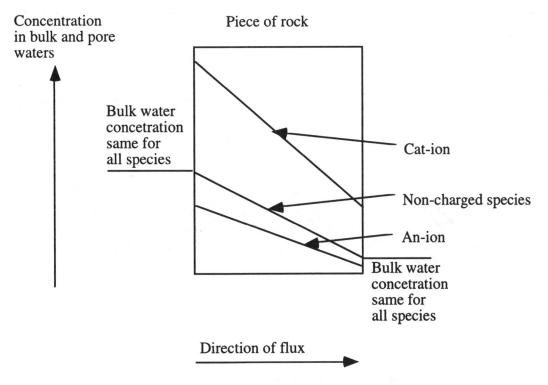

Figure 2 Concentrations in pore and bulk waters of neutral species, an-ions and cat-ions

These effects are clearly seen for diffusion in bentonite clay which has similar mineral surfaces as those of granites and gneisses.

A recent compilation of diffusivities in compacted bentonite has been made by Yu and Neretnieks [11]. For low salinity waters it was found that the an-ions iodide, chloride, carbonate and pertechnetate had one to three orders of magnitude smaller effective diffusivities due to ion exclusion than comparable non-influenced nuclides. Surface diffusion was found for cesium, protactinium, radium and strontium. One to two orders of magnitude increase in the effective diffusivity was noted. Other cationic nuclides were not found to have a clearly noticeable surface diffusion component. The ion exclusion as well as the surface diffusion effects disappeared at higher ionic strength pore waters. We mean by high ionic strength or high salinity a salt concentration of 10 000 mg/l or more. With low salinity waters we mean salt concentrations around 1 000 mg/l or less. These are rather arbitrary values but they were chosen because a group of important data that were used had approximately these salinities. There is a definite need to study these effects further.

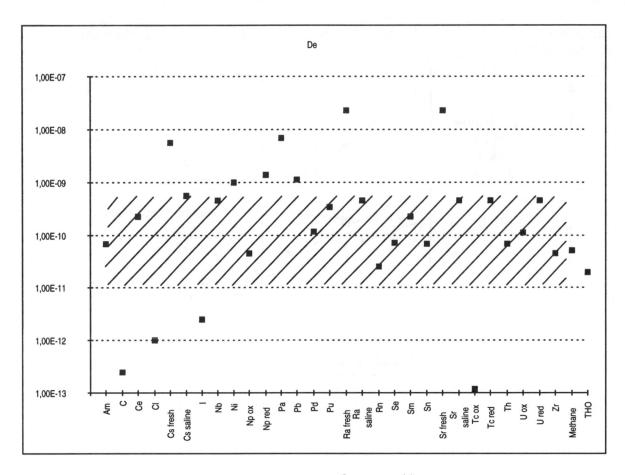

Figure 3. Effective diffusivities. Dotted lines at 10^{-9} and 10^{-11} indicate span where neither ion exclusion effects nor surface diffusion effects are important.

Figure 3 shows a compilation of the diffusivity data in compacted bentonite. The variability of these data are nearly two orders of magnitude if one excludes the nuclides that either have ion exclusion or surface diffusion effects. A visual estimate of the standard deviation of these data is $\sigma=0.5$ on the ^{10}log scale. Part of this variance is due to the differences between the nuclides so the variability for a given nuclide is less.

It has not been possible to make a similar compilation of diffusion data for the different nuclides in granite. In the earlier cited papers [8,9,10] where electrical conductivity measurements and diffusion measurements using iodide and an-ionic dyes were used similar variabilities were found for different rock pieces. A visual estimate of the standard deviation of these data is $\sigma=0.2$-0.3 in the ^{10}log scale.

For use in a performance assessment for a known site we have recommended the following procedure[12].

Site specific data on the formation factor are determined for such a number of samples that a mean value and a standard deviation can be obtained. The formation factor can be measured using electrical conductivity measurements or direct steady state measurements of the diffusive flux through the samples. This is done for the different rock types that may be encountered by the nuclides. The water chemistry and the ionic strength is determined for at the different locations. A table showing the calculation procedure can then be constructed. In the second column the bulk diffusivities of the nuclide is entered. This multiplied by the formation factor will give the effective diffusivity in column 4 if there are no ion exclusion or surface diffusion effects. This is for high ionic strength. For low ionic strength waters these effects must be accounted for and we suggest a decrease of a factor ten for those nuclides that are subject to ion exclusion and an increase by a factor of ten for those nuclides that are known to exhibit surface diffusion. This factor is given in column 7. There is however, a large uncertainty in the factor 10. it could well be a factor 100 as seen in Figure 3 above. The effective diffusivity is then entered in column 8.

For use in ongoing safety analysis performed by SKB in Sweden we have suggested the use of formation factor of $4.2*10^{-5}$. The effective diffusivities for the nuclides that will be used in the sample calculations are shown in Table 1.

Specie	D_w	K_d	$D_e=D_w*F_f$	$D_a=D_e/(\varepsilon+\rho K_d)$	K_d	f	$D_e=D_w*F_f*f$	$D_a=D_e/(\varepsilon+\rho Kd)$
				High ionic strength			Low ionic strength	
HTO						1		
IE						0.1		
SD						10		
other						1		

where IE denotes "ion exclusion" i.e. anions, SD denotes surface diffusion i.e. ion exchanging cations.

Table 1.   Calculation schedule to obtain D_e and D_a when D_w, K_d and F_f are known

Sorption data

Carbol and Engkvist ([13] have recently compiled sorption data for a large number of nuclides for the same types of waters as were used to estimate the diffusion data. The waters are reducing. with an Eh <200 mV and have a pH=7. They are typical of the Gideå (low salinity) and Äspö (high salinity) site respectively.

Their data show that the sorption coefficient of mono- and di- valent cat-ions decrease by a factor 5 to 10 in the higher saline waters compared to low salinity waters. They are in the range 0.1 m^3/kg for Co(II), Ni(II), Pd(II), Ra(II) and Cd (II), 0.01 m^3/kg for Sr(II) and 0.5 m^3/kg for Cs(I) and Ag(I) for the low salinity waters. For tri- to penta-valent nuclides the sorption coefficients are much higher and are not influenced by the salinity of the water. The sorption coefficients range from 1-5 m^3/kg with the higher value, 5, for the actinides Th(IV), U(IV), Np(IV) and Pu(IV), the value 3 for Ac(III), Am(III), Cm(III), the value 2 for Sm(III), Eu(III) and Ho(III). Zr(IV), Nb(V), Tc(IV) and Pa(IV,V) have K_d=1 m^3/kg. The anions I, Cl have K_d=0 m^3/kg. Carbonate has a K_d of 0.001 m^3/kg.

The uncertainty interval has been estimated to be a factor between 4 and 50, for the different nuclides. A typical value for the uncertainty interval is 10 between the highest and the lowest value in the span.

Uncertainty interval and distribution function

In very few instances a formal statistical analysis has been presented and there is little information on the form of distribution function and on the standard deviation of the distribution.

For illustration purposes it is assumed in this paper that the distribution is log-normal. The term uncertainty interval is not defined. Interviews with several colleagues working in this and related fields indicate that most of them agree to the following statement. "The uncertainty interval means that very few data points will be found outside the interval". When pressed somewhat more on how many point would be permitted to lie outside the interval the answers range from about 1 to no more than 10%.

If 5 % lie outside the interval this will mean that the interval spans four standard deviations on the logarithmic scale. Thus for an interval of a factor of ten the standard deviation is $\sigma = 0.25$ on the ^{10}log scale. For intervals 4, 20 and 50 they are 0.15, 0.33 and 0.43 respectively.

Often the proposed uncertainty intervals are asymmetric around the suggested value. This does not seem to indicate that the distribution is asymmetric, only that the use of the expressions is imprecise and blunt. It is unclear in this case what is uncertainty and what is variability.

In the examples a value of $\sigma=0.3$ will be used for D_e and K_d and an uncertainty interval of 3 up and down.

Specific flow wetted surface and conductivity distribution

The entity that is most poorly known is the flow wetted surface. There is no well tested and universally accepted method to determine this entity. There are also very few field measurements that have been used to determine the FWS in the form of the specific FWS a_r.

Abelin et al. [14] used the non recovery of a part of the tracer mass injected in a long time experiment in Stripa to estimate the FWS. Values ranges from 0.5 to more than 20 for five different flowpaths if virgin rock matrix properties were assumed. If the matrix diffusion had mostly taken place in somewhat altered rock adjacent to the fracture surfaces the values ranged from 0.2 to 2.7 m^2/m^3.

Birgersson et al. [15] obtained a_r values ranging from 5 to 24 m^2/m^3 from the non-recovery of tracers in experiments in a fracture zone in the Stripa experimental mine.

Bore hole fluid injection measurements in the Äspö rock laboratory on fracture frequencies and conductivities in the observed fracture zones gave a value of a_r 1.2 m^2/m^3. A $\sigma = 1.6$ on the ^{10}log scale was found for the local conductivities. Gylling et al. [16] and Moreno and Neretnieks [4] analysed data from three different sites and found σ for the conductivities to be about 1.6-2-4. This shows that there will be a very large range of flowrates in different locations. It was found in simulation using the Channel Network Model that the standard deviation for local flowrates is approximately 0.85 of that for the conductivities.

From a geometrical point of view the flow wetted surface can be approximately estimated from the flowing fracture frequency in boreholes by the following expression [17].

$$a_r = \frac{\pi^2}{2} \cdot \frac{1}{H}$$

(8)

H is the average distance between conducting fractures in the borehole. There are two difficulties in using this method. One is that of cut off in the measurement. The low flow channels are not detected. Moreno et al try to compensate for this by extrapolating the observed higher flow data to the low flow region by assuming a log normal distribution of flow rates.

The other difficulty is that the spacial resolution in the measurements is often very coarse. Commonly packer distances of more that 10 m are used in boreholes in the good rock. This means that one can in principle know if there is more than one conducting feature in that interval, not how many more. This may seriously underestimate the FWS.

There are not enough experiments where the FWS has been obtained using the geometric method and other methods that observe the impact of the FWS on the retardation or non recovery of tracers. In the few instances where this has been attempted the differences seem to be about an order of magnitude.

The data on the magnitude of the FWS are thus very uncertain. The variability in the form of the standard deviation of the local FWS/flowrate is perhaps better known. It is at present used in the Channel Network Model. The simulation results for a tracer are much more influenced by the total FWS encountered along the flowpath than by the magnitude of the standard deviation.

For a standard deviation the ^{10}log scale varies of 1.6, the difference between the low and high values is more than one million. The low and high values are then taken at -2σ and $+2\sigma$. This span contains 95 % of all values.

In the examples a value of $a_r = 0.4$ m^2/m^3 will be used with a low value of 0.1.

Other data

A water flux in the rock at repository depth of 0.1 l/m^2·year is used. This would be obtained in a rock with a hydraulic conductivity of 10^{-9} m/s and a hydraulic gradient of 0.3%.

For use in Equation 1 for the single fracture description the fracture frequency is 1 per 5 meters for the chosen specific FWS. This gives a flowrate Q of 0.5 l/year in a fracture 1 m wide.

In the example we study a fracture or streamtube somewhere near a degraded canister and assume it is 50 m long before reaching a region with different properties e.g. a fracture zone. This would illustrate a near field fracture. The water may thereafter emerge into a region with high flux.

The fracture aperture is assumed to be 0.01 mm. This choice is not critical. Even very large variations do not influence the results. These values give a water residence time of 0.1 year. If the fracture aperture would be 1 mm for a loosely gouge filled fracture the residence time would be on the order of 10 years.

Summary of data used in the examples

Nuclide	$D_e *10^{13} m^2/s$	K_d m^3/kg	$^{10}Log\ \sigma_D$*	$^{10}Log\ \sigma_K$*
I Low salinity	0.08	0	0.3	0.3
I high salinity	0.8	0	0.3	0.3
Cs Low salinity	4	0.5	0.3	0.3
Cs high salinity	0.4	0.05	0.3	0.3
Pu and Np both salinities	0.4	5	0.3	0.3

*Standard deviations are examples for use in the examples. They are not based on statistical analysis

In the sample calculations it is assumed that the uncertainties in diffusivity and sorption data is a factor of three up and down.

	Entity	Value	Uncertainty span*
a_r.	Specific flow wetted surface in good rock	0.4 m^2/m^3 rock	0.1-5
2b	Fracture aperture	0.01 mm	0.01-1¤
L	Flow path length	50 m	
u_o	Water flux at repository depth	0.1 l/ m^2	-
d	Thickness of porous alteration material	0 mm	0-1#
	Thickness of porous gouge material in fracture	0 mm	0-2b#
ρ	Density of rock	2700 kg/ m^3	none

*Subjective values. This spans a conductive fracture distance of 1-50 meters
¤ Subjective values
Sorption coefficients are taken to be as for food rock

Nuclide	Halflife of nuclides years
^{137}Cs	30
^{239}Pu	24 400
^{237}Np	$2.1*10^6$

Table 2. **Effective diffusivities, sorption coefficients and estimated standard deviations of some nuclides. Reducing water.**

Influence of variability along the flowpath

It was demonstrated earlier, Equation (7) that variations along the flowpath can be accounted for along the flowpath as shown in the expression $\int_0^z \frac{A_o(z)a_r(z)\sqrt{D_e K_v}}{Q}dz$. In the streamtube, or we prefer to think of the flow as taking place in a succession of channels, Equation (8), the flowrate Q is the same all along the flowpath. It may be noted that $\int_0^z A_o(z)a_r(z)dz$ is the total FWS along the flowpath contacted by the water flowrate Q. The flowpath may have traversed good rock, fracture zones and other features. In principle it can be assessed by following the streamlines obtained from

206

the hydrological modelling and from knowledge of specific FWS a_r in the different regions. Provided there is no correlation between a_r and the sorption coefficient and the effective diffusivity the integration can be made as just described. At present there is no information available on any correlation between the entities and thus this assumption is necessary at present. There is also no known correlation between D_e and K_v. The integration can also here be made independently.

The average of the square root of D_e and K_v differs from the square root of the average of the entity. For the assumed log normal distribution and the value of $\sigma = 0.3$ the difference is only a few percent. However, the expected value of this the entity is nearly three times that of the logarithmic mean in the distribution. There seems therefore to be a need to analyse the form of the distributions and what is actually meant by the values that are recommended for the different entities.

Although a part of the variability in the data are caused by natural stochastic variations that will be evened out along the pathway, some sample calculations will be made assuming that the variability is due to uncertainties.

Sample calculations

Influence of uncertainties in sorption, diffusion and flow wetted surface

For ^{137}Cs the effluent concentration after 50 m travel distance has decayed to insignificance, $<< 10\text{-}9$, for the central case. Only when all the variables are given the most unfavourable values simultaneously the concentration will become noticeable and be at most a few % of that at the inlet of the streamtube. Then the travel distance is 50 m, the diffusivity and the sorption coefficients are a factor 3 lower than the central value, FWS is at the lowest value and the water is saline. For a non saline water even the worst case combination will give a totally insignificant concentration.

For ^{239}Pu the effluent concentrations will decay to insignificance for the central case. A decrease of both the sorption coefficient and the diffusivity by the uncertainty factor 3. still gives no significant concentration. The same applies if only the FWS is given its lowest value. An increase of the matrix diffusion group $\int_0^z \dfrac{A_o(z)a_r(z)\sqrt{D_eK_v}}{Q}dz$ by a factor of 10 gives noticeable concnetrations. Any combination of variables that give factor of 10 increase will have the same effect. This includes also an increase of the flowrate Q.

Log Concentration

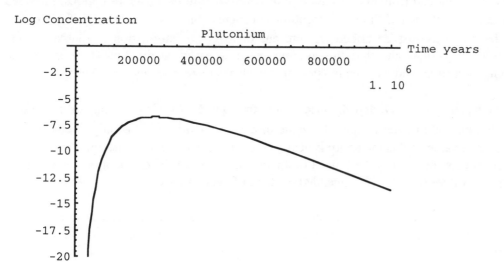

Figure 3 . Outlet concentration for plutonium for a combined uncertainty of a factor 10 increase.

For ^{237}Np the breakthrough curves are shown in Figure 4 for a factor of 3, lower curve, and factor of 10, upper curve, uncertainty in the group $\int_{0}^{z}\dfrac{A_{o}(z)a_{r}(z)\sqrt{D_{e}K_{v}}}{Q}dz$,

Log Concentration

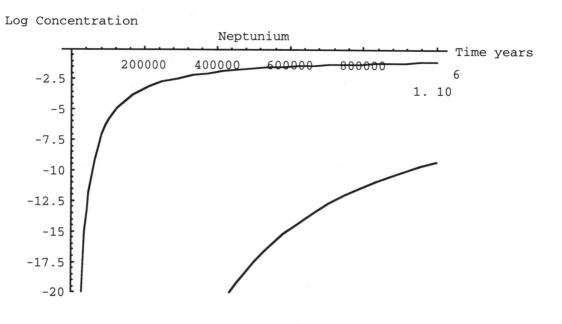

Figure 4.   Outlet concentration for Neptunium. upper curve factor 10, lower, factor 3 uncertainty.

Influence of uncertainties of the presence of alteration and filling materials

Below is an example of the impact of fracture filling and fracture alteration material. The fracture filling material fills out all the fracture aperture, 0.01 mm and the alteration material is 1 mm thick. Both have the same volumetric sorption coefficients as the good rock and are instantly equilibrated. It is the fracture alteration material that causes 99% of the effect in this case.

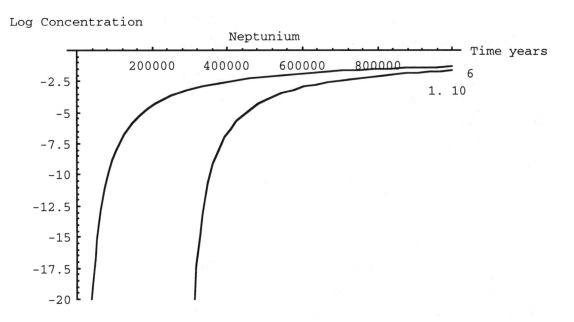

Figure 3 . Outlet concentration for neptunium. Both curves factor 10 uncertainty for the matrix diffusion group b. Right hand curve has 0.01 mm fracture filling and 1mm alteration material.

Discussion and conclusions

There is at present very little information on the stcochastic properties of most of the entities that influence the migration of radionuclides in fractured rock. Much of the obderved variability in reported data on sortion and diffusion must at present be assumed to be uncertainties. This may be due to systematic differences between different measurement techniques. The stochastic variability of these entities can be accounted for by integrating the properties along the flowpath as has been demonstrated.

The impact of systematic errors and uncertainties in these entities is somewhat diminished becuase they enter the expression by a square root. The local flowrate of water, Q, per Flow Wetted Surface, FWS, has a stonger inpact. Furthermore this entity is much less studied and the magnitude of this entity is more uncertain. Fortunately theory indicates that the variability of this entity along the flowpath can be accounted for in a simple manner.

The by far most uncertain entity is the magnigtude of the specific flow wetted surface. It also has the highest impact . It is recommended that this entity be studied further and that new techniques are developed to measure flowrate distributions in boreholes with a very high spacial resolution. Present day practice of using packer distances of 10 or more meters give data that can strongly underestimate the FWS.

In fractured rocks with fracture filling and porous fracture alteration materials of more than a fraction of a mm, such material can significantly retard the nuclides in addition to the matrix diffusion effect.

Notation

a_r	Specific flow wetted surface	m^2/m^3 rock
A_o	Cross section of stream tube	m^2
b	Half aperture of fracture	m
c	Concentration	mol/m^3
d	Thickness of alteration material	m
D_a	Apparent diffusivity	m^2/s
D_e	Effective diffusivity	m^2/s
F_f	Formation factor	-
H	Average distance between open fractures in borehol	m
K_a	Surface sorption coefficient for rock surface	m
K_d	Mass sorption coefficient for rock	m^3/kg
K_{df}	Mass sorption coefficient for gouge	m^3/kg
K_{da}	Mass sorption coefficient for alteration material	m^3/kg
K_v	Volumetric sorption coefficient for rock	m^3/m^3
K_{vf}	Volumetric sorption coefficient for gouge	m^3/m^3
K_{va}	Volumetric sorption coefficient foralteration material	m^3/m^3
L	Channel length	m
Q	Flowrate of water	m^3/s
R_a	Retardation factor	-
t	Time	s
t_o	Travel time of nuclide for plug flow	s
t_w	Travel time of water for plug flow	s
u_o	Water flux	$m^3/m^3 \cdot s$
W	Channel width	m
z	Flow path length	m
ε	Rock matrix porosity	-
λ	Decay constant for nuclide	1/s
σ	Standard deviation	-
ρ	density of rock	kg/m^3

Literature

1. Neretnieks, I., Diffusion in the rock matrix: An important factor in radionuclide retardation? J. Geophys. Res. 85, p 4379-4397, 1980.
2. Carslaw H.S., Jaeger J.C. Conduction of heat in solids, 2nd ed. Oxford University press 1959.

3. Moreno L. and I. Neretnieks, Flow and nuclide transport in fractured media. The importance of the flow wetted surface for radionuclide migration, In J.I. Kim and G. de Marsily (Editors), Chemistry and Migration of Actinides and Fission Products, Journal of Contaminant Hydrology, Vol 13, 49-71, 1993a.

4. Moreno L. and I. Neretnieks, Fluid flow and solute transport in a network of channels. J. Contaminant Hydrology 14, 163- 192, 1993b.

5. Tullborg Eva-Lena, Terralogica. Personal communication March 1997.

6. Neretnieks I. A note on the potential impact of fracture filling and alteration materials. Report prepared for SKB. Dept. Chemical Engineering and Technology, Royal Institute of technology, Stockholm Sweden. Feb 1997.

7. Ohlsson Y. and Neretnieks I. Literature Survey of Matrix Diffusion Theory and of Experiments and Data Including Natural Analogues. SKB TR 95-12.

8. Skagius, K., and I. Neretnieks, Porosities of and diffusivities of some non-sorbing species in crystalline rocks. Water Resources Res. 22, p 389-398, 1986a.

9. Skagius, K., and I. Neretnieks, Diffusivity measurements and electrical resistivity measurements in rock samples under mechanical stress. Water Resources Research 22 (4), p 570-580, 1986b.

10. Birgersson, L., and I. Neretnieks, Diffusion in the matrix of granitic rock. Field test in the Stripa mine. Water Resources Research 26 p 2833-2841, 1990.

11. Ji-Wei Yu and Neretnieks I., Diffusion and sorption properties of radionuclides in compacted bentonite, SKB TR 97-xx, in print.

12. Ohlsson Y. and Neretnieks I. Diffusion data in granite for SR 97, Recommended values SKB TR 97-xx, in print.

13. Carbol P., Engkvist I. Compilation of radionuclide sorption coefficients for performance assessment. SKB TR 97-xx, in print 14. Abelin H., Birgersson L., Moreno L., Widén H., Ågren T., Neretnieks I. A Large Scale Flow and Tracer Experiment in Granite II. Results and interpretation. Water Resources Research, 27(12), p 3119-3135, 1991.

14. Abelin H., Birgersson L., Moreno L., Widén H., Ågren T., Neretnieks I. A Large Scale Flow and Tracer Experiment in Granite II. Results and interpretation. Water Resources Research, 27(12), p 3119-3135, 1991.

15. Birgersson, L., L. Moreno, I. Neretnieks, H. Widén, and T. Ågren, A tracer migration experiment in a small fracture zone in granite, Water Resour. Res., 29(12), 3867-3878, 1993.

16. Gylling B., Moreno L., and Neretnieks I. A channel-network-model for radionuclide transport in fractured rock.- Testing against field data, Paper presented at the MRS Meeting on Scientific basis for nuclear waste management XVIII, Kyoto, Oct 24-26, 1994, Proceedings vol 353, p 395-402.

17. Gylling B., Moreno L., and Neretnieks I. Modelling of solute transport in fractured media using a channel network model. Submitted to ICEM 1997.

What can be Learnt from Other Research Fields and Applications

Gh. de Marsily
University Paris VI, France

Abstract

As a preliminary example, the case of the contamination of the Rhine alluvial aquifer in the vicinity of Mulhouse, France, was briefly presented. In the early 1900's, the potash mines (Mines de Potasse d'Alsace) disposed of the waste salt (NaCl) extracted from the mine together with the potash as "salt dumps" on the surface of the alluvial plain. Since then, the salt has been leached by rainfall and has slowly propagated in the aquifer downstream. The movement of the salt plume has been monitored for several tens of years, and was, some years ago, nearing the well fields of the city of Mulhouse, threatening to make them unfit for water supply. To prevent this, the city decided to find a new location for its well field, and decided to move it some kilometres to the west, away from the direction followed by the salt plume. To everybody's great surprise, the first well drilled in this new site appeared to be already inside a salt-contaminated area ! There had been indeed another plume, starting from the same salt dump, in another direction, which had been totally undetected, and had transported salt even further downstream than the main one ! This example tends to show that heterogeneity in the flow and transport of solutes can (i) remain largely undetected, and (ii) transport solutes at areas that had been expected to be out of reach from the salt. The contamination has now been controlled at the source, to remediate this situation.

It seems therefore appropriate to derive methods to detect anomalies in the continuity of confining strata, to avoid the type of undetected migration that occurred in the previous example. The effort can follow at least three paths : (i) geophysical methods; (ii) hydraulic testing; (iii) in situ natural tracers. For the geophysical techniques, there is a great need to develop techniques that can remotely detect fractures and discontinuities in all the geologic formations which are being considered to host wastes. This is not the topic of this paper, however it is clear that when a repository is being built, e.g. in a granite or even clay environment, one will assume that the emplaced wastes is away by a given minimum distance from any flowing fracture (e.g. 100 m in the KBS-3 concept). Since it is not possible to drill in every direction to make sure that there is no such fractures at some distance from each wastes package, it is necessary that a geophysical technique (that can be applied in underground tunnels) can be developed, which would be used systematically. Ground penetrating radar have been suggested, but still needs to be improved.

Hydraulic testing is a very efficient way to test the hydraulic properties of pervious media. In fractured systems, the "Generalised Radial Flow" interpretation method developed by Barker (1988) makes it possible to determine the apparent dimension of the medium in which the test is performed. Dimensions tending towards one indicate a highly channelised medium, where anomalous short circuits may exist, whereas dimensions tending towards three indicate a medium having rather spatially homogeneous distribution of hydraulic conductivity, with much less channelling.

Application of this method has been very successful for the SKI Site-94 Project (Geier et al, 1996). Herweijer (1997) also used pumping tests in alluvial aquifers to detect the presence of buried channels of high conductivity, by comparing the drawdown in different wells at increasing distances from the pumped well. He found that, using "classical" interpretation methods such as the Theis type curve method, produced values of hydraulic conductivity which were similar for all wells, but storage coefficients that varied with the distance, indicating thus the presence of anomalous conduits, or high permeability channels, biasing the apparent value of the storage coefficient. However, such methods are not effective in media with very low permeability, because the influence of the hydraulic test is very slow to propagate in the medium, and therefore the size of the medium affected by the test is very small, and is not sufficient to detect the presence of anomalies at some distance.

The third method that can be considered is the use of natural tracers, which have migrated over very long periods of time in the medium. One such tracer is heat, and temperature anomalies have sometimes been used to detect the presence of conduits or channels. For instance, within the seabed programme (NEA, 1988), temperature profiles were measured on the see bottom (by cylindrical penetrators that were dropped into the sediments, and measured the temperature at three different depths under the sea bottom). If these profiles showed a linear increase of the temperature with depth, it indicated that thermal conduction was the major heat transport mechanism; however some points showed a more or less parabolic profile, indicating the effect of fluid advection on the profile, thus the presence at short distance of a conductive anomaly, as the permeability of theses sediments is too low to permit significant advective transport of heat, unless there is a fracture. Burrus (1997) also used thermal profiles in wells in deep sedimentary basins to detect the presence of advective heat transport in several layers of these basins. He showed, for the case of the Paris Basin, that the present-day temperature profiles cannot be interpreted as being in steady state, because, for large depth, the effect of the climate changes that occurred at the end of the last glaciation are not yet fully dissipated, and the present-day profile still bears the imprint of the increase in temperature that occurred around 18,000 to 7,000 years ago.

Another type of tracers that have been used with great success in sedimentary basins is the rare gases. Castro et al (1998a,b) have used helium, argon and neon to analyse the flow of fluids in the Paris basin. They showed that these tracers can be used to understand the movement of water in the permeable aquifer and in the impervious aquitards as well. Radiogenic helium is mainly produced in the deep upper crust by radioactive decay of uranium and thorium, and transferred to the bottom of the sediment series. Helium is also produced in the sediments, but in much less quantity. Primordial helium can also come from the degassing of the mantle. By using the ^{4}He and ^{3}He ratio, it is possible to understand the origin of helium (atmospheric, radiogenic or mantellic) and to study its rate of transfer in the sediments. Used in conjunction with argon, it was found that helium seems to be transported by diffusion through the impervious sediments, whereas argon, which has a smaller diffusion coefficient, is mainly transported by advection, and is thus sensitive to the permeability of the clayey aquitards. The presence of water conducting faults can thus probably be detected by argon anomalies (ratio to helium), if these faults are pervious. Such faults are known to exist in the Paris basin, but could not explicitly be detected, for lack of sufficient sampling. If the rare gases are indicators of diffusion transport, they may be quite useful for studying the mechanism of matrix diffusion around a fracture, by comparing their ratios at different depth along a vertical fracture.

To interpret the rare gases measurements, it is important to use a numerical model representing the various transport mechanisms and the sources, for each gas, and then calculating their ratios, to be compared with the observations. See Castro et al, 1998b.

References

Barker, J.A. (1988) A generalized radial flow model for hydraulic tests in fractured rock. Water Resour. Res. 24, 10, 1796-1804

Castro, M.C., Jambon, A., Marsily, G. de, Schlosser, P. (1998) Noble gases as natural tracers of water circulation in the Paris basin. Part 1 : Measurements and discussion of their origin and mechanisms of vertical transport in the basin. Submitted (2nd draft) to Water Resour. Res.

Castro, M.C., Goblet, G., Ledoux, E., Violette, S. Marsily, G. de (1998) Noble gases as natural tracers of water circulation in the Paris basin. Part 2 : Calibration of a groundwater flow model using noble gas isotope data. Submitted (2nd draft) to Water Resour. Res.

Geier, J.E., Doe, T.W., Benabderrahman, A., Hässler, L. (1996) Generalized radial flow interpretation of well tests for the Site-94 project. SKI Report, Stockholm, 96:4, 178 p.

Herweijer, J.C. (1997) Sedimentary Heterogeneity and Flow Towards a Well. Assessment of flow through heterogeneous formations. Ph.D. Dissertation, Free University Amsterdam.

NEA-OECD (1988) Feasibility of sub-seabed disposal of high-level radioactive waste. OECD Press, Paris.

SESSION IV

What is Wanted and What is Feasible:
Views and Future Plans in Selected Waste Management Organisations

Chairmen: A. Gautschi (Nagra, Switzerland)

Modelling Strategies and Usage of Site Specific Data for Performance Assessment

A. Hautojärvi
Posiva Oy, Finland

J. Andersson
Golder Associates AB, Sweden

H. Ahokas
Fintact Oy, Finland

L. Koskinen and A. Poteri
VTT Energy, Finland

A. Niemi
Royal Institute of Technology, Hydraulic Engineering, Sweden

Abstract

The presented strategy plan builds on an evaluation of different approaches to modelling groundwater flow in crystalline basement rock, the abundance of data collected in the site investigation programme in Finland, and the modelling methodology so far developed in the programme. The whole system is anticipated to be modelled using nested models, where larger scale models provide boundary conditions for the smaller ones. Furthermore, it seems motivated to adopt different conceptual models for the different scales studied. Equivalent porous medium, stochastic continuum and discrete fracture network models are suggested to describe groundwater flow for performance assessment needs.

INTRODUCTION

Posiva Oy manages the R&D programme for spent nuclear fuel disposal in Finland. Currently Posiva is exploring four investigation sites as potential hosts for a deep repository. According to the schedule set by the authorities the site for the repository has to be selected by the end of year 2000. The site selection will be supported by a site-specific performance analysis.

This paper outlines a strategy plan for groundwater flow and transport modelling to be considered for the site specific performance assessment analysis of spent nuclear fuel disposal. Such an analysis, in addition to showing that the proposed repository is safe, needs to strengthen the link between field data, groundwater flow modelling and derivation of safety assessment parameters, and needs to assess uncertainty and variability.

The groundwater flow used as input to the source term modelling should be within the ranges calculated from the groundwater flow modelling. Regarding far-field migration there is a need to strengthen the estimation of the ratio between channel width and groundwater flow from actual field data, to evaluate different migration paths and not only the most pessimistic ones and to study the effect of variability in the fracture plane. The groundwater flow modelling should discuss implications for the biosphere in terms of dilution in the geosphere and distribution of release points.

Given the geometrical detail at the canister scale it seems that the most appropriate model would be a discrete fracture network description. It is suggested to limit the size of the discrete model to small blocks contained by larger size fractures. Outside this domain a stochastic continuum model would be applied, but with hydraulic and transport properties derived from a discrete network model.

In a regional and repository scale, adoption of the equivalent porous medium concept seems the most appropriate. The regional scale takes the regional fracture zones and watersheds into account whereas the repository scale model is more focused to the repository area and describes the local fracture zones and deterministic structures found in the site. These models provide boundary conditions to more detailed scale models, but do not resolve the variability of the hydraulic properties.

GROUNDWATER FLOW AS INPUT TO SAFETY ASSESSMENT

In order to work out a strategy for the groundwater flow modelling it is necessary to assess how the groundwater flow modelling may be used in the quantitative safety assessment as well as to assess other specific uses of groundwater flow modelling in the safety assessment. Especially the variability of groundwater flow at various scales due to irregular appearance and heterogeneity of water conducting pathways in crystalline rock is an essential difficulty and challenge for a modeller. The groundwater flow information needed in performance assessment is viewed briefly here.

GROUNDWATER FLOW IN THE SCALE OF DEPOSITION HOLES AND TUNNELS

The near-field release model used, REPCOM (Nordman and Vieno, 1994) is a compartment model that can handle different release pathways. In the recently updated safety analysis in Finland, TILA-96 (Vieno and Nordman, 1996), the following pathways are included:

- Q_F the "fissure pathway" into a fracture which directly intersects the deposition hole – this release will depend on the darcy velocity in the fracture and the fracture geometry,

- Q_{DZ} the pathway from the top of the deposition hole into the excavation damaged zone – this release will depend on the geometry and flow rate in the damaged zone,

- Q_T the pathway from the tunnel into the geosphere – this release will depend upon the flow in the rock and the hydraulic properties of the tunnel.

Apparently the groundwater flow (and the geometry of the flow) affects the release in all considered pathways. However, in TILA-96 the approach taken is not to estimate the groundwater flow parameters in the near-field model from the groundwater flow modelling. Instead sensitivity analyses are performed to show that the release is robust (i.e. quite insensitive) to a wide range of groundwater flow conditions. In coming assessments it appears that description of flows would be warranted.

In order to show that the assessment really concerns the analysed sites (i.e. is site specific) it seems appropriate to show that the groundwater flows calculated within the groundwater flow modelling are within the ranges considered in the source term modelling.

There are prospects to show that the flow ranges now considered, indeed, are too wide – at least for a majority of canister positions. Even if this information is of little value in a pessimistic single canister failure analysis, it appears to be a positive feature of a site if it contains many canister positions with properties much more favourable than the ones assumed in the pessimistic assessment case. The fact that the release is robust to a large variety of flows indicate that there is little "risk" to go for a realistic modelling of the flow – the safety arguments could only be strengthened.

On the other hand there also are some practical constraints to consider:

- The results will partly depend upon the technical properties of the tunnel backfill, efficiency of seals etc. and there is a need for the safety assessment people to make their own variations about these without having to go back to groundwater flow modelling all the time. Consequently, it would be necessary to develop a baseline input from groundwater flow modelling, which then can be used by safety assessment as a basis for different variation cases.

- The flow input to the source term model ranges from the deposition hole scale to the tunnel scale. In providing the flow input to the source term these scales have to be considered, the proper conceptual model for this scale needs to be selected and the proper representation of the repository component (by a plate or individual tunnels) has to be selected.

In conclusion, the idea is to strengthen the link between source term input and results of the groundwater flow modelling, while still addressing the above constraints.

FAR-FIELD MIGRATION

The far-field migration model used in TILA-96 evaluates transport along a single one-dimensional migration path. The retention is essentially governed by the ratio WL/q, where W is the total width of channels per unit area of rock cross section, L is the transport distance and q the Darcy velocity in the rock.

An attempt is made to make the analysis site specific by analysing flow routes from the groundwater flow simulations of Kivetty and Romuvaara respectively. For each site q and L for segments of a fast route and a slow route are presented. Estimates of W for each segment is taken from TVO-92 (Vieno *et al.* 1992).

In TVO-92 the effects of conductive fracture frequency were estimated. It was noted that low frequencies and high channelling give most conservative results. For different rock parts (intact rock, disturbed rock zone and different fracture zones) it was simply assumed that the flow was equally divided over the chosen number of parallel fractures in each rock part. In the intact rock it was further assumed that the channels cover 10% of the width. The assumption that in the disturbed rock zone and in the various fracture zones the flow is divided over the given numbers of fractures can be discussed. Along a path with varying W, L and q the effective transport resistance is simply the sum of the WL/q for the individual segments. This important relation is not explicitly stated in Vieno *et al.* (1992) although it was used and is also evident from the mathematical derivation.

It could be noted that Selroos and Cvetkovic (1996) demonstrate that the sum of WL/q can be extended to an integral along the pathway. It may also be of interest to note that Andersson *et al.* (1997) suggest how to provide effective parameters for far-field migration in SKB SR 97 using procedures quite similar to the ones applied already in TVO-92.

Even if the TILA-96 approach attempts at a site specific migration analysis there is certainly a need for improvements, which is also noted in the report. The following questions, in particular, could be raised:

- There is a need to strengthen the conceptualisation and estimation of flow distribution in the bedrock and derivation of the W/q estimates. In the simplified analyses the estimates of the conductive fracture frequency could clearly be made more explicit and the flow rates and flow distributions over fracture systems in various rock parts should be assessed carefully and thoroughly. In a rock block with several conductive flow paths it is likely that the flow will vary considerably between different paths resulting in a distribution of W/q values representing the different paths. Furthermore, it is only the paths where migration actually takes place that are important. This will affect how to sample the W/q distribution into effective parameters.

- A key aspect when determining W/q is the effect of flow variability in the fracture plane. A small portion of the fracture area may account for a large portion of the flow (see e.g. Tsang *et al.*, 1991). A real fracture may be thought of to be divided into several parallel (potential) migration paths (see Figure 1), although only some of these may be "used" depending on the spread of the migration plume. If flow is evenly distributed over the plane of the fractures, the individual W_i/q_i in each such path would equal the total width of the fracture divided by the total flow across this width, but in general the flow will vary, resulting in a distribution of W_i/q_i. An important issue is to what extent free water diffusion may even out the effects of the flow variability, which would make the estimation of the W/q and its coupling to the fracture geometry more robust. Some initial studies on this subject were made already in TVO-92 (Vieno *et al.*, 1992), but the issues seem to require further studies. A potentially promising approach would be migration simulations with heterogeneous (as well as homogeneous) discrete fracture networks.

- Allowing for a detailed groundwater flow modelling which includes the effect of small scale variability (i.e. stochastic continuum or discrete networks) may result in much faster pathways (and lower WL/q) than presently results from the equivalent homogeneous porous medium approach adopted. In TVO-92 and in TILA-96, there is an attempt to compensate the homogeneous model by selecting extreme values. Furthermore, the flow paths usually end up very quickly in the fracture zones. However, the approach of selecting the worst possible migration paths may thereby not disclose that the majority of migration paths would have much *more favourable* migration characteristics. Evaluation of different paths (in at least a semi-stochastic sense) seems to be needed.

- The present migration modelling does not imply a route dispersion term. Multiple path analyses could provide means of substantiating the dispersion effects and thus be used to validate the conclusions from the sensitivity analysis conducted within TILA-96. Coupled to this is the questions whether the one-dimensional flow tube in the transport model is an oversimplification. Clearly, the one-dimensional migration model is used in a majority of safety assessment produced so far (NEA, 1997). Furthermore, it is a misunderstanding that three-dimensional advection-dispersion equation should represent reality *a priori* better than one-dimensional one. In fractured rock it is also clear that transport takes place along single flow paths often with very little mixing between paths. One could also claim that given the small size of the source and the very limited contact between flow paths at fracture intersections the release will seldom branch into multiple paths, thus validating the single flow path model. However, given the lack of international consensus on these issues, the question needs to be addressed and possibly illustrated with a few numerical examples.

In conclusions, it is evident that groundwater flow modelling has a very important role in describing uncertainty and variability in migration parameters. Determination and uncertainty of WL/q in relation to available field information is in fact the main challenge for the far-field flow analysis. It is recommended that these problems will be studied with some detail by the DFN and detailed SC approaches.

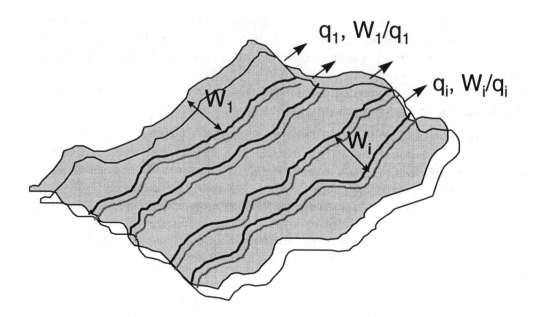

Figure 1: *A plane of a fracture may be thought to be divided into several parallel migration paths with different W_i/q_i. The proper equivalent value to use for the entire fracture will depend upon the flow variability in the fracture plane, the effect of free water diffusion and the size of the source term. These aspects can be studied in detailed discrete fracture network analyses.*

BIOSPHERE RELEASE POINTS AND DILUTION

The TILA-96 biosphere model is based on the WELL-96 scenario, which is an update of WELL-94 (Vieno, 1994). However, too much reliance on well doses may be challenged and questions regarding bio-accumulation may be asked. In such a case, the question of spread of release points as well as geosphere dilution will be of great importance for the biosphere modelling. Such estimates could possibly be obtained from groundwater flow modelling. Consequently, it is recommended that the groundwater flow modelling also discusses implications for the biosphere in terms of geosphere dilution and distribution of release points.

Regional and repository scale groundwater flow modelling will provide a general view of the distribution of potential release points. The potential release points will basically be confined to a few fracture zones. There is probably no need (nor foundation) for describing the heterogeneity of fracture zones - the present assumption of constant transmissivity fracture zones is considered as sufficient. If needed, stochastic continuum modelling could be applied to demonstrate the effect of the variability of flow paths and to illustrate whether there is a possibility for spread between several fracture zones and release points. However, such a study would be rather generic.

HYDROGEOLOGICAL MODELLING APPROACHES

This chapter discusses, in general, different approaches to modelling groundwater flow, in crystalline basement rock, to be used as input to radionuclide migration modelling and other safety assessment relevant uses. The approaches discussed are generally those applied or discussed in safety assessments performed in programs in other countries.

GROUNDWATER MODELLING PHILOSOPHY

In general, it should be emphasised that a model is formulated for a specific purpose and is not the same as reality. Among other things this implies:

- A model only includes the processes and interactions that were deemed necessary – it cannot represent all phenomena observed in the field.

- An important component in making a hydrogeological modelling credible is consistency, which calls for a rigorous quality assessment of the data and methods used, regardless of how flow and transport in the rock will be conceptualised.

- In a heterogeneous environment a model is formulated for certain scale - this means that model predictions do not necessarily match field measurements made at a scale different from the model, although there may be means of up-scaling detailed data.

- The choice of processes and scale has to be motivated from the purpose of the model – one means of testing the appropriateness of choices made is to compare with more complex models.

Among other things these general observations apply to the questions of selection of conceptual models for the groundwater flow, evaluation of uncertainty and to the degree of calibration necessary. Furthermore, the available possibilities need to be contrasted against the actual safety assessment needs.

CONCEPTUAL MODELS FOR GROUNDWATER FLOW

Conceptual models for groundwater flow concern selection of physical processes governing the flow and the selection of the proper geometrical framework for the models. For the latter case there exists conceptually different structure models, such as homogeneous porous medium, stochastic continuum, discrete fracture network and channel networks.

Processes Governing Groundwater Flow

The rate of groundwater flow through a rock mass depends on two main factors: the resistance to flow, determined by the characteristics of the constricted pathways through which the water flows; and the driving force. The representation of the resistance to flow is connected to the representation of the rock structures, for which various conceptual models have been formulated. In contrast, it is well established that the main driving force is gravity through topographical differences, although there are possibilities for coupled phenomena such as hydro-mechanical couplings, thermal buoyancy, density effects due to salinity and concentration differences as well as clogging or weathering of migration paths.

The density variations due to the heat production from the waste is fairly limited. In cases where the natural hydraulic gradient is significant, which usually is the case, Thunvik and Braester (1980) show that the thermally induced driving force can be neglected. However, the heat driving force may need to be considered when there are no other drivning forces, which could be the case for a very deep repository in saline water (Claeson and Probert, 1996).

Modelling of hydro-mechanical couplings is of a research character for example within the framework of the international DECOVALEX-project (see e.g. Stephansson *et al.* 1996). Hydromechanical effects may be important in the vicinity of tunnels and may need to be considered for extreme events during glaciations. However, in most of the rock mass the added affect of hydromechanical changes, is probably minor, at least before a glaciation event.

Large density variations, for example due to highly saline waters, need to be incorporated, at least in large scale modelling. When considering density dependent flow it is usually not sufficient to assume steady-state flow (see e.g. Voss and Andersson, 1993).

Deterministic Porous Medium Models

When modelling groundwater flow in crystalline rock at the large scale the most common approach is to represent the fractured rock by an Equivalent Porous Medium including deterministic regions representing large scale fractures or fracture zones. In fact, most recently published performance assessments concerning crystalline rock (SKB-91, Kristallin-I, AECL, Nirex-95, SKI SITE-94) take this approach for the large scale (see e.g. NEA, 1997). In these applications there are essentially two uses of the model

- to provide boundary conditions to smaller scale models,

- to evaluate the large scale effects of boundary conditions changes (e.g. due to climate changes.

In earlier performance assessments (such as KBS-3, SKBF, 1983), but also in TVO-92 (Vieno *et al.*, 1992) the porous medium approach was the only model concept applied also for providing flow and migration paths. Given the pronounced spatial variability of the flow, caused by the fact that it takes place in the discrete fractures of the crystalline rock, it appears that the (deterministic) porous medium approach is not fully appropriate for these latter applications. (In TVO-92 this problem was at least partly resolved by including deterministic features down to relatively detailed scale combined with the conservative assumption of placing the completely failed canister in the "worst possible location").

In the model the location of the large scale regions is obtained from a geological structure model and the hydraulic data in each feature is obtained by averaging small scale permeability measurement in the different hydraulic regions. Clearly both the structure model and the permeabilities used are subject to uncertainty.

The importance of the structural model uncertainty is problem dependent. It is clear that for many applications the flow in the repository region is not very sensitive to the exact location of the fracture zones. On the other hand, it is probably very important to be able to exclude the existence of a major zone directly in the repository area. A straightforward approach to explore the relevance of the uncertainty in the structure model is to carry out variational cases.

The proper way of establishing the permeabilites to the different hydraulic units is not straight forward. Usually measurements exist from injection tests or other bore hole well tests, performed at a much smaller scale than model block sizes. Within SR 97 semi-empirical scaling rules, originally developed within the Äspö project (Rhén *et al.*, 1997), are used. As long as equivalent porous medium models are used in the large scale and mainly for establishing gradients this scaling uncertainty may be of relative small significance. In the more detailed scale, however, the scaling problem is one of the reasons for the decrease of the applicability of the equivalent porous medium model.

Stochastic Continuum Models

In the stochastic continuum (SC) approach (e.g. Matheron, 1973 or Neuman, 1987 followed by many others) it is assumed that the permeable features of the rock mass may be described as a continuous, but spatially varying hydraulic conductivity field. In repository performance assessments the stochastic continuum model has mainly been used in Sweden; i.e. in SKB-91, as the main alternative in SR 97 and as one alternative in SKI SITE-94. The main uses of the stochastic continuum model have been to:

- derive migration paths and distribution of groundwater flow in these paths

- to derive a distribution of the near-field darcy velocitites

Conceived advantages of the stochastic continuum concept are that the models can be applied to relatively large regions, that they, through multiple realisations, provide a means to handle uncertainty and spatial variability and that the models can be conditioned on point data. However, there are also question marks regarding their applicability. One should also note that the stochastic continuum models do not directly provide information on the correlation between flow and flow wetted surface.

A key property of a stochastic continuum model is the correlation structure, which will determine the effective connectivity of the modelled rock mass. Different correlation structure models exist (Gómez-Hernández and Srivastava, 1990), for example Gaussian, non-Gaussian, and fractal. There is little direct experimental evidence to support any particular correlation structure. In addition, simulation results (e.g. Tsang *et al.*, 1996) also show that the flow distribution and the migration paths strongly depend on the model actually selected.

In theory the model parameters, at the desired support scale, can be taken directly from field measurements. However, a permeability (or a transmissivity) profile estimated along a bore hole based on single hole hydraulic tests, which is the typical data available, will have a varying support scale depending on the measured values (a high transmissivity value represents a much larger rock portion than a small value). Consequently, a much better approach is to determine the parameters through simulations of the hydraulic tests directly - for example by using a discrete network model (Niemi *et al.*, 1997). The advantage of the latter is a more realistic description of possible anisotropic properties of the conductivity field. Estimates of safety assessment related parameters can also be determined more realistically if fracture networks are used for determining the support scale parameters.

Discrete Fracture Network Models

In the discrete fracture network (DFN) approach the rock mass is represented by a network of discrete fractures (usually discs) of different size, orientations, locations, and hydraulic properties. All these properties are regarded as stochastic parameters.

The DFN approach has for example been applied at the detailed scale in the repository performance assessment exercises Nirex-95, SKI SITE-94 (see e.g. Geier, 1996) and will be one of the alternatives in SKB, SR 97 (Dershowitz *et al.*, 1997). The DFN-approach has also been used to interpret one candidate site in Finland (Poteri and Laitinen, 1997), the LPT2 experiment at Äspö (Dershowitz *et al.*, 1996) and is considered a main alternative in the current TRUE block scale experiment (Winberg, 1996). The main conceived advantages of the DFN model are that

- it actually tries to describe flow as it can be observed from individual bore holes and tunnels and thus lend itself to a coordinated interpretation of structure and hydraulic data,

- for transport applications DFN has the intuitive benefit of actually describing the interaction between flow and the rock itself, and can thus provide direct estimates of the correlation between flow and the flow wetted surface,

- DFN approaches can be adopted to relatively large scale applications (see e.g. Geier, 1996), but then it is necessary to exclude the details of the fracture network (i.e. only include the large features) in the outer regions of the model domain,

- DFN models are good for small scale modelling, and can also be used successfully to take local data and produce appropriate effective parameters for large scale modelling as discussed previously.

The individual fracture network parameters can be estimated from geometrical information, such as trace maps and hydraulic information. Some of the remaining uncertainties can be quite significant. Critical to the properties of fracture network is the effective connectivity which would be a combination of size, density and transmissivity distributions. There are further conceptual variants in defining the local structure, and in addressing variability in hydraulic conductivity within a single fracture. These difficulties need to be considered and addressed when performing discrete fracture network modelling.

Channel Network Models

Another alternative for the detailed hydrogeological (and transport) modelling is the Channel Network concept (see e.g. Gylling *et al.*, 1994 or Moreno *et al.*, 1996). In the channel network model flow and transport takes place in a network of one-dimensional channels. The model is characterised by the conductivity distribution of the channels, transport properties and frequency of channels. However, in contrast to the discrete fracture network concepts the channel network models do not use structural information explicitly.

The channel network model has been applied to the LPT2 experiment at Äspö (Gylling *et al.*, 1994) and is considered as one alternative in the SKB SR 97 assessment. However, one could argue that conceptually the channel network model is quite similar to an integrated finite difference model, but where the explicit channels facilitate particle tracking algorithms. As with both the stochastic continuum and the discrete network approach the migration path characteristics will depend on the effective connectivity (i.e. a function of channel permeability and frequency distribution).

EVALUATION OF UNCERTAINTY AND AVAILABLE DATA

Handling uncertainty and variability is a major concern of all assessments of flow in fractured rocks. Variability can be treated through an appropriate model, but the fact that there is a choice of conceptual models leads to uncertainty, as does incomplete knowledge of the parameters that characterise the chosen model. A particular concern is that the various alternative conceptual models for the details of flow in heterogeneous rock, namely Stochastic Continuum, Discrete Fracture Network, and Channel Networks models all capture local variability but scale up differently. Unfortunately, limited data exist for validating their large scale behaviour

Where the model is stochastic, various realisations can be produced, and the range of results can be taken to represent the effect of uncertainty. However, the approximations inherent in each of the model choices is a further source of uncertainty.

There is significant variability in the small scale, particularly in the hydraulic properties of fractures. There is also uncertainty in how these structures interconnect. Different conceptual variants give different answers and there is not always direct field data to select the appropriate model

At a small scale, channels in fractures affect the flow but their location is uncertain. Details of the structure at this scale is of less importance for flow modelling than it is for migration modelling, where it is probably the most important source of uncertainty.

In order to express uncertainties of the detailed groundwater flow a stochastic approach needs to be contemplated and the existence of alternative models need at least be addressed. This does not necessarily mean that the multitude of approaches tried in e.g. SKI SITE-94 (or in SKB SR 97) need to be re-applied. A more careful motivation on when a certain modelling concept really would be applicable would possibly diminish the need for a lot of parallel approaches - in particular as in many cases the selection of the parameters affecting the effective connectivity could be more important than the choice of overall conceptual model. (This latter observation is also made in SKI SITE-94).

CONCLUSIONS

The following conclusions could be drawn from the work done and discussion in this paper:

With regard to driving forces there does not appear to be a need to include the thermal effect as a rule when usual hydraulic gradients exist. With very low hydraulic gradients and in layered repository lay-outs the thermal effects may become significant. There is, however, no harm in analysing thermal effects for presentation purposes.

For the coastal sites at least the regional modelling needs to include the effect of density dependent flow and changing boundary conditions due to land up-lift and coming sea-level changes due to permafrost and glaciation. Permafrost and glaciation, in turn, imply major changes of the boundary conditions.

In the regional scale the equivalent porous medium approach seems appropriate, although there is a need to address the uncertainty in the structure models. In fact it would be very good to show that model results are not too sensitive to the details of a structure model.

At the detailed scale (for source term and migration) it is necessary to represent the pronounced spatial variability at the scale of migration paths. There is also a need to represent uncertainty with a stochastic approach and at least discuss the applicability of different conceptual models. However, as also noted in SKI SITE-94 there is sometimes more uncertainty concerning the general hydraulic conductivity (which can be expressed both by stochastic continuum and discrete network models) than concerning the actual selection of the conceptual model.

For radionuclide transport the discrete network approach seems to be the only possibility to make reasonable modelling work in search of defensible migration parameters.

The data estimation problems and the issue of calibration need to be addressed taking into account the nature of the data (e.g. point like) and the scale in question.

REFERENCES

Andersson J.,M. Elert, L. Moreno, B.Gylling, J.-O. Selroos. Derivation and treatment of the flow wetted surface and other geosphere parameters in the models FARF31 and COMP23 for use in safety assessment, SKB Report (in progress), 1997.

Claeson J., and T. Probert, Temperature field due to time-dependent heat sources in a large rectangular grid - Derivation of analytical solution, Swedish Nuclear Fuel and Waste Managemnet Co., SKB TR 96-12, 1996.

Derschowitz W., A. Thomas, R. Busse, Discrete fracture analysis in support of the Äspö Tracer Retention Understanding Experiment (TRUE-1), ICR 96-05, Swedish Nuclear and Fuel Waste Management Co., 1996.

Dershowitz W., Follin S., Eiben T. and J. Andersson, Discrete fracture modeling for the SR 97 performance assessment project: Aberg, SKB Report (in progress), 1997.

Geier, J., SITE-94, Discrete feature modelling of the Äspö Site. 3 Predictions of hydrogeological parameters for performance assessment, Swedish Nuclear Power Inspectorate, SKI Report 96:7, 1996.

Geier J, and A. Thomas, Discrete feature modelling of the Äspö Site 1, Discrete fracture network models for the repository scale, Swedish Nuclear Power Inspectorate, SKI Report 96:5, 1996.

Gómez-Hernandéz, J.J. and R.M. Srivastava, ISIM3D: an ANSI-C three-dimensional multiple indicator conditional simulation program, Computers and Geosciences, vol 16(4), 395-440, 1990.

Gylling B, Moreno L and Neretnieks I and L. Birgesson, Analysis of LPT2 using the Channel Network model, the Swedish Nuclear Fuel and Waste Management Company, SKB ICR 94-05, 1994.

Matheron, G., The intrinsic random functions and their applications, Advances in Applied Probability, 5, 438-468, 1973.

Moreno L., Gylling B., and Neretknieks, I., Solute transport in fractured media. The importance mechanisms for performance assessment, Journal of Contaminant Hydrology (in press), 1996.

NEA, Working party on the Integrated Performance Assessments of Deep Repositories. Lessons learnt from Phase-1 activities (1995-1996), Final report, NEA/IPAG/DOC(97)1, OECD Nuclear Energy Agency, 1997.

Neuman S.P., Stochastic continuum representation of fractured rock permeability as an alternative to the REV and fracture network concepts. In: Farmer I.W. *et al.* (eds) *Proc. 28th U.S. Symp. Rock Mech.*, 533-561, Balkema, Rotterdam, 1987.

Niemi et al.: Stochastic continuum modeling with block conductivities based on hydraulically calibrated fracture networks - description of the methodology and application to Romuvaara data. (Work in progresss, to be published as Posiva Technical Report), 1997.

Nordman H and T Vieno, Near-field model REPCOM, Helsinki, Nuclear Waste Commission of Finnish Power Companies, Report YJT-94-12, 1994.

Poteri A. and Laitinen M., Fracture network model of the groundwater flow in the Romuvaara site. Posiva Oy, Report POSIVA-96-26., 1997.

Rhén I. (ed), Gustafson G., Stanfors R., and Wikberg P., Äspö HRL - Geoscientific evaluation 1997/5. Models based on site characterization 1986-1995, SKB TR 97-06, Swedish Nuclear Fuel and Waste Management Co., 1997.

Selroos J.O. and V Cvetkovic, On the characterization of retention mechanisms in rock fractures, Swedish Nuclear Fuel and Waste Management Co., SKB TR 96-20, 1996.

Stephansson O., L. Jing and C.-F. Tsang (eds), Coupled Thermo-Hydro-Mechanical Processes of Fractured Media. Mathematical and experimental studies, Recent developments of Decovalex project for radioactive waste repositories, Development in Geotechnical Engineering 79, Elsevier, Amsterdam, 1996.

Thunvik R., and C. Braester, Hydrothermal conditions around a radioactive waste repository, SKBF/KBS TR 80-19, Swedish Nuclear Fuel Supply Co., 1980.

Tsang C.F., Tsang Y.W. and Hale F.V., Tracer transport in fractures: Analysis of field data based on a variable-aperture channel model, Water Resources Research, 27(12), 3095-3106, 1991.

Tsang Y.W., C.F. Tsang, F.V. Hale and B. Dverstorp, Tracer transport in a stochastic continuum model of fractured media, WRR 32(10), 3077-3092, 1996.

Vieno, T., A. Hautojärvi, L. Koskinen, H. Nordman, TVO-92 safety analysis of spent fuel disposal. Nuclear Waste Commission of Finnish Power Companies, Report YJT-92-33E, 1992.

Vieno T., Well-94 A stylized well scenario for indicative dose assessment of deep repositories, Nuclear Waste Commission of Finnish Power Companies, Report YJT-94-19, 1994.

Vieno T., and Nordman H., Interim report on safety assessment of spent fuel disposal TILA-96, POSIVA-96-17, Posiva Oy, 1996.

Voss C., and Andersson J., Regional flow in the baltic shield during holocene coastal regression, Ground Water Vol. 31, 6, 989-1006, 1993.

Winberg A., Tracer Retention Understanding Experiments (TRUE). Test plan for the TRU Block Scale Experiment, Swedish Nuclear Fuel and Waste Management Co., Äspö ICR 97-02, 1996.

Training for Incorporating Spatial Variability in Performance Assessment Approach

Lionel Dewière, Michel de Franco
Andra, France

Emmanuel Mouche
CEA/DMT, France

Executive summary

Performance assessment is faced to a number of uncertainties inducing discrepancies in modelling results. As emphasized in INTRAVAL conclusions [1], sedimentary rocks and fractured media have to be considered separately.

In sedimentary rocks, physical processes are well understood, at least well studied, and we know that the proposed models are representative of the real world. This enables us to propose simple or complex models taking into account most of the observations. There comes a certain level of confidence in performance assessment predictions. Nevertheless, spatial variability and scale effect, induce uncertainties in extrapolation both in space and time.

The fractured media seems different. Some physical processes are not caught yet, and this is mainly due to the spatial variability at different scales preventing us to trap the phenomenon. Physicists are still debating on the Poiseuille approximation, or in other words on the cubic law, strongly linked with the microscopic description of the walls of the fracture. Hydrogeologists can develop a lot of theories on fractures zones, or intersections of fractures, but cannot propose the relations between fracturing and hydraulic properties. Geologists try to describe and to predict the location of fractures.

As uncertainties lowers the confidence level, regulators demand the waste management organizations to provide proof of their ability to deal with the dispersion of the parameters. So a lot of technics have been developed using geostatistics, fractals, stochastic direct or indirect modelling, and applied both to sedimentary and fractured media.

Let us focus on cases where the Representative Elementary Volume (REV) exists, i.e. at the considered scale, the flux is proportional to the hydraulic gradient, as in sedimentary layers or, under a set of hypothesis in single fractures or network of fractures, depending of the scale: so no uncertainty occurs from the conceptual model.

Within the framework of research program developed for the underground repository, the challenge is to provide predictions from scarce measurements which of course have to be completed during siting. Decisions on new investigation arise at specific steps of the demonstration approach in an iterative approach involving performance and safety assessment. As outlined during the DECOVALEX project [2], these uncertainties must be incorporated in the safety assessment calculations, in order to estimate their level of acceptability from the safety point of view. There are two possible scenarios :

1   The safety is not affected by this uncertainty ; i.e. the integration of the component tested for describing the system is good enough. However, we need to maintain progress in taking into account this component. If necessary, the assessment should be repeated.

2   The uncertainties concerning the geological medium and its impact on safety assessment are unacceptable : then these uncertainties must be reduced.

To reduce uncertainties, we can :

–modify the design parameters of the disposal concept
–improve the characterization of the medium,
–acquire further data on the intrinsic parameters of constitutive laws or models used.

In this iterative approach, the role of safety assessment is to determine the area of acceptability, while performance assessment is to determine the limits of possibilities and range of uncertainties in predictions. The final product is a disposal concept involving a series of safeguards that mutually reinforce each other until the system reaches a level of risk [3] that is called "As Low As Reasonably Possible".

Within this framework we will examine what we did, what we do and finally what we will do? Also which tools can be used and what are the challenges for future program of investigation?

Hydrologists are asked to provide models, part of the safety assessment, based on a set of data. Analysing this data, they choose the conceptual model and then include hard and soft data by using direct or inverse modelling.

In case of direct modelling (try and error method), final results are evaluated in safety assessment by choosing averaged values : the uncertainty represented by minimum and maximum plausible values. The expert has the right to decide how far from the real world he is. At that time, ANDRA made it for subsurface Low Level Waste disposal. After construction of a two dimensional model for the upper aquifer, safety assessment has been done using a one dimension model. The main flowpath has been determined as the shortest one and travel time deduced for a range of permeability and gradient. Uncertainty was judged unacceptable and experts were asked to go back on their models, so hypothesis were revisited and new field data were added, then another run began.

If inverse methods are selected, the quality of which depend on the data density, we can obtain an idea of variability and elaborate models for parameter distribution in terms of statistical moments giving us the opportunity to use them in stochastic models as developed by Gelhar [4], Dagan [5]. Such a stochastic model [6] has been developed and then applied for training on a subsurface disposal facility in France. The training was to simulate the uncertainty coming from a given model for variability. That model was defined analysing the data available from an expert point of view. The results showed us the ability of the tools to answering questions related to uncertainties, at least those due to the spatial permeability distribution. The output of the model enables us to quantify the uncertainty associated with a given model for variability.

Such an approach can be chosen for the geological disposal project. The use of stochastic modelling can help us both at the previous stage of siting to define the need in characterization and then in performance assessement. In the preliminary characterization stage, the lack of data prevents us to make statistical descriptions of the heterogeneity but, one key question is how to reach the parameter we need. Nevertheless, we can test the impact on performance-safety assessment of some reasonable laws for spatial variability based on geological and geophysical data. This set of hypothesis has then to be verified on the site, if needed :

– In the case of acceptability, meaning that the uncertainty does not affect the safety of the storage, the risk can be judged "As Low As Reasonably Possible", and then no additional data is needed. However, for another purpose, it can be decided to increase the knowledge selecting new data location.

– In case of unacceptability, the next step might be to gather additional data, of which the location should come from the suspected structure of the permeability field. So, two possibilities arise :

Unfortunately, the hypothesis is confirmed, so the concept must be adapted.

The hypothesis is wrong, we reach an objective element for the demonstration, increasing confidence in the concept proposed.

[1]   "The International INTRAVAL Project - Final Results", 1996, Nuclear Energy Agency, Organisation For Economic Co-operation and Development.

[2]   Dewière, L., Plas, F., Tsang C.F., Lessons learned From DECOVALEX, in Coupled Thermo-Hydro-Mechanical Processes of Fractured Media, 1996, Elsevier, p.495.

[3]   Devillers, C., Repository System integration and Overall Safety, IAEA seminar on Requirements for the Safe Management of Radioactive Waste, Vienna, 28-31 August 1995.

[4]   Dagan, G., Flow and Transport in Porous Formations, Springer Verlag, 1989.

[5]   Gelhar, L., Stochastic Subsurface Hydrology, Prentice Hall, 1993.

[6]   Mouche, E., Treille, E., Dewière, L., An Investigation of Validity of Stochastic Flow Models in Nonuniform Average Flow, Computational Mechanics Publications, Proceedings of the eleventh international conference on computational methods in water resources, 1996.

SKB's Approach to Treatment of Spatial Variability in PA Modelling

Jan-Olof Selroos and Anders Ström
SKB, Sweden

Abstract

This paper discusses how the Swedish Nuclear Fuel and Waste Management Company (SKB) treats issues concerning spatial variability in rock properties. The discussion emphasizes Performance Assessment (PA) applications.

In an ongoing safety assessment exercise (SR 97), SKB is for the first time applying three different conceptual models of the spatially variable geosphere. A conceptual model here implies the manner in which the spatially variable rock mass is conceptualized (pictured) and described mathematically rather than how the geologic structures are located in the rock mass. The models employed are i) a stochastic continuum model (HYDRASTAR), ii) a discrete feature network model (FracMan/MAFIC), and iii) a channel network model (CHAN3D). HYDRASTAR is a code developed and used within SKB in previous safety assessment studies. The main reason for applying different models is to address conceptual model uncertainty. A stronger confidence in the results is foreseen as a consequence of the use of multiple conceptualizations. Furthermore, a regulatory recommendation states that the effect of different conceptualizations should be studied in the ongoing PA exercise.

The present paper discusses how the three models are used in SR 97 and what specific aspects are being investigated using these models. A short discussion follows on still unresolved issues within SKB's approach to the modelling of spatial variability. Finally, the issue of spatial variability is put into perspective relative to other issues with possibly high priority for future PA development.

1   Introduction

In Performance Assessment (PA) studies, the Swedish Nuclear Fuel and Waste Management Company (SKB) has traditionally used fairly sophisticated models for the description of spatial variability in primarily hydraulic properties. Continuum and stochastic continuum models on both regional and local scales have dominated in the PA and safety assessment studies such as KBS-3 [1] and SKB 91 [2]. Discrete models have also been adopted, but mainly in more research oriented projects.

The issue of whether fractured rock should be conceptualized and described as a continuum (porous media model) or as discrete permeable features (discrete fracture model) has received some attention during recent years (e.g., [3] and [4]). The choice is a conceptual one - how do we picture water flow and transport of solutes to take place in the bedrock? The subsequent mathematical statement of the problem dependends on this conceptual choice. In SKB terminology, models based on various conceptual approaches are referred to as different *conceptual models*. In other contexts this term is sometimes used to denote different geologic structural interpretations; however, we denote those models as descriptive structural models and reserve the term conceptual model for the above introduced usage.

The current approach by SKB to the treatment of spatial variability in PA applications will be exemplified below by discussing some aspects of ongoing projects and analyses.

2   SR 97 and the Alternative Models Project

At the present time, SKB is involved in producing a new safety assessment study called SR 97 (Safety Report 97) for a deep repository for spent nuclear fuel. SR 97, which is part of a series of activities scrutinizing all pertinent aspects of deep-rock nuclear waste disposal, is primarily aimed at analyzing the long-term safety of a deep-rock repository in Sweden. New features of SR 97 as compared to previous PA analyses are that more emphasis is placed on the canister integrity (i.e., the canister is given a more realistic life span) and on the completeness and assessment of uncertainties in the analysis. Furthermore, three separate locations in Sweden are analyzed as possible disposal candidates. The objective is to illustrate the importance of site-specific conditions. The safety report, which is to be delivered mid 1998, is a site-specific application of SR 95 [5] which was a framework indicating how a complete safety assessment study should be performed.

A subproject within SR 97 is the so-called Alternative Models Project (AMP), aimed at quantifying conceptual model uncertainty. The AMP can be seen as part of the strive for completeness regarding uncertainty within SR 97; moreover, in recommendations by pertinent regulatory agencies the effect of different conceptualizations has been identified as an issue to be studied in the ongoing PA exercise. The specific objectives of AMP are thus to *illustrate rock barrier performance* using different conceptual models and to *assess model robustness* in terms of relevant far-field performance measures.

Three different conceptual models of the spatially variable geosphere are used. These are a stochastic continuum model, a discrete fracture network model, and a channel network model. The different conceptualizations of the rock mass are illustrated in Figure 1. The stochastic continuum model used is HYDRASTAR, a code developed and extensively used within SKB (e.g., [6] and [7]). In HYDRASTAR the hydraulic conductivity, which is the spatially variable property, can be conditioned on both measured conductivity and head values. Furthermore, deterministic zones can be defined independently. Conservative transport is modelled through particle tracking.

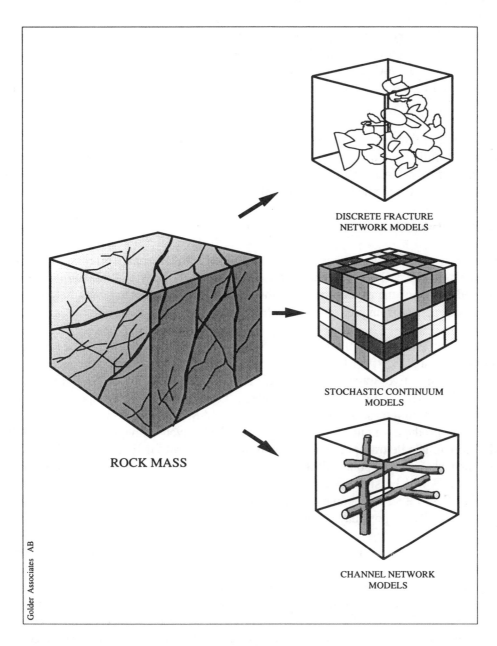

Golder Associates AB

Figure 1. The three conceptual models for groundwater flow in fractured rock. The figure is taken from [8].

The channel network model is called CHAN3D and is developed through full financial support from SKB. In CHAN3D a stochastic network of water conducting channels, rather than discrete fractures, is generated. Individual channel segments are assumed to have constant conductance, volume and parameters governing potential mass transfer reactions. Field data from boreholes is used to obtain these parameters. Discrete zones can be defined deterministically. Transport is modelled using a particle following technique. The channel network model has been developed as a result of field observations, specifically in tunnels, of highly channelized flow in fractured rocks. CHAN3D has been used in both experimental and PA applications (e.g., [9] and [10]).

The discrete feature model employed is FracMan/MAFIC, a commercially available code which has only partially been supported by SKB through studies within the Stripa exercise and at the Äspö Hard Rock Laboratory (e.g., [11] and [12]). Flow is solved for in the stochastically generated fracture network, and transport is modelled by a pathway analysis approach where the main pathways through the connected network are identified by a graph theory search algorithm [13]. Effective transport properties for use in subsequent transport modelling are calculated along the identified pathways.

The choice of these three models in the AMP exercise is primarily based on the fact that all models have successfully been used in previous analysis of field-scale flow and transport experiments in Sweden. Thus, it is of interest to evaluate if models that all explain field data will result in significant differences when applied in PA applications.

3   Project Strategy, Results and Experiences

In order to enable conditions which permit a quantitative comparison between the different conceptual models, it was early recognized that a clear project strategy and modelling specifications had to be defined. Some of the aspects defining the AMP will be discussed below before proceeding to some experiences gained from preliminary results.

In the model chain currently utilized by SKB in PA applications, the darcy flow at canister locations obtained from HYDRASTAR is fed into a near-field transport code denoted COMP23 [14], and conservative (nonreactive) travel times from HYDRASTAR are fed into a far-field transport code denoted FARF31 [15]. The far-field code incorporates advective-dispersive transport, matrix diffusion combined with equilibrium sorption, and chain decay. In AMP it was decided that all conceptual models were to produce input for COMP23/FARF31 even though both CHAN3D and FracMan/MAFIC can handle transport directly both in the near and far field zones. The decision was based on an emphasis on conceptual differences in flow related descriptions. It is believed that the analysis within AMP will be simplified by only comparing transport related parameters for use in FARF31 and COMP23 rather than comparing final results from individual transport codes. However, in future stages of AMP transport may be simulated directly by all three models.

It was also decided to use site-specific data from the Äspö Hard Rock Laboratory within AMP to the greatest possible extent. The site-specific data used pertains to discrete zones with known transmissivity and conductivity statistics for the rock mass in between. Since parts of the geological data are only known in a statistical sense, multiple realizations of the spatially variable rock domain are to be employed. Furthermore, the boundary conditions used were obtained from a regional model for Äspö, and the canister locations coincide with parts of a hypothetical repository placed at Äspö within the SR 97 safety assessment study. A strict formalism has been developed for the modelling teams on how to refer to the different databases and reports available; this has been done in order to safeguard data extractability. Moreover, the assumptions in the models are also to be reported in a similarly strict manner.

A strong emphasis in AMP has been put on the definition of output entities to be calculated and subsequent performance measures for comparison of the output. The main parameters for use in the transport codes are the darcy velocities at canister scale and the groundwater (conservative) travel times from the canisters to the geosphere/biosphere interface. Additional input parameters for FARF31 to be estimated are the flow wetted surface and Pe-number along the flow path. The stochastic continuum model HYDRASTAR can only produce the first two output parameters; estimates of flow wetted surface and Pe-value have to be obtained from separate analyses. The other two models in AMP can, however, provide direct estimates of the flow wetted surface along flow paths. A fifth output entity defined is the so-called F-factor (e.g., [16]) which is the product of the flow wetted surface (per volume of water) and the groundwater travel time. The F-factor determines to a large extent the amount of matrix diffusion that will occur.

A number of performance measures quantifying various aspects of the output have been formulated. In short, the distributions of the individual output entities are to be calculated within each realization and over all realizations. The distributions may also be characterized by moments and/or percentiles. The distribution of e.g. the groundwater residence time within a realization is directly related to the breakthrough curve and is a measure of the spatial variability and canister location extent (i.e., source size). This distribution may subsequently be characterized by its moments or percentiles. By analyzing the ensemble of realizations, the resulting distributions of these moments/percentiles may be obtained. These distributions now reflect the uncertainty in the moments/percentiles of the breakthrough curve. Apart from the distributions and moments/percentiles, a plot of the release locations in the ensemble of realizations is also to be presented.

The AMP exercise is still ongoing and no final results are available at the present moment (August 1997). However, a few preliminary findings and conclusions may be reported. Our experience is that even though strict modelling specifications are defined, it is hard to safeguard that the different conceptual models actually address the same identical problem. This is primarily due to the fact that not every single aspect of the problem can be defined in the specifications, thus leaving room for individual judgment and decisions, which may be biased by the possibilities offered by the conceptual model utilized. Furthermore, the results obtained so far indicate that it may be hard to quantitatively compare e.g. absolute values of the travel times resulting from different models due to the different assumptions invoked. Rather, the span of travel times and corresponding uncertainties are more relevant for comparison.

An outcome of the AMP which was not originally foreseen is that possible inconsistencies in e.g. structural model, site-specific data or inferred boundary conditions from the regional model are easier detected and identified when several independent modelling groups analyze the same problem and iteratively compare results and update models. The increased understanding of the site resulting from this collective effort thus seems greater than what could have been expected from strictly individual undertakings. It is finally noted that the AMP may be extended compared to its present form and may e.g. incorporate other sites than Äspö, incorporate transport and subsequent dose calculations by the individual models, and/or incorporate different variants (e.g., different underlying descriptive structural models) within each conceptual model. However, for such an extension even stricter modelling specifications may have to be introduced.

4   Conclusions and Future Action

The AMP is a subproject within the ongoing PA analysis SR 97 and will not affect the bulk calculations of dose levels in the biosphere where the existing model chain will probably be utilized as the single calculation tool. However, the results obtained in AMP may have some consequences on SR 97 in general and on future PA analyses specifically. Some of the implications of AMP on future PA analyses are summarized below, followed by still unresolved issues related to the description of spatial variability in PA applications.

The main reason for applying different models is to address conceptual model uncertainty. A stronger confidence in the results is foreseen as a consequence of the use of multiple conceptualizations. It is emphasized that in the future, spatial variability in other than hydraulic parameters (e.g., parameters affecting radionuclide retention) may also be incorporated in PA analyses. Both theoretical model development and experimental work (at the Äspö Hard Rock Laboratory) are presently being performed in order to address such issues.

In future PA analyses it may turn out necessary to retain all conceptualizations of the geosphere for subsequent radionuclide transport and dose calculations unless it can be shown possible to discriminate between the different models. As a consequence, the worst model in terms of biosphere impact may also have to be considered in the final analysis unless the model can be proven wrong on structural or hydraulic grounds. Furthermore, the use of multiple conceptualizations in the whole chain of a regular PA analysis may turn out computationally prohibitive. However, our belief is that there is not a need for such a strategy; the consequences of different geosphere conceptualizations can be addressed without propagating the uncertainty to the dose level.

Since the use of several conceptual models is believed to render a stronger confidence in the overall results, the different models should be seen as complementary rather than competing. Thus, the AMP is not designed to identify the one conceptual model to be used in future PA studies, but rather designed as a scheme for how several models can be utilized simultaneously.

Still unresolved issues within SKB's approach to the modelling of spatial variability are presently believed to be related to data acquisition/abstraction and to scaling effects when moving up from experimental scales to scales of interest for PA applications. Future priorities thus include the continuing development and use of alternative conceptual models, resolving issues related to the data abstraction process, and studying the influence of spatial variability (in hydraulic and other parameters) on the site selection process. The last issue is related to the question of how much data is needed to make site-specific models on a PA scale and to discriminate between different sites characterized by spatial variability and other uncertainties.

The issue of alternative conceptual models for the description of spatial variability has to be put into perspective relative to other issues with possibly high priority for future development. Such issues might include the understanding and description of mass transfer and retention processes affecting radionuclide transport in the geosphere.

5   References

[1]   SKB, Final storage of spent fuel – KBS-3, 1983, Swedish Nuclear Fuel and Waste Management Company, Stockholm, Sweden.

[2]   SKB, SKB 91 – Final disposal of spent nuclear fuel. Importance of the bedrock for safety, 1992, SKB TR 92-20, Swedish Nuclear Fuel and Waste Management Company, Stockholm, Sweden.

[3]   Savage, D. (ed.), The Scientific and Regulatory Basis for the Geological Disposal of Radioactive Waste, 1995, John Wiley & Sons.

[4]   Bear, J., Tsang, C-F., and de Marsily G. (ed.), Flow and Contaminant Transport in Fractured Rock, 1993, Academic Press Inc.

[5]   SKB, SR 95 - Template for safety reports with descriptive example, 1995, SKB TR 96-05, Swedish Nuclear Fuel and Waste Management Company, Stockholm, Sweden.

[6]   Norman, S., HYDRASTAR - a code for stochastic simulation of groundwater flow, 1992, SKB TR 92-12, Swedish Nuclear Fuel and Waste Management Company, Stockholm, Sweden.

[7]   Walker, D., Eriksson, L., and Lovius L., Analysis of the Äspö LPT2 pumping test via simulation and inverse modelling with HYDRASTAR, 1996, SKB TR 96-23, Swedish Nuclear Fuel and Waste Management Company, Stockholm, Sweden.

[8]   Geier, J. E., Axelsson, C-L., Hässler, L., Benabderrahmane, A., 1992, Discrete fracture modelling of the Finnsjön rock mass: Phase 2, SKB TR 92-07, Swedish Nuclear Fuel and Waste Management Company, Stockholm, Sweden.

[9]   Gylling, B., Moreno, L., Neretnieks, I., and Birgersson, L., Analysis of LPT2 using the Channel Network Model, 1994, SKB HRL ICR 94-05, Swedish Nuclear Fuel and Waste Management Company, Stockholm, Sweden.

[10]   Gylling, B., Romero L., Moreno, L., and Neretnieks, I., Coupling of the near field release to the far field transport - using NUCTRAN and CHAN3D, 1996, SKB PR U-96-02, Swedish Nuclear Fuel and Waste Management Company, Stockholm, Sweden.

[11]   Dershowitz W., Wallman P., Discrete-fracture modelling for the Stripa tracer validation experiment predictions, 1992, Stripa Project TR 92-15, Swedish Nuclear Fuel and Waste Management Company, Stockholm, Sweden.

[12]   Uchida M., Doe T., Dershowitz W., Thomas A., Wallman P., Sawada A., Discrete-fracture modelling of the Äspö LPT-2, large-scale pumping and tracer test, 1994, SKB HRL ICR 94-09.

[13]   Dershowitz, W., Wallmann P., Shuttle, D., and Follin, S., Canister and far-field demonstration of the discrete fracture analysis approach for performance assessment, 1996, SKB PR U-96-41, Swedish Nuclear Fuel and Waste Management Company, Stockholm, Sweden.

[14]   Romero L., Moreno, L., and Neretnieks, I., Model validity document - NUCTRAN: A computer program to calculate radionuclide transport in the near field of a repository, 1995, SKB Arbetsrapport AR 95-14, Swedish Nuclear Fuel and Waste Management Company, Stockholm, Sweden.

[15]   Norman, S., and Kjellbert N, FARF31 - A far field radionuclide migration code for use with the PROPER package, 1990, SKB TR 90-01, Swedish Nuclear Fuel and Waste Management Company, Stockholm, Sweden.

[16]   SKI, SKI SITE-94, Deep Repository Performance Assessment Project, Summary, 1997, SKI Report 97:5, Swedish Nuclear Power Inspectorate, Stockholm, Sweden.

Methods to Incorporate Different Data Types in the Characterization Process

J. Jaime Gómez-Hernández
Polytechnic University of Valencia, Spain

Jesús Carrera and Agustín Medina
Polytechnic University of Catalonia, Spain

Abstract

That spatial variability of the hydrodynamic parameters controlling radionuclide transport causes large uncertainties in the predictions is known and accepted. Methods have been devised to analyze spatial variability of these parameters and to model the uncertainty of the predictions. However, the final use given to large portions of the total data collected is minimal. Many data are simply disregarded. Data that are used is, in many cases, accepted without scrutiny and used blindly in the prediction models. In the quest to extract as much information as possible from the available data, we have developed and implemented techniques with the aim of incorporating all types of data in the characterization of the spatial variability of conductivity/transmissivity. This serves to reduce the uncertainty in the predictions and to increase the confidence in the model. More importantly, this makes the model less dependent on unspoken assumptions, which are often implicit and sometimes unknown to the modeler. Types of data used in models include: geometric information (relevant to connectivities commonly not included in stochastic analyses), transmissivity data (either after interpretation of a slug/pump test, or raw as given by the drawdown curves in response to long pump tests), piezometric data (both steady-state and transient), geological/geophysical information (as soft data) tracer test concentration data (relevant for transport parameters), and isotopic data (informative of residence times and transport conditions over long time periods and small fluxes, more relevant for repository performance than short term tests)

1. Introduction

The characterization of the formation parameters that are relevant to fluid flow and mass transport in the geological barrier must be addressed with a stochastic approach; reasons being the inherent spatial variability of those parameters and their sparse sampling. Random function modeling, as any other stochastic approach, should be understood as a tool that has been proven efficient for the purpose of modeling the heterogeneity of the parameters of interest, never as a direct representation of reality. As soon as a random function model approach is adopted, reality is replaced by an ensemble of alternative "realities" that are build to satisfy a number a constraints generally expressed in statistical terms. (For example, all components of the ensemble should have the same overall spatial average, or, at a given location, all components should reproduce a known measured value.) Randomness is not a property of reality but of the random function model used to describe it. In this respect, the type of knowledge one can gain from analyzing the random function model results is not an image of what is happening in reality, but it helps in learning about it.

After accepting the use of random function models in the characterization process, our aim should be to attempt to make the realities of the ensemble, hereafter referred to as realizations, as similar to each other as possible. The larger the similarity, the more confident we will be in the predictions based on this model. The key to attain similarity across realizations is conditioning. If the realizations are drawn from a conditional random function, they will not only be consistent with some global statistical parameters but each one of them will reproduce the measured data at the sampling locations.

Conditioning is not a foreign concept in stochastic groundwater modeling, however in most of the applications found in the literature, the conditioning data is limited to conductivity or transmissivity measurements. In the following a number of approaches to achieve conditioning to the large amount of information that is generally available and not often used are discussed.

The paper is just a recollection of techniques available to incorporate information regarding geology, geophysics, piezometric heads, concentrations and water ages in the characterization of the spatial distribution of conductivity fields.

2. Conditioning to geology and geophysics

By conditioning to geology and geophysics we imply that any small correlation that could be deduced between conductivity and geological or geophysical attributes should be used to restrict the degree of variability of conductivity at specific locations. Typical examples are seismic surveys from which measured attributes such as impedance or wave velocities can be correlated to conductivity or to porosity. Three-dimensional seismic surveys provide exhaustive information in the horizontal plane, which, even if it is weakly correlated to conductivity, can help in reducing the uncertainty on conductivity and thus on groundwater model predictions. Conditioning to geology can be achieved by using the geological maps to set bound intervals on the conductivity values at given locations, or, even better, to specify likelihood functions for conductivity according to the different facies.

Two approaches that can be used to generate conductivity realizations conditional to geophysical or geological information are non-parametric geostatistics and simulated annealing. Non-parametric geostatistics can incorporate data of various types and degrees of correlation with conductivity, through the common coding of information in series of 0's and 1's. Each datum is transformed into a, possibly incomplete, series of indicator variables, which are later updated to produce the final realization. Simulated annealing works by minimizing an objective function that measures the mismatch of the realization with respect to the pursued conditioning constraints. The interested reader is referred to the books by Deutsch and Journel [3] and Goovaerts [5] for the necessary explanations.

An example of the impact of including secondary information in the characterization of the spatial variability of conductivity is shown next. It is based on the Walker Lake exhaustive data set [6]. Figure 1 shows a grayscale map of a subarea from the Walker Lake data set representative of a hypothetical aquifer 2.0 km by 1.2 km in size and discretized by 6000 cells. The primary variable represents transmissivity and has been sampled only at 20 locations, the secondary variable represents a geophysical attribute that, although with a poor linear correlation with the primary variables displays the same large scale conductivity patterns. The secondary variable is assumed to come from a 3-D seismic survey and it is exhaustively available.

Two Monte-Carlo analyses were carried out, both consisting of: (i) generation of conditional transmissivity realizations, (ii) solution of single-phase saturated flow with prescribed head boundaries along the lateral edges and no flow boundaries at the top and bottom edges and (iii) advective particle tracking of a contaminant plume with source at the center of the upstream (left) boundary. The difference between the two analyses is the amount of information used in step (i) for conditioning. In the first case, only the small sample of primary data are used, in the second case, both the primary and the secondary data are used. The resulting transmissivity, piezometric head and pathlines corresponding to three of the realizations are shown in Figure 2. It is evident that incorporating the secondary information helps in reducing the variability across realizations, therefore reducing the uncertainty on the predictions.

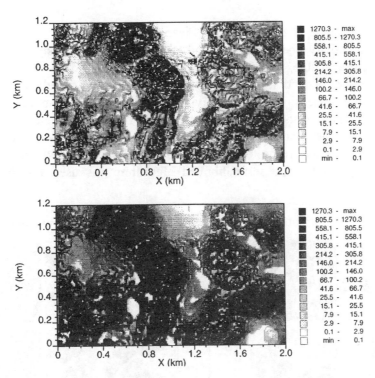

Figure 1. A subarea of the Walker Lake data set. On top, the primary variable representing transmissivity, sampled for 20 regularly spaced data; on bottom, the secondary variable representing a seismic attribute, exhaustively sampled.

These conclusions are also evident from Figure 3 which shows histograms of the arrival times and arrival positions of the particles to the right edge compiled from the 100 realizations generated. These histograms can be used to provide a quantitative measure of the uncertainty on the predictions about these variables.

3. Conditioning to piezometric head data

By conditioning to piezometric head data we mean that the solution of the flow equation on the conductivity/transmissivity realizations matches the measured piezometric heads (either steady-state or transient). Note that some authors refer to piezometric head conditioning only on a statistical sense such as the reproduction of the cross-covariance between conductivity and heads; this reproduction alone does not ensure actual reproduction of the piezometric head values in the conductivity fields so generated, particularly if the range of variability of conductivity is large.

Conditioning to piezometric head requires incorporating the groundwater flow state-equation into the stochastic generation process with the problems associated to the non-linear dependence of transmissivity on heads and the need to specify some boundary conditions for the groundwater flow solution.

Figure 4 shows an example of conditioning t Piezometric head values are generally always more abundant than conductivity measurements and are also less expensive to obtain. When few conductivity data are available, piezometric head information could be very valuable on detecting large scale trends on the conductivity fields, partly due to the longer correlation length of piezometric heads.o heads of a two-dimensional transmissivity field. The data available in this experiment were five transmissivity measurements towards the center of the field, plus the steady-state heads measured at the time of drilling (when undisturbed conditions were supposed to be present) plus the transient heads resulting from an interference test in which one well was pumped and the head decline was monitored in all five wells. The self-calibrated algorithm was used for the generation of the conditional realizations [1,4].

No secondary information   With secondary information

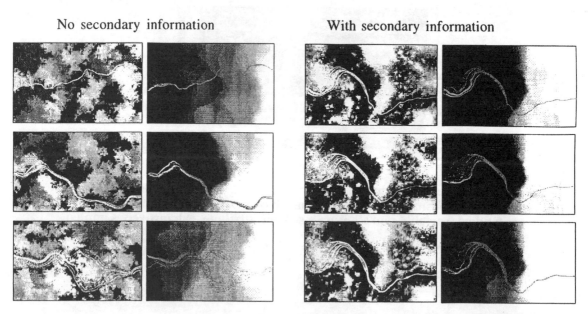

Figure 2. Three realizations from each of the two Mote-Carlo exercises performed. On the left, only the samples from the primary variable are used; on the right, both the primary and the secondary data are used. In both cases, the transmissivity maps, the piezometric head solution of the groundwater flow equation and the particle path lines are shown (the release is from the center of the left border).

Conditioning to only five transmissivity measurements using a variogram with a range about one tenth of the field size produces the field on the left of Figure 4, in which, no spatial trends are noticeable. This field is a typical realization of a stationary multi-Gaussian random function. However, when the piezometric head information is included in the conditioning, the specific response patterns of the different wells reveal the presence of a U-shaped low transmissivity barrier in the center of the field. This feature repeats itself consistently in all the realizations. Figure 5 shows the ensemble mean transmissivity field corroborating this aspect. Figure 5 also shows the ensemble variance on logtransmissivity which can be used as an indication of the degree of uncertainty on logtransmissivity after conditioning.

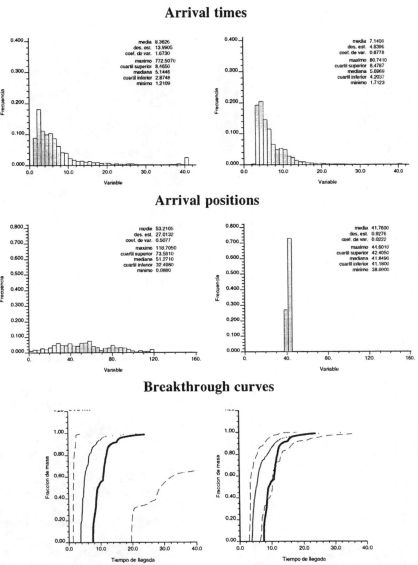

Figure 3. Uncertainty quantification (all graphs summarize the results from 200 realizations). Top, histogram of particle arrival times; middle, histogram of arrival positions; bottom, median breakthrough curve (dotted line), and a 95% confidence interval (dashed lines) compared to the breakthrough curve computed in the exhaustive field of Figure 1 (thick solid line).

4. Conditioning to tracer data

While in the previous section all realizations are conditioned to piezometric head data, in the sense that the solution of the flow equation reproduces the measured piezometric head values, it seems natural to attempt to incorporate also the information about tracer data in order to constrain further the different realizations [7]. The problem of conditioning to tracer data although conceptually similar to the problem of conditioning to piezometric head (there is the need to incorporate an additional state-equation relating conductivity to concentration) is much more complicated from a formulation point of view and from an implementation point of view.

Initial logT field   **Final logT field**

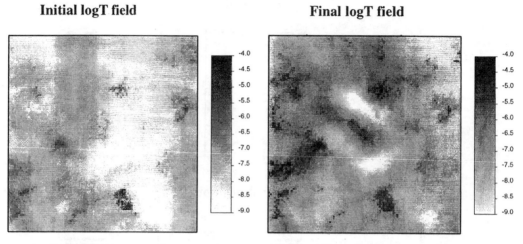

Figure 4. A realization of a 2-D transmissivity field. On the left, the logT field is only conditional to the five transmissivity measurement; on the right, the logT field is conditonal to transmissivity and steady-state and transient piezometric heads.

Ensemble mean   **Ensemble variance**

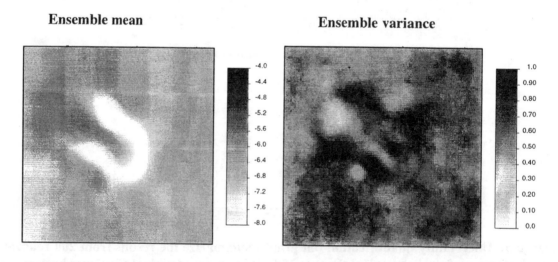

Figure 5. Ensemble mean and ensemble variance of logtransmissivity, steady-state head and transient head data.

Figure 6 shows an example of how different types of conditioning affects the characterization of a transmissivity field. A synthetic transmissivity field discretized in 500 rectangular cells is taken as reality. In this synthetic field the groundwater flow equation has been solved under steady-state flow conditions, and for the aquifer response to an interference test; the advection-dispersion transport equation has also been solved under to types of boundary conditions, in one case there is a background concentration that enters the aquifer along the boundary and in the other case, the concentration corresponds to four tracer tests carried out within the aquifer.

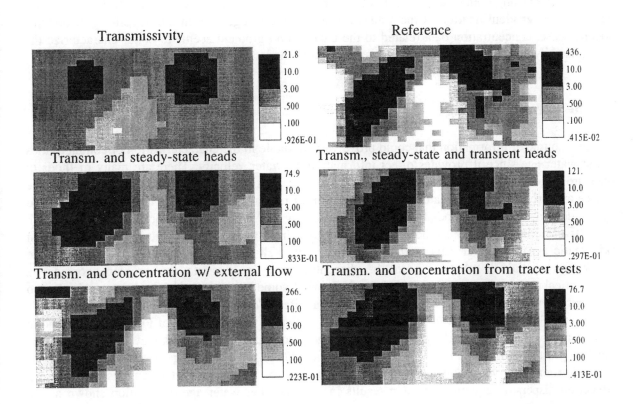

Figure 6. Impact of conditioning to different amounts of information. Top right is the reference, the remaining five fields are conditional estimations of the reference field using different types of data as indicated by the image title

The synthetic reality was sampled at four locations for transmissivity values, piezometric head data (both steady-state and transient) and tracer data. Then, conditional estimates of the reference transmissivity field were obtained. First, using only the four transmissivity measurements (top left of Figure 6); the resulting field is extremely smooth, the two zones of highest transmissivity are only guessed. Then, steady-state piezometric head data are included (middle left of Figure 6), the definition of the two high transmissivity zones and the low transmissivity zone in-between is better. This definition further improves when both steady-state and transient heads are considered (middle right of Figure 6), the central area, that concentrates most of the information, is well identified whereas the zones towards the edges are less. The bottom row in Figure 6 shows the impact of using concentration data. The two types of concentration data (on the left, breakthrough curves of contamination entering with the background flow from the left edge; on the right, breakthrough curves resulting from four radial tracer tests) provide different degrees of information with regard to transmissivity characterization. In both cases, the resulting estimated maps are comparable to that obtained when using both steady-state and transient heads, however, in the bottom left of Figure 6 we can see a better identification of the values towards the left edge than in the bottom right. For this specific case, concentration data related to the background gradient seems to better characterize the transmissivity field than tracer test data, maybe because the latter has a more limited spatial extent.

5. The use of water age

Another source of information that could prove very valuable in the characterization of conductivities is the water age since it provides direct information on water velocity. There are many methods to measure water age: radioactive isotope decay, isotopic relations or unbalance between radioactive source and products. Unfortunately, the development of measurement methods has not been followed by the development of methods for global understanding of water ages. The most common approach consists of defining the age as the time taken to travel from the source point to the point of interest (kinematic age). However, this method does not work well in heterogeneous media, in which the water age spatial distribution could be discontinuous. To illustrate the problem consider a hypothetical aquifer composed by a uniform facies with some high-transmissivity lenses as in the top of Figure 7 which shows the flownet in such an aquifer. Computing the kinematic age using this flownet results in the water age spatial distribution shown in the center row of Figure 7. The chaotic appearance of this field and the presence of high age gradients induce to think that dilution processes must be present. An alternative definition of water age [2] as the average age of the particles that are present around a given point leads to a state equation for water age that resembles the standard advection-dispersion equation and that results in the smoother water age distribution shown at the bottom of Figure 7,

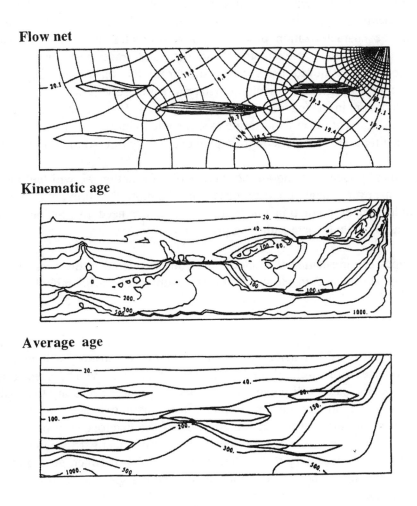

Flow net

Kinematic age

Average age

Figure 7. **An example calculation of water age.** **Top, flow net commonly used for the calculation of the kinematic age, i.e., the age related to the travel time needed to reach a point (shown in the middle).** **Bottom, water age when computed with a state equation similar to the advection-dispersion equation in which water age is defined as the average age of the particles around the point considered.** **Notice how much smoother is the bottom field with respect to the middle one**

No example is shown on how to incorporate this information in the characterization process, but, if water age is modeled using an equation similar to that of mass transport, the same techniques used to incorporate concentration data could be used to incorporate water age data.

6. Summary

There may be more information related to conductivities that is generally used. The inherent problem of lack of measurements on hydraulic conductivities could be alleviated by incorporating other types of information in the characterization process. A few examples have been shown of successful integration of geophysical data, piezometric head data and tracer test data. The integration of geological data could be achieved using tools similar to the ones used for the integration of geophysical data, and the integration of isotopic data, more precisely information regarding water age could be accounted for in a manner like the one used for integrating concentration data. Unfortunately, at this moment, there is no single tool capable to integrate all types of information, and some of the available tools are still difficult to implement in fully three-dimensional models.

There is no doubt that by incorporating all these types of information we will feel more confident on the model predictions, even though, on some occasions, the more data brought in will not necessarily entail a direct reduction of prediction uncertainty due to the need to re-evaluate the underlying conceptual model along with the data gathering process.

There are other types of information that should also be considered in the characterization process, such is the case of geochemical data. However, this is a problem of considerably larger proportions.

7. Afterthoughts

Characterization of the geological barrier around an underground repository will not be based on large amounts of data collected from wells drilled in the repository vicinity. The need to minimize man-made potential fast release paths will prevent it. For this reason, it is necessary to consider and devise techniques that can bring all the information possible, qualitative and quantitative, directly or indirectly related, hard or soft into the characterization of the heterogeneity of parameters, such as conductivity, on which the performance assessment of the geological barrier relies more directly. To achieve this objective, it is necessary to ease the communication between geologists, geophysicists, hydrologists and geochemists, who, on occasions, tend to be too specialized as to end up speaking in terms difficult to understand across disciplines.

8. References

[1]   Capilla, J. E., J. J. Gómez-Hernández and A. Sahuquillo, Stochastic Simulation of Transmissivity Fields Conditional to Both Transmissivity and Piezometric Data, 2. Demonstration on a Synthetic Aquifer, Journal of Hydrology, 1998, accepted for publication.

[2]   Carrera, J., M. R. Varni, and M. W. Saaltink, Una ecuación en derivadas parciales para la edad del agua subterránea, in Hidrología y Recursos Hidráulicos, Asociación Española de Hidrología Subterránea, 1995, volume XIX, part 1, 325-334.

[3]   Deutsch C. V. and A. G. Journel, GSLIB, Geostatistical Software Library and User's Guide, 2nd Edition, Oxford University Press, 1998.

[4]   Gómez-Hernández, J. J., A. Sahuquillo and J. E. Capilla, Stochastic Simulation of Transmissivity Fields Conditional to Both Transmissivity and Piezometric Data, 1. Theory, Journal of Hydrology, 1998, accepted for publication.

[5]   Goovaerts, P., Geostatistics for Natural Resources Evaluation, Oxford University Press, 1997.

[6]   Isaaks, E. H. and R. M. Srivastava, An Introduction to Applied Geostatistics, Oxford University Press, 1989.

[7]   Medina, A. and J. Carrera, Couple Estimation of Flow and Transport Parameters, Water Resources Research,, 1996, 32(10), 3063-3076.

POSTER SESSION

Chemical Modelling Studies on the Impact of Small Scale Mineralogical Changes on Radionuclide Migration.

A. T. Emrén

Dept. Nucl. Chemistry, Chalmers Univ. Techn., S-41296 Göteborg, Sweden.

Abstract

In the literature, one finds several models for control of redox properties in groundwater. The proposals for redox controlling substances include iron oxides, chlorites, methane, pyrite and polysulphides. The CRACKER program, has been developed to model groundwater formation in crystalline rock that usually shows a great degree of small scale spatial variability from the chemical point of view. The program has been used to model observed Äspö groundwaters. With reasonable mineral sets, the modelled and observed groundwater properties have been found to be similar.

The program has been has been used to investigate the influence of several redox control models on the modelled properties of present and possible future Äspö groundwaters. In the simulations, one or more of the possible redox reactions have been prevented from occurring. The groundwater has then been assumed to react with minerals distributed in the fracture walls. Due to the discreteness of mineral grains, a certain amount of fluctuations in groundwater properties are occurring. The process of sampling water for measurement has been simulated by letting about 900 waters from different locations mix.

It has been found that some of the models have difficulties in explaining other properties than the pE-pH behaviour (properties like element concentrations), while other models perform quite well. In the figure, pE-pH results are shown for a model consisting of some thirty minerals and a high salinity groundwater at two temperatures. The redox properties have been assumed to be controlled by several redox reactions occurring simultaneously. The most obvious feature is the decrease in pH at a higher temperature. The difference in behaviour is caused by the fact that the solubility increases with temperature for some minerals, while it decreases for other minerals.

Further, it has been found that modelled retardation of radionuclides is lower if the mineral distribution shows a spatial variability at a length scale of a few millimetres rather than being homogeneous at such length scales.

INTRODUCTION

Wastes, which like used nuclear fuel, is dangerous if handled in the wrong way has to be taken care of in a way that prevents damage to people or environment. Among the important properties of the repository - groundwater -rock system, pH and redox potential of the groundwater as well as mineralogy of the rock and fracture systems are assumed to be of major interest. The reason is that an unfavourable combination of properties may cause artificial and natural barriers to become inefficient. As examples may be mentioned that canisters may corrode rapidly, dangerous radionuclides may have high solubilities and sorption of dissolved radionuclides may be too weak to prevent them from reaching the surface.

Spatial variation of the chemical properties in the system has often been overlooked in models for the behaviour of radionuclides released from a leaking canister. As will be discussed below, the spatial variability gives rise to three main effects:

1. No minerals are in global equilibrium with the groundwater. The water is forced to perform some kind random walk in the composition space. The sampling process causes several kinds of water to mix.

2. Lowering of reaction rates in mineral - groundwater reactions. Due to the spatial variability of the chemistry, the stoichiometry is locally constrained and thus, the reactions are not able to aim at the global equilibrium (if such exists) but rather to local sub-equilibria.

3. Lowering of efficiency in retardation of radionuclides by sorption since some particle tracks being able to avoid the sorbing minerals.

The CRACKER program and spatial variability of rock composition

The CRACKER program [1] has been designed to handle properties like spatial variability and phase rule violations in the rock - groundwater system. In this program, minerals are distributed randomly across fracture surfaces in accordance with observed or reasonable abundances. Minerals forming solid solutions, ion exchangers or intermediate forms as well as ordinary minerals are handled by the program.

As the water propagates through the fracture it reacts with the minerals that are within diffusion distance. Normally, the water is allowed to reach equilibrium locally (although kinetic effects may be included). Results from each time step are stored in a result file containing information about position (x and y), water composition, pH, pE and minerals present in each diffusion cell.

A key concept is that of diffusion cells. That is, a region in which diffusion is able to make the water approximately homogeneous from a chemical point of view. Different diffusion cells are assumed to be chemically independent of each other, while reactions may take place between water, dissolved species and minerals in the same diffusion cell. In each time step, the water moves from one diffusion cell to the next.

To understand the concept of diffusion cells, one has to realise that the time required for diffusion to make a water packet homogeneous is proportional to the linear size squared. Thus, if reactions with a mineral grain change the water composition, the change in composition spreads through the water with a velocity that is initially great, but rapidly decreases. Sooner or later, the velocity becomes less than the flow velocity of the water. At that time, the region influenced has reached a certain size, defining a diffusion cell. Any water package greater than that size will be unable to reach equilibrium during the time it is contact with a mineral grain.

Reduced reaction rates due to spatial variabilities

In studies on mineral - groundwater reactions, it is generally found that reactions rates occuring in nature are two or more orders of magnitude lower than the rates measured in laboratories. One major reason appear to be that due to the spatial variability of the chemistry, the stoichiometry is locally constrained and thus, the reactions are not able to aim at the global equilibrium (if such exists) but rather to local sub-equilibria.

The spatial variability of chemical properties causes the possible reactions to be constrained for stoichiometric reasons. In a diagram of free enthalpy vs. concentrations of components, chemical reactions are represented by straight lines. Since they rarely pass the global minimum, the reaction proceeds towards a local equilibrium (Fig. 1). Since this is shallower than the global minimum, the driving force is less.

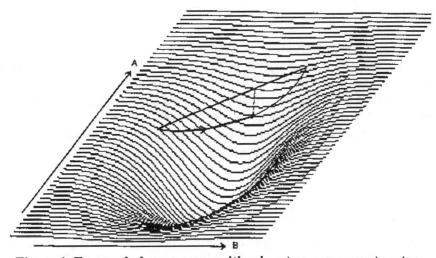

Figure 1. Free enthalpy vs. composition in a two component system.

When the water moves to another environment, the stoichiometry changes and the system is able to further approach the global equilibrium. Such behaviour can be simulated with CRACKER as is seen in Fig. 2. In this figure, merely three solid phases are used and instantaneous local equilibrium is assumed. Nevertheless, the system requires several months to approach equilibrium.

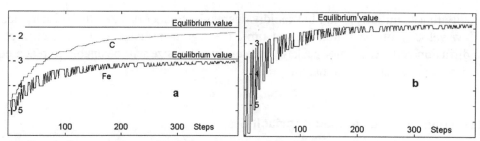

Figure 2. Convergence towards equilibrium in a one dimensional CRACKER simulation with a rock consisting of the three minerals Goethite, Siderite and Pyrite.
a. Logarithms of Carbonate and Iron concentrations.   b. PE values.

REDOX CONTROLLING REACTIONS

The redox potential

Closely connected to global disequilibrium caused by spatial variability is the question of redox properties of the groundwater. Obviously, a proper understanding of the mechanisms controlling the redox potential is desirable in the safety assessment of a proposed repository for nuclear wastes.

When discussing the redox properties of ground water, one has to bear in mind that the redox potential of a groundwater sample generally is not a unique quantity. The primary reason is that several reactions are too slow to reach equilibrium. This would not prevent a unique metastable redox potential to be established if the water were able to contain free electrons. Since this is not the case, the redox potential has to be defined indirectly. Groundwater contains several redox couples and for each couple there is a certain set of species concentrations. Such a set of concentrations establishes a redox potential by means of a half cell reaction like the following

$$aA + bB \rightarrow cC + dD + e^- \tag{1}$$

In principle, this redox potential can be measured if an electrode is found, which selectively catalyses the corresponding reaction. A different route to determination of the redox potential is to measure the concentrations of the different reactants and products of the reaction. Knowing the equilibrium constant (standard potential) of the reaction and the species concentrations, the corresponding redox potential may be determined.

pH and redox properties in groundwater

In attempts to find mechanisms controlling the redox properties of ground waters, the behaviour of redox potential versus pH is often used. This is possible only if there are considerable variations in pH values in the region considered. Further, although the pH value usually is a unique quantity, the redox potential as mentioned above may have several values depending on how it is measured.

It is often found that protons or hydroxyl ions are involved in redox reactions and this fact constitutes a relation between pH and the redox potential. Let us assume the following reaction to take place:

$$aA \rightarrow bB + n\,H^+ + e^- \tag{2}$$

The corresponding equilibrium equation may be written as

$$\frac{\{B\}^b\{H^+\}^n\{e^-\}}{\{A\}^a} = K \tag{3}$$

Converting to logarithms and making use of Nernst's equation gives

$$\log(\{B\}^b/\{A\}^a) - n\,pH + \log\{e^-\} = E^0\,F\,\ln 10\,/(RT) \tag{4}$$

Rearranging (4) and making use of the pE concept is used rather than the redox potential, a more convenient expression may be derived.

$$pE = -n\,pH + pE^0 + \log(\{B\}^b/\{A\}^a) \tag{5}$$

Besides being simpler than (4), the temperature dependence has disappeared in (6). Further, the slope gives a direct measure of the ratio of protons to electrons in the corresponding reaction.

Since no free electrons are able to exist in water, oxidation reactions like (1) always have to be coupled to reduction reactions in natural waters. In this case, relations like (5) are describing the net ratio of protons to electrons in the assembly of reactions governing the redox properties of the water.

Since the pH values measured in natural groundwaters often are within the range 6 - 9, the proton concentrations typically are less than 10^{-6} M. As far as the concentrations of important redox couples are more than that, changes in pH have a small impact on the concentrations of redox couples. Thus, the last term of (5) often is approximately constant. Then plots of pE vs. pH give approximate information of the net ratio of protons to electrons in the redox controlling reactions of the groundwater.

Proposed redox controlling reactions

Several reactions for redox control have been proposed in the literature. Allard [2] Grenthe et al. [3] have proposed Fe(II) and/or Fe(III) minerals and the corresponding redox couple in the water to be responsible for the redox control of deep granitic groundwaters.

Iron as the key element for redox control may involve other kinds of minerals than oxi/hydroxides and carbonates. Thus Arthur [4] has proposed mineral sets including Epidote or Daphnite to control the redox properties of ground water.

A quite different approach has been taken by Nordstrom et al. [5] who have proposed polysulphides to control the redox properties of the groundwater.

A third possibility is the methane - carbonate reaction. Although it is assumed to have a negligible reaction rate at normal temperatures, it has been observed [6] that the methane concentrations in natural water often are fairly close to the equilibrium values of the methane carbonate reaction. The explanation might be microbial activity and/or catalysis by minerals. Further, it has been reported in the literature [7] that semiconducting minerals, like Pyrite, are able to catalyse reactions in which organic molecules are involved.

The sulphide - sulphate reaction generally is assumed to be too slow to influence the redox properties in ground water. It has been pointed out by Glynn [8] that in some groundwaters, the sulphide / sulphate ratios are fairly close to the values expected at equilibrium. This might have explanations similar to those mentioned above in connection with the methane - carbonate reactions.

Finally, the redox properties are controlled by many reactions occurring more or less simultaneously. The situation is complicated by the fact that mineral sets found in the nature are too large to fulfil Gibb's phase rule [9]. A consequence is that no global equilibrium exists until the groundwater is sampled and looses its contact with the minerals. Consequently, one would expect groundwater a spatial variability in the groundwater composition even at a small scale (centimetres).

CRACKER SIMULATIONS OF EXISTING GROUNDWATERS

In the simulations, a typical groundwater found at a depth of about 500 m at Äspö has been used. Äspö is situated at the Swedish east coast close to the town Oskarshamn. In the first simulation, all observed minerals have been used and all redox reactions except nitrogen reactions have been allowed to occur. Then identical simulations have been performed with modified databases. The modified data bases are blocking some of the redox reactions by formally treating different oxidation states of an element as different elements. In certain cases, some minerals have been blocked from influencing the water properties by assuming them to be inert. A special case has been simulations with the hypothetical mineral proposed by Grenthe et al. [3]. Here all redox reactions except those involving Fe (and water oxidation/reduction) have been blocked out. Further, all iron minerals have been replaced by the hypothetical mineral.

PROPERTIES OF MODELLED CURRENT GROUNDWATERS

The deep groundwaters are moving quite slowly (~1 m/year) and thus (although equilibrium does not exist) they could be expected to be close to a steady state with respect to minerals in the surrounding fracture walls. It has been shown [10] that the CRACKER program is successful for simulating present day groundwaters from known mineral sets and initial waters.

In the simulations, the observed water has been used as input to simulations in which observed minerals have been allowed to react with the water. Although the water locally may deviate considerably in properties (as illustrated in figure 3), the mixing that occurs during the sampling process causes the fluctuations to disappear.

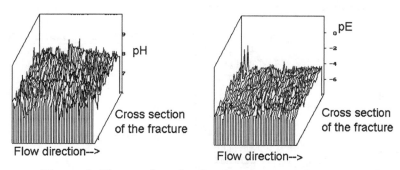

Figure 3. Fluctuations in simulated properties across a fracture.

In spite of the fluctuations found during simulations, the observed water is essentially a steady state water when used as input to the CRACKER program. This is seen in Fig. 4 where element concentrations from the simulated steady state are compared with those of the observed water.

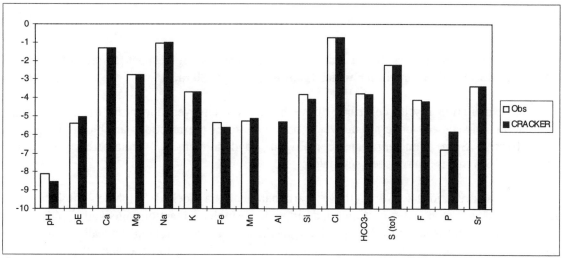

Figure 4. Observed and simulated steady state properties of the studied water (sample from Äspö at a depth of about 500 m). Al-concentrations are not measured at Äspö.

Since the mineral sets found in nature normally contain incompatible minerals (Gibb's phase rule is violated), global equilibrium rarely is possible. This is illustrated in figure 5. Although a mineral as calcite is very common and is in local equilibrium with the groundwater, the sampled water is slightly oversaturated both in the simulations and in the measurements. CRACKER makes use of the well-known program PHREEQE [11] to calculate speciation as well as results from mixing, dissolution and precipitation reactions. PHREEQE has also been used to calculate saturation states of the observed waters.

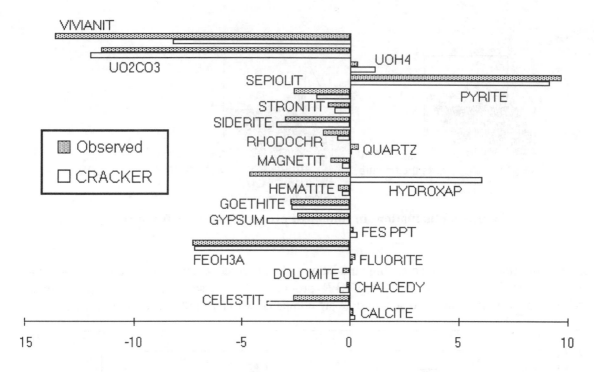

Figure 5. Saturation indices for a number of minerals with respect to the observed and CRACKER-simulated end member water from KAS 03 at 130 m.

The simulation results described above have been achieved under the assumption that all redox reactions discussed above are active. If one or more of the reactions are blocked, the picture looks different. Some concentrations remain in good agreement, while other are changed considerably. Seven cases have been simulated:

All redox reactions active, Methane - carbonate reaction blocked, Sulphide - sulphate reaction blocked, Methane - carbonate reaction and sulphide - sulphate reactions blocked, Fe(II) - Fe(III) reactions blocked, all iron minerals replaced with the hypothetical mineral and all redox reactions allowed, and finally all iron minerals replaced with the hypothetical mineral and all redox reactions except Fe(II) - Fe(III) blocked.

Diagrams of simulated values of pE vs. pH found at different simulated locations should give information on the net ratio of electrons to protons in the reactions controlling the redox properties in the simulations. Some typical results are shown in Fig. 6. Each plot consists of some 3000 values sampled from a simulation.

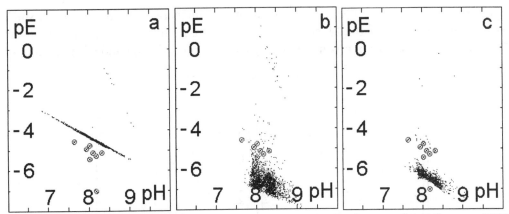

Figure 6. Plots of pE vs. pH in the simulations. Each plot consists of some 3000 values taken from a simulation with a particular set of redox controlling reactions. The symbol ⊗ is used to denote observed values.
a. All redox reactions active.
b. Methane - carbonate reaction and sulphide - sulphate reactions blocked.
c. All iron minerals replaced with the hypothetical mineral and all redox reactions except Fe(II) - Fe(III) blocked.

It is obvious that more reactions than the Fe(II) - Fe(III) reactions are needed to make the simulated properties similar to the observed properties. One should note, however, that the iron concentrations in Äspö groundwaters are anomalously low. At other places, reactions involving iron might be much more important.

Since the Swedish concept of a repository for spent nuclear fuel includes the use of copper canisters, the corrosive properties of the groundwater with respect to copper are of interest. At a temperature of 15 °C, oxidation of Cu to Cu_2O is negligible ($< 10^{-13}$ g Cu per litre of groundwater) in all the cases studied. The phase $CuFeO_2$ on the other hand is thermodynimacally more stable than Cu with respect to the simulated low temperature groundwaters except cases d and g for which Cu is virtually inert. The formation of this compound means that Cu and possibly Fe(II) too are oxidised and thus something has to be reduced. Essentially, the iron content of the groundwater is the limiting factor in production of the compound. The corrosion capacities in the cases discussed above are about 0.1 mg per litre of groundwater for cases a, b, c and e, while it is about 0.2 μg in case f.

If, on the other hand some iron source, as goethite is available in excess quantities the corrosion capacity is limited by the redox buffering capacity of the water, which in turn depends upon which redox reactions is active. In cases d, e and g, the contribution is negligible, while it is quite substantial in the remaining cases, a, b and f: 3 g/l, c: 60 mg/l.

PROPERTIES OF MODELLED FUTURE GROUNDWATERS

In the near field of a repository, the ground water chemistry could be expected to deviate significantly for several reasons: Changes in mineralogy, presence of clay, cement, canisters and oxidised rock. Further, the temperature will be increased to more than 50 °C during some 10000 years. Thus, one could expect the groundwater entering the repository to differ more or less from the present properties.

267

Most macroscopic properties remain unchanged as the simulated temperature is increased. A small number of properties change significantly. Most important is a decrease in the pH and an increase in the pE value. Further, the Si concentration is increased by half an order of magnitude. The pattern of deviations for the cases b and c found at 15 °C is appearing again at 80 °C.

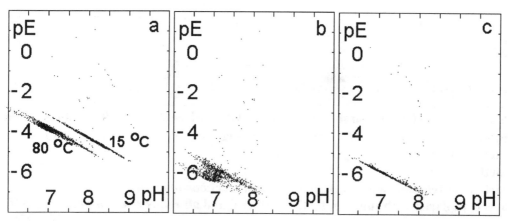

Figure 7. Simulated pE vs. pH for different cases at 80 °C. Since this property is not measured at 80 °C, comparison with observed values is not meaningful.
a. All redox reactions active.
b. Methane - carbonate reaction and sulphide - sulphate reactions blocked.
c. All iron minerals replaced with the hypothetical mineral and all redox reactions except Fe(II) - Fe(III) blocked.

The decrease in pH of about one unit does not appear to be significant at all. On the other hand, some of the cases give rise to pE - pH combinations outside the stability region of Cu. Each point in the diagrams represents one simulated water. The mixture of some 900 randomly chosen such waters represents one simulated sample (one set of bars in Fig. 6).

At 80 °C, oxidation of Cu to Cu_2O is no longer negligible except in cases c, d and g. In cases a, b, e and f, the capacity of the water to corrode Cu to Cu_2O is about 0.3 g/l. The phase $CuFeO_2$ is much more stable in this case too if a suitable source of Fe, as goethite is available. In this case, between 8 and 15 g/l may be corroded in the cases a, b, c and f.

Influence of spatial variability upon retardation
An important barrier for preventing released radionuclides from reaching the biosphere is constituted by the rock itself. Sorption and matrix diffusion are assumed to serve as major mechanisms for the retardation. In most modelling work no spatial variability is considered along the sorbing surfaces.

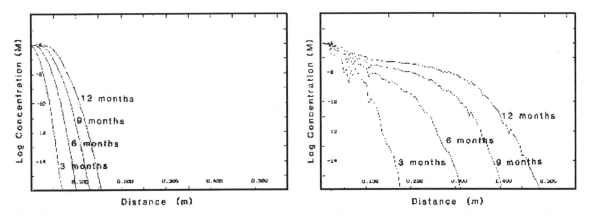

Figure 8. Simulated log concentrations of Np in groundwater. In the first case (a) spatial variability of chemical properties is absent while in the second (b) it has been taken into account. From [12]

As has been shown [12] this may lead to an overestimate of the capacity of fracture walls to retard moving radionuclides. The CRACKER program is designed to take the variability into account. A typical fracture in granitic rock has been used to simulate the concentration of Np as a function of distance with and without spatial variability. The result is illustrated in figure 8. Obviously, the radionuclide spreads considerably faster if there is a spatial variability in the chemical properties of the rock. The lowering of sorption efficiency is a consequence of certain particle tracks being able to avoid the sorbing minerals. Further, there is an "interest upon interest" effect that also decreases the sorption efficiency.

CONCLUSION

The CRACKER program has been found to give a good agreement between observed and simulated properties of deep groundwater at Äspö and with the use of minerals observed at the place. To make the simulated water similar to the observed water, either the methane - carbonate reaction or the sulphide - sulphate reaction or both have to be active.

The spatial variability considered in CRACKER is needed to get a satisfactory agreement between observed and simulated values of element concentrations and saturation properties of observed minerals at the Äspö site.

From the simulations it is not possible to choose one reaction that mainly is responsible for the redox properties. Should the redox control work as discussed above, with several reactions involved, the temperature rise close to a repository may cause the groundwater to turn agressive with respect to copper. Oxidation to Cu_2O may then be an important reaction for assessment of the canister endurance.

Further, the spatial variability of chemical properties is leads to a considerable decrease in the reaction rates and to a decrease in the efficiency of sorption in retarding radionuclide leaching from a failed canister in a repository for nuclear wastes.

ACKNOWLEDGEMENT
This work has been supported by the Swedish Nuclear Power Inspectorate (SKI)

REFERENCES

[1] Emrén A. T., "CRACKER a Program Coupling Inhomogeneous Chemistry and Transport" Computers & Geosciences, To appear (1997)

[2] Allard B., Larson S. Å., Tullborg E-L. and Wikberg P., "Chemistry of Deep Groundwaters from Granitic Bedrock", SKBF KBS TR 83-59, Svensk Kärnbränsleförsörjning AB, Stockholm (1983), 11 pp

[3] Grenthe I., Stumm W., Laaksoharju M., Nilsson A. C., Wikberg P., "Redox potentials and redox reactions in deep groundwater systems", Chemical Geology, **98** (1992), 131

[4] Arthur R., Emrén A.T., Glynn P. and McMurry J., "The modeler's influence on results for calculated Plutonium(IV) Hydroxide Solubility in Hypothetical Environments at the Äspö Hard-Rock Laboratory", SKI TR under preparation, SKI, Stockholm (1996)

[5] Nordstrom K. and Puigdomenech I., "Redox chemistry of deep groundwaters in Sweden", SKB TR 86-03, Svensk Kärnbränsleförsörjning AB, Stockholm (1986), 30 pp

[6] Shock E. L. "Catalysing methane production", Nature **368** (1994), 499-500

[7] Mango F. D., Hightower J. W., James A. T., "Role of transition-metal catalysis in the formation of natural gas", Nature **368** (1994) 536 - 538

[8] Glynn P., U. S. Geological Survey, personal communication.

[9] Emrén A. T., "Geochemical Modelling: CRACKER Improvements, Groundwater formation and Temperature Gradients", SKI TR 90:13, SKI, Stockholm (1991), 31 pp

[10] Emrén A.T., "Äspö groundwater modelling with the CRACKER program", Submitted to J. Appl Geochemistry (1996)

[11] Parkhurst D. L., Thorstenson D. C. and Plummer L N, "PHREEQE - a computer program for geochemical calculations", U. S. Geological Survey, Water Resources Investigations, report USGS/WRI-80-96 (1980, 1985 revision)

[12] Emrén A.T., "The influence of heterogeneous rock chemistry on the sorption of radionuclides in flowing groundwater", Journal of Contaminant Hydrology, **13** (1993) p 131-141.

Upscaling Conductivities and Dispersivities

J. Jaime Gómez-Hernández and Xian-Huan Wen
Polytechnic University of Valencia, Spain

Abstract

Data are never measured at the scale at which they are later needed for in models. In most cases the measurement scale is smaller than the model-gridblock scale, although, there are cases in which the opposite is true. In order to use measured data, we must account for the change of scale.

We only focus on the process of upscaling, that is, how to incorporate in the model, data measured at a scale smaller than the model gridblocks. More precisely, we analyze the upscaling of conductivity derived from cores, transmissivity derived from slug tests and dispersivities derived in the laboratory or from short-scale tracer experiments.

For conductivity, we have developed "selective upscaling", a process in which we start with a very fine grid of conductivity values generated at the scale of the measurement data, which is later upscaled in a selective manner so as not to average out the fast flow velocity areas in the formation (those flow channels that are most consequential from a regulatory point of view).

Selective upscaling proceeds as follows. First, a conductivity realization is generated using a uniform, very fine grid with pixels at the scale of the measurements. Second, a uniform coarser grid, with as many gridblocks as considered tractable by the flow and transport models, is overlaid on the fine scale realization and a rough upscaling is performed (for instance by taking the geometric mean of the pixel values within each block). Third, using the flow velocities derived after solving the groundwater flow equation in the coarse model, apply an elastic deformation of the initial coarse grid so that the variances of flow velocities within blocks are minimized. With the new resulting gridblock geometry, gridblock values are recomputed and flow is solved again at the coarse scale and the third is repeated. The process continues iteratively until the coarse grid geometry stabilizes. The result is a final coarse grid in which the solution of the flow equation displays the same type of fast flow channels as observed in the fine scale model.

For dispersivities, we have analyzed the validity of the macrodispersivity values predicted by first-order analytical solutions to the flow and transport equations to represent macrodispersivities at the block scale. We compare the analytically-derived macrodispersivities to macrodispersivities obtained after the numerical solution of the flow and transport equations in heterogeneous 2-D transmissivity fields for multi-Gaussian and non-multi-Gaussian distributions.

The results show that analytically-derived macrodispersivities are valid for multi-Gaussian related transmissivity fields for variances of log-transmissivities as large as 1.0. However, these macrodispersivities are not applicable when transmissivities are not multi-Gaussian related, especially, if the connectivity of high or low values is larger than the one intrinsic to multi-Gaussian models. For transmissivity fields in which the covariance of log-transmissivity is isotropic but the high or low values are preferentially connected in a given direction, the macrodispersivities resulting after the numerical upscaling are anisotropic and may be up to an order of magnitude different from the analytically-derived ones depending on the size of the gridblock.

271

1. SELECTIVE UPSCALING

We propose an approach for the upscaling of heterogeneous conductivities that considers intimately the definition of the block geometries and the block upscaled conductivity tensor. The objective of the upscaling is to produce a new conductivity field with a number of blocks up to two orders of magnitude smaller and able to reproduce the fastest flow paths observed in the fine scale field. The method proposed is compared to a naïve upscaling approach by regular partitioning of the fine scale field, and to an elastic upscaling approach which is based only on the conductivity field values.

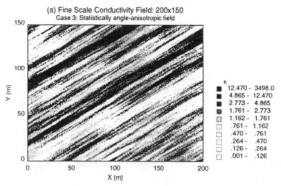

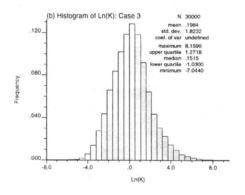

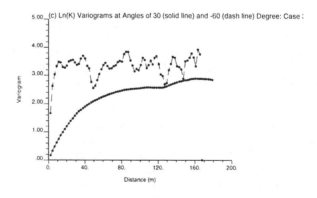

Figure 1. Base case. Spatial image, histogram and variogram of small scale conductivity.

The base case (Figure 1) in which the comparisons are carried out is a 200 by 150 cell grid of spatially anisotropic conductivities with direction of major continuity angled at 30 degrees with

272

respect to the horizontal axis. The solution of groundwater flow and particle advection at the fine scale for no flow boundaries at top and bottom, and prescribed heads, constant in each vertical boundary, is shown in Figure 2 as velocity fields, particle pathlines, the breakthrough curve and the histogram of arrival positions to the right edge.

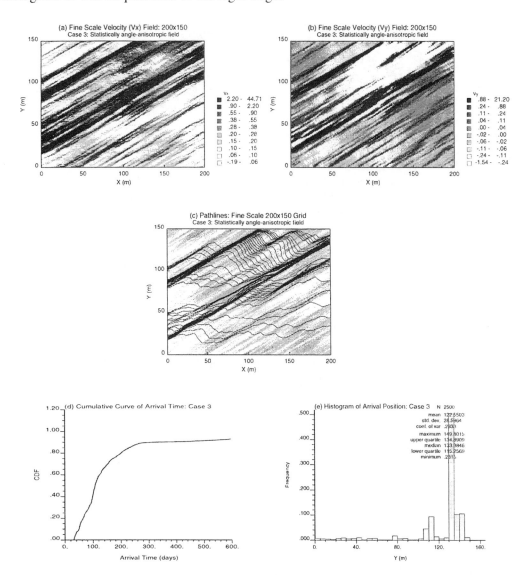

Figure 2. Base case. Top. Flow velocity components; middle, transport pathlines; bottom, transport results.

The grid geometries corresponding to the three upscaling approaches are shown in the first column of Figure 3. The two approaches that use an elastic grid, apply a minimization algorithm to displace the gridblock corners within the flow domain to obtain a minimum energy state [1]. In the middle row, the energy is expressed in terms of intra-cell conductivity variances, in the bottom row, the one we propose, this energy is defined in terms of intra-cell total flux variance. The advantage of the proposed approach is that the grid will be distorted in those areas where the fluxes are higher, therefore reducing the smoothing effect of any upscaling approach. For the specific case presented here, the presence of a few very high conductivity values near the top makes the first elastic grid too coarse in the bottom part.

Once the geometry is defined, the block conductivity is computed by isolating the block, solving the flow equation at the small scale and determining the block conductivity tensor that is able to reproduce the average flow crossing the block for the same average gradient [2].

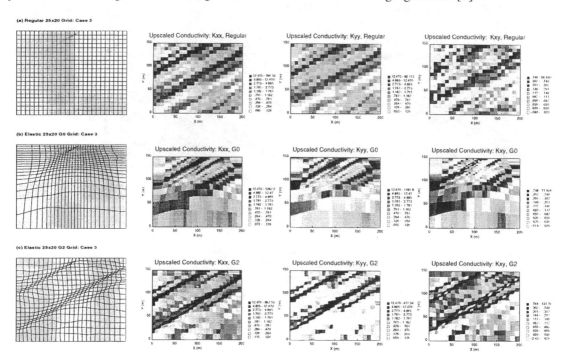

Figure 3. Three upscaling grids. Top, regular grid; middle, Garcia's grid, and bottom proposed grid. The geometry and the corresponding values of the components of the conductivity tensor are shown.

Because the proposed approach requires to know the flow distribution in order to determine the geometry, it is necessary to use an iterative approach, whereby, the flow velocities are first estimated using a regular grid upscaling, then a first estimate of the grid geometry is estimated, which is used to re-estimate the flow velocities and so on.

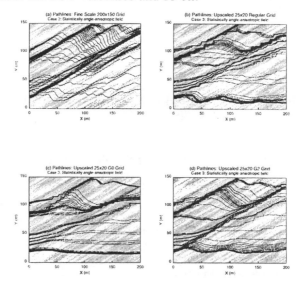

Figure 4. Comparison of transport pathlines. (a) base case, (b) regular grid, (c) Garcia's grid and (c) proposed grid.

The resulting upscaled block conductivity tensors and the solutions to flow and advective transport problems for the three upscaling approaches show the good performance of the proposed method, especially in the reproduction of early breakthroughs and channeled transport as shown in Figures 4 and 5.

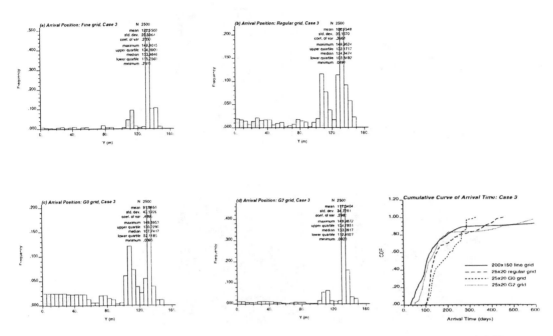

Figure 5. Comparison of arrival locations and breakthrough curves. (a) Base case, (b) regular grid, (c) Garcia's (G0) grid, (d) proposed (G2) grid.

2. MACRODISPERSIVITIES IN NON-GAUSSIAN FIELDS

Conductivity fields that display connectivities in the extreme values should not be modeled using multi-Gaussian random functions because of their inherent lack of connectivity at extreme values [3]. Since most of current stochastic hydrogeology results are modeled using a multi-Gaussian random function, or using first-order approximations to the flow and transport equations (an approximation specially congenial to multi-Gaussian random functions), we decided to contrast block macrodispersivities as predicted analytically by first-order theory to macrodispersivities numerically derived from the heterogeneous conductivity fields according to four different random function models, model A is a multi-lognormal one, model B displays high connectivity at high conductivity values, model C displays high connectivity at low conductivity values, and model D displays high connectivity at both high and low conductivity values. All models share the same log-normal histogram of zero mean and variance 2.0. All models have the same isotropic log-conductivity covariance. The models differ only on their indicator variograms at extreme thresholds. Figures 6 and 7 show typical realizations and pathlines for each model.

The numerical experiments consist in computing the average macrodispersivities at different distances from a source line by analyzing the temporal moments of the breakthrough curves. Two flow situations were considered, one parallel to the direction of connectivity of the extreme values and one orthogonal to that (referred to as horizontal and vertical flow conditions). In addition, the analytical results obtained by first-order theory using the mean, variance and isotropic covariance shared by all four models are also computed. The numerical results correspond to average values computed after 100 unconditional log-conductivity realizations. The breakthrough curves are integrated values along the entire cross-section. The results show the similarity between multi-Gaussian based results and first-order analytical ones.

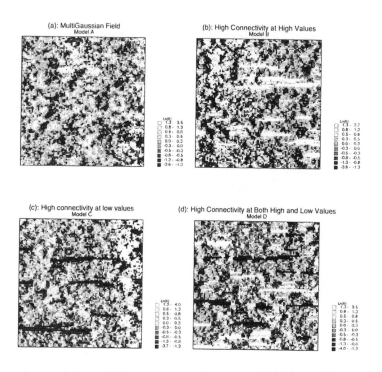

Figure 6. A typical realization of conductivities for the four models.

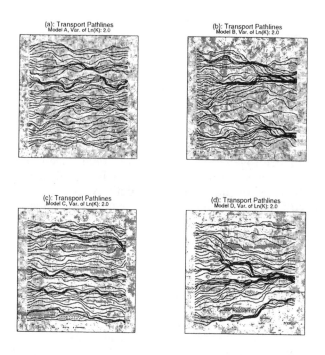

Figure 7. Transport pathlines for horizontal flow conditions from left to right in the realizations of Figure 6.

The residence time distributions represented by the breakthrough curves (Figure 8) show significant differences between the four models considered, especially under horizontal flow conditions. With respect to model A (the multi-Gaussian one), model B shows earlier arrival of particles and shorter tails, model C shows later arrival and longer tails and model D shows earlier arrival and longer tails. That is, high connectivity of highs leads to early arrivals, and high connectivity of lows leads to long tails. For vertical flow conditions the differences are smaller.

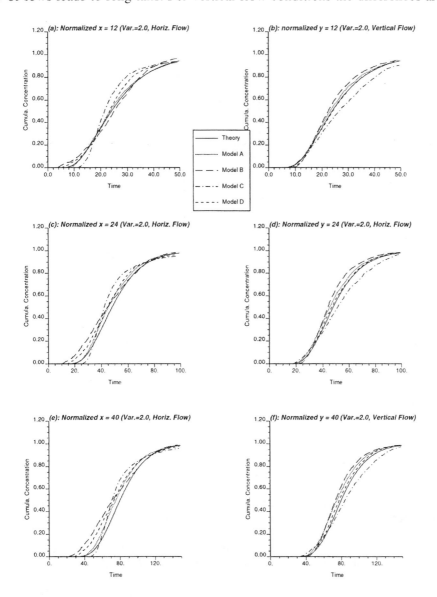

Figure 8. Residence time distributions at distances of 12, 24 and 40 correlation scales from the source line. Left column for horizontal flow, right column for vertical flow.

As it can be seen in Figure 9, for all four models, the pattern of variability of macrodispersivities with distance is the same: longitudinal macrodispersivities increase with distance towards some asymptotic value and transverse macrodispersivities reach a maximum at a certain distance from the source then decrease (apparently asymptotically towards zero). However, in absolute terms, the values vary significantly among the four models. Under horizontal flow conditions, the longitudinal macrodispersivities for models B and D are larger than for the multi-Gaussian model. Transverse macrodispersivities for model B are much larger than for the multi-Gaussian model, whereas model C results in smaller values. For vertical flow conditions, an anisotropic behavior is observed in all models except for the multi-Gaussian one, even though all models share the same isotropic log-conductivity covariance.

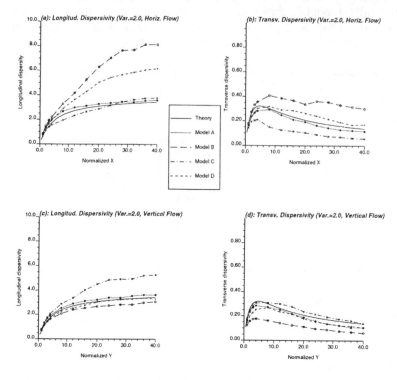

Figure 9. Apparent macrodispersivities as a function of distance from the release plane. Top, horizontal flow; bottom, vertical flow. Left, longitudinal macrodispersivities; right, transverse macrodispersivities

3. REFERENCES

[1] Garcia, M.H., A.G. Journel and K. Aziz, Automatic grid generation for modeling reservoir heterogeneities, SPE Reservoir Engineering 1992, 278-284.

[2] Durlofsky, L.J., Numerical calculation of equivalent grid block permeability tensors for heterogeneous porous media, Water Resources Research, 1991, 27(5), 699-708.

[3] Gómez-Hernández, J. J. and X.-H. Wen, To be or not to be multi-Gaussian? A reflection on stochastic hydrology, Advances in Water Resources, 1998 21(1), 47-61.

Channel Network Model for Flow and Radionuclide Migration

Björn Gylling
Luis Moreno
Ivars Neretnieks

Department of Chemical Engineering and Technology
Royal Institute of Technology
Stockholm, Sweden

Abstract

The Channel Network model has been developed for simulations of flow and solute transport in fractured rock. The model, and its computer implementation CHAN3D, has been used to simulate and analyse field experiments performed in Sweden. In the performance assessment of repositories for nuclear waste a tool was needed which could handle the release from the near field and simulate the radionuclide migration in the far field. In order to develop the tool the Channel Network model was integrated with a model which calculates the near field release in detail.

The integrated model concept which may be used for performance assessment of a nuclear repository is presented. The tool is developed by coupling of two models, one near field and one far field model. A compartment model, NUCTRAN, is used to calculate the near field release from a damaged canister. The far field transport through fractured rock is simulated by using CHAN3D, based on a three-dimensional stochastic channel network concept. The near field release depends on the local hydraulic properties of the far field. The transport in the far field in turn depends on where the damaged canister(s) is located. The very large heterogeneities in the rock mass makes it necessary to study both the near field release properties and the location of release at the same time. In order to demonstrate the capabilities of the coupled model concept it is applied on a hypothetical repository located at the Hard Rock Laboratory in Äspö, Sweden. Two main items were studied; the location of a damaged canister in relation to fracture zones and the barrier function of the host rock. In the study of the far field rock as a transport barrier the retardation and the maximum release of the radionuclides was addressed.

1. INTRODUCTION AND BACKGROUND

The Channel Network model has been developed for simulations of flow and solute transport in fractured rock. The model, and its computer implementation CHAN3D, has been used to simulate and analyse field experiments in Sweden, e.g., the long-term pumping and tracer test, LPT2 and the tracer retention understanding experiment, TRUE. However, in the performance assessment of a repository for nuclear waste a tool was needed to calculate the release in the near field and simulate the radionuclide transport in the far field. In order to develop the tool the Channel Network model was integrated with a model which describes the near field release in detail.

The Swedish repository for spent nuclear fuel is planned to be located at large depth in crystalline rock. In the repository, radionuclides escaping from a damaged canister will spread into the backfill material surrounding the canister and then migrate through different pathways into the mobile water in the rock fractures. The transport through the near field, occurring mainly by diffusion in the bentonite surrounding the canister and the sand-bentonite mixture in the tunnel, is modelled by using NUCTRAN [1,2]. The far field transport of the escaping radionuclides is then simulated by using the Channel Network model [3]. In addition to advection in the channels, the model can account for diffusion into the rock matrix and sorption within the matrix, which are the by far most important mechanisms retarding the nuclides.

The aim of this paper is to describe how the Channel Network model may be used to simulate the migration of radionuclides in the far field using the near field model as an advanced source term. The output from the source depends on the local flow rates around the repository calculated by CHAN3D. The integration of the source for radionuclides into the far field model makes it possible to simulate release and transport over the entire distance from a damage in a canister, into and through the backfill by various simultaneous paths, through the geosphere to the biosphere and still consider the detailed scale related to the near field, and also model the most important mechanisms in the far field. The concept may provide a useful tool in the performance assessment of a nuclear waste repository. To demonstrate the concept the integrated models are applied on a hypothetical repository layout [4]. Among the addressed items, the influence of the position of the initially damaged canisters and the importance of the rock as a barrier for release of escaping nuclides were studied.

2. THE USED MODELS FOR THE NEAR AND FAR FIELD

Nuclides escaping from the near field are transported by the flowing water through fractures in the rock mass. This is calculated by the far field model CHAN3D using the temporal release from the repository calculated by the near field model NUCTRAN as input. The calculated local flow around the repository, calculated by CHAN3D, is needed as input to NUCTRAN. In that sense the models are integrated. The used models in the performance assessment calculations are described below. In the next section the integration of the models is described.

2.1 NUCTRAN: A COMPARTMENT MODEL FOR THE NEAR FIELD

In the repository the canisters will be deposited in vertical boreholes below the floor of the tunnels. The deposition holes are backfilled with compacted bentonite and the tunnels with a mixture of sand and bentonite. Some canisters may be initially defective or later damaged by mechanical or corrosion processes. Nuclides will then dissolve in the intruding water and leak out of the canisters. The clay and the rock matrix have very low hydraulic conductivities. Since there is no water flow the nuclide transport through them is only by diffusion. Advection is restricted to the fractures in the rock surrounding the repository. Nuclides escaping from the canisters may diffuse into fractures intersecting the deposition holes and/or diffuse into the altered rock in the disturbed zone around the deposition tunnels. Figure 1 shows four considered pathways: path Q_1 directly to the water in a fracture intersecting the repository hole, path Q_2 up to the disturbed zone around the tunnel, path Q_3 into the tunnel backfill and further to a fracture (zone) intersecting the tunnel and path Q_4 through the rock to a nearby fracture or fracture zone. The release of nuclides from the canisters is calculated by using a compartment model.

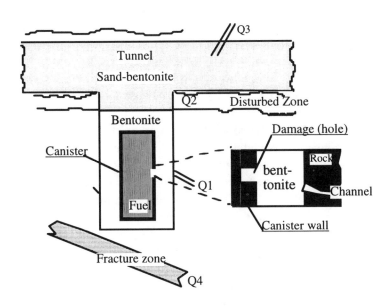

Figure 1. Schematic view of the used repository design, showing the small hole in the canister and the location of the various escape routes.

Radionuclides leaking from a damaged canister diffuse into the backfill material surrounding the canister and then migrate through different pathways into water-bearing fractures in the rock surrounding the repository. In the model, the processes governing the nuclide transport in the near field repository are dissolution, diffusion, advection, and sorption.

For the purpose of numerical simulations the model is formulated in terms of integrated finite differences [5], where the concept of "compartments" is introduced to define the discretization of the system. Average properties are associated over these compartments with representative nodal points within the compartment. From the theoretical point of view the compartments may have any shape, but are of the same material. The compartments are defined by their diffusion length(s), contact area(s) to other compartments and their volume. These dimensions together with the transport properties such as diffusion and sorption will define the capacity and the diffusion resistance(s) characterizing the compartments. The compartments can be arranged in a three-dimensional structure. For the fuel dissolution two approaches may be considered: a) the concentration of a nuclide in the canister is controlled by its own solubility and b) congruent dissolution where the release rate of the nuclides is determined by the dissolution rate of the spent fuel matrix.

The model uses a rather straightforward discretization process into compartments considering the geometry of the system and the materials. To avoid a fine discretization at regions with very narrow pathways, analytical or semi-analytical solutions are introduced in these zones. This kind of approach is used, for example, when nuclides are released through the small hole at the wall of the damaged canister into the bentonite. This concept is also used when the nuclides migrate from the bentonite surrounding the canister into a fracture intersecting the deposition hole. Finally, at the boundaries of the repository, the diffusive transport into the flowing water in the fractures is accounted for by using the notion of the equivalent flow rate which is a fictitious flow rate of water that carries with it a concentration equal to that at the contact interface. The magnitude of the equivalent flow rate is assessed using the boundary layer theory [1]. The data needed for NUCTRAN are the canister inventory, nuclide half-lives, diffusion coefficients for nuclides in bentonite, sand-bentonite mixtures and rock, and properties for the materials involved. The output from NUCTRAN is the transient release through the different paths. Details about the model formulation and the analytical or semi-analytical expressions used may be found in [1] and [2].

2.2   CHAN3D: A CHANNEL NETWORK CONCEPT FOR THE FAR FIELD

In the far field the nuclides escaping from the near field are transported by the flowing water through fractures in the rock mass. In our concept the water flows through channels mainly located in fractures and fracture zones. The fractures intersect in the rock and they form a three-dimensional network of interconnected channels. Solute transport through the far field is calculated by using the code CHAN3D based on a channel network model [3]. The model may accommodate the transport of interacting and hypothetically non-interacting solutes. For interacting solutes diffusion into the rock matrix and sorption on the interior of the matrix are considered. These are very important processes when assessing retardation of radionuclides that may escape from repositories for nuclear waste.

Fluid flow and solute transport through the far field are calculated by using the code CHAN3D based on the Channel Network model [3]. For visualization purposes, the network is depicted as a rectangular grid. A fraction of the grid is shown in Figure 2. The channel length is not explicitly used in the model, it is implicitly included in the conductance and volume of the channels and the flow-wetted surface. For these reasons, the exact form of the grid is not significant.

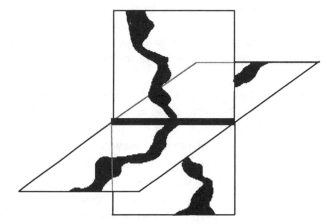

Figure 2. A fraction of the grid in the model showing the channel in the fracture intersection and the channels in the fracture planes.

To generate the grid a hydraulic conductance is assigned to each channel of the network. This is the only entity needed to calculate the flow, if the pressure field is known. The conductance is defined in analogy with electrical networks where it is the reciprocal resistance. Here, it becomes the ratio between the flow in a channel and the pressure difference between its ends. The conductances of the channels are assumed to be log-normally distributed and not correlated in space. For laminar conditions, the flow, Q, through a channel is proportional to the pressure gradient:

$$Q_{ij} = C_{ij}\left(P_i - P_j\right)$$ (1)

where C_{ij} is the conductance of the channel connecting the nodes "i" and "j". P_i and P_j are the pressure at these nodes. The pressure field is calculated by writing the mass balance at each intersection point:

$$\sum_j Q_{ij} = 0 \quad \text{for all i}$$ (2)

The dominating water flow paths are determined by the major fracture zones. These fracture zones are responsible for most of the water flow. In the model, we have assumed that the fracture zones may be described by two parallel planes delimiting the zone with high permeability. Channels present between these two limiting planes have higher conductances than channels within the rock mass. Conductances are obtained from a distribution with a mean value that matches the experimentally estimated transmissivity of the fracture zone.

If the residence time for noninteracting solutes is to be calculated, then the volume of the channels is needed as well. Owing to lack of data, it is assumed that the volume of the channel is proportional to the third root of the conductance derived from the cubic law. This is important for noninteracting species but not for those that sorb and/or diffuse in the rock matrix. If sorption onto the fracture surface or diffusion into the matrix will be included in the model, the surface area of the flow-wetted surface must also be included. Then some properties of the rock are needed, such as rock matrix porosity, diffusivity, and sorption capacity for sorbing species.

The solute transport is simulated by using a particle-following technique [6,7]. Many particles are introduced, one at a time, into the known flow field at one or more locations. Particles arriving at an intersection are distributed in the outlet channels with a probability proportional to their flow rates. This is equivalent to assuming total mixing at the intersections. Each individual particle is followed through the network. The residence time for noninteracting tracers in a given channel is determined by the flow through the channel and by its volume. For sorbing solutes matrix diffusion and sorption are included in addition to the advection. The residence time of an individual particle along the whole path is the sum of residence times in every channel that the particle has traversed. The residence time distribution is then obtained from the residence times of a multitude of individual particle runs. Hydrodynamic dispersion in individual channels is considered to be negligible. The dispersion in the system is caused by the uneven flow distribution due to the heterogeneity of the rock and the interaction mechanisms. The travel time for each particle in a channel is determined by choosing a uniform random number in the interval [0,1] and the travel time is then calculated by solving for t in Equation (3) [8]:

$$[R]_0^1 = \text{erfc} \left[\frac{LW}{Q} \left(\frac{K_d D_e \rho_p}{t - t_w} \right)^{0.5} \right] \tag{3}$$

3.   TOOL DEVELOPMENT

To calculate the release from the near field, the number of canisters initially damaged and their locations are determined by a random process. The data needed for NUCTRAN are the canister design and information about the nuclide properties. Data for CHAN3D may be obtained from hydraulic tests and field observations [9]. For transport calculations with CHAN3D information about the interactions between the rock and the nuclides are also needed. Sorption and matrix diffusion data may be obtained from laboratory measurements. The magnitude of the flow-wetted surface and hydraulic conductivity distribution are obtained from borehole measurements [8]. The hydraulic model is solved using the corresponding boundary conditions. The water flow rates obtained in these calculations are then used to determine the advective transport through the channel network. Once the water flow rate is determined by using the flow part of the model (CHAN3D-flow), the release from the near field may be addressed. Flow rates in the disturbed zone and in the fracture/fractures which intersect the deposition hole are needed to calculate the release from the near field. Once the release from the near field has been determined, the transport part of the model (CHAN3D-transport) is used to calculate the transport of radionuclides through the far field. The model integration is shown schematically in Figure 2.

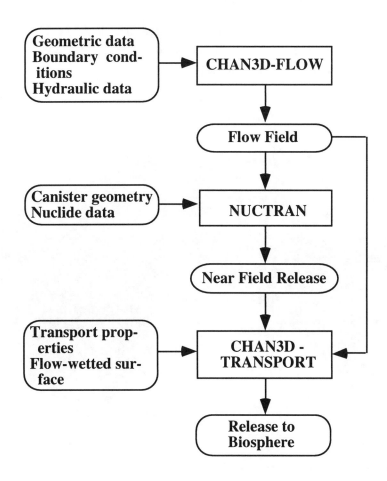

Figure 3. Flow-sheet over the model integration.

4. APPLICATION OF THE TOOL TO A SITE

As the location of the repository for nuclear waste is not yet decided, there is no detailed repository layout and for this reason a hypothetical repository located at the Äspö HRL site [4] is used in this scenario study. In principle, the model repository consists of a few central tunnels and from these tunnels several horizontal deposition tunnels are excavated, where the deposition holes will be drilled. Due to the fracture zone geometry the hypothetical repository is made up of three sections. We have chosen to study one of these sections.

4.1 SCENARIO FOR A HYPOTHETICAL REPOSITORY

The dominating water flow paths are determined by the major fracture zones which then are responsible for most of the water flow. The most important conductive fracture zones at Äspö HRL that have been used and the repository location are shown in Figure 3. In addition to the fracture zones the access and deposition tunnels may also act as water conducting features. The boundary conditions are taken from a regional study [10] calculated by using the code NAMMU. In the studied repository section, 25 deposition tunnels are located, which together may contain about 2 100 canisters. A distance between deposition holes (canister) of 6 m and a distance of 25 m between deposition tunnels are used. The deposition tunnels are located at a depth of 450 m below the sea level. It is assumed that the probability that a canister is initially damaged is 0.1% and that the initial, 1 mm^2, hole in the canister will increase in size after 10 000 years.

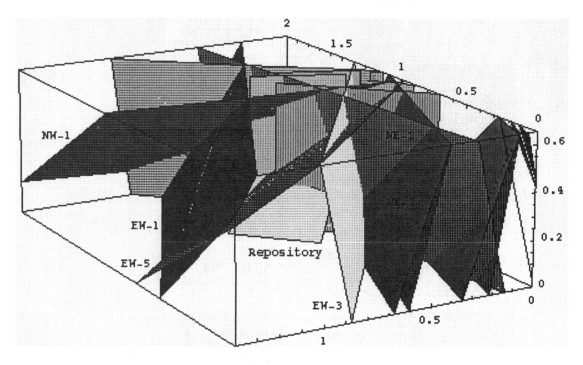

Figure 4. **A section of a hypothetical repository located in the rock mass between the fracture zones EW-1, EW-3, EW-5, NE-1, NE-2 and NW-1.**

5. RESULTS

In the calculations, the radionuclides C-14, I-129, Cs-135, Cs-137, U-238, Pu-240, and Ra-226 were chosen for our study. Realizations of the channel network were generated and the flow field was calculated. Then three different realizations were made for transport calculations. The number of damaged canisters in each run depends on the probability for an initial damage. The expected value for each run is 2.1 damaged canisters.

The release from the far field to the biosphere for the nuclide Cs-135 is shown in Figure 5. The figure show the total release for three runs. In each run, the total release is the sum of the release from two or three canister. Run R1 shows the largest retardation, since no damaged canister is found near a fracture zone. Run R2 shows the smallest retardation, because a damaged canister is located near a fracture zone.

The maximum release rates and the corresponding times of I-129 and Cs-135 from the far field are shown in Table 1. The release of C-14 was on the same order of magnitude as for I-129. Due to decay in the far field no release of Cs-137 was obtained. For U-238, Pu-240 and Ra-226 very small release rates were obtained.

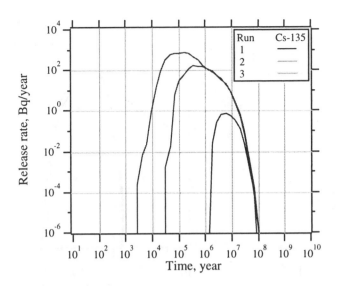

Figure 5. The far field release of Cs-135 ($K_d = 0.15$ m^3/kg) from the three performed runs.

Nuclide	Run	Maximum Release Rate [Bq/year]	Time for Maximum [years]
I-129	1	$2.9 \cdot 10^3$	$4.7 \cdot 10^4$
	2	$7.0 \cdot 10^3$	$3.2 \cdot 10^4$
	3	$3.6 \cdot 10^3$	$3.2 \cdot 10^4$
Cs-135	1	$7.4 \cdot 10^{-1}$	$6.1 \cdot 10^6$
	2	$7.5 \cdot 10^2$	$1.6 \cdot 10^5$
	3	$1.6 \cdot 10^2$	$3.6 \cdot 10^5$

Table 1. The maximum release rates from the far field and the corresponding times for the maximum release for the studied nuclides.

In order to investigate the retardation during the transport through the far field, we have plotted the release from the near field and the far field in the same figure. We have chosen to show the release from run R1 with the largest retardation and run R2 with the smallest retardation. The retardation of Cs-135 is shown in Figure 5. The rate of transport increases considerably when the damage has grown after 10 000 years. In run R1 the damaged canisters are located in favourable locations and in run R2 one canister is near a fracture zone. It has been found that the release from a certain run depends mainly on the locations of the damaged canisters.

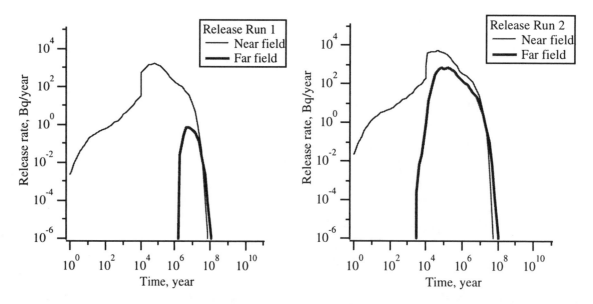

Figure 6. Near field and far field release of Cs-135 from the runs R1 and R2. In run R2 there is a damaged canister located in a region strongly connected to a fracture zone.

To illustrate the paths from the canisters only two particles were used for each release path from the near field simulations. In Figure 6 four nuclide traces are shown from one canister. As can be seen it may be a fairly large difference in paths between for the nuclides which are released in fractures intersecting the deposition hole and the disturbed zone. There may also be a difference in the paths for the individual nuclides released through the same near field path.

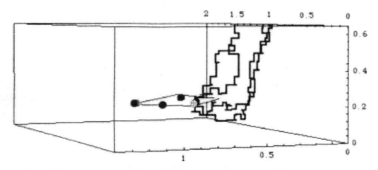

Figure 7. The paths from one canister. The two traces to the left are from the nearfield path Q$_1$ and the other two are from the path Q$_2$.

6.   DISCUSSION AND CONCLUSIONS

In the performance assessment of a planned repository the concept using a near field model integrated into a far field model may be very useful. Using this concept the detailed calculations of the near field may be used in addition to the far field modelling of the most important mechanisms. The coupled models can be used as an efficient tool to study the impact of various factors including the location of the damaged canister(s) and the importance of the host rock as a barrier against release of potentially escaping nuclides. Here, the effect of the retardation in the far field was studied.

From the results it may be seen that the location of the canister with an initial damage is of vital importance. The retardation in the far field depends mainly on the water flow rate and the flow-wetted surface of the channels in addition to the sorption and diffusion of the nuclides in the rock. Therefore, the retardation of nuclides that are transported by large flow rates to the surface can be very small. The detailed release from the near field model had an impact on which paths the nuclides traversed in the far field.

In the near field calculations it is assumed that the damage is very small initially but grows abruptly to a large hole after 10 000 years. In these calculations 5% of the Cesium is assumed to have migrated out of the uranium oxide matrix and is available for direct dissolution and transport. The slow build up of the Cs-135 concentration initially is due to the large transport resistance through the initially small damage. The rate of transport increases considerably when the damage has grown after 10 000 years. Then the rate of transport is controlled by the diffusion into the flowing water in the release paths. The transport through the far field is considerably retarded. When the Cs-135 is released in a region connected to a fracture zone the retardation is considerably smaller.

7.   REFERENCES

[1]   Romero L., Moreno L. and Neretnieks I. Fast multiple path model to calculate radionuclide release from the near filed of a repository, Nuclear Technology, 1995, Vol. 112,1 pp 89-98.

[2]   Romero L., Moreno L. and Neretnieks I. The fast multiple path model "Nuctran" – Calculating the radionuclide release from a repository, Nuclear Technology, 1995, Vol. 112,1 pp 99-107.

[3]   Moreno L. and Neretnieks I. Fluid flow and solute transport in a network of channels, Journal of Contaminant Hydrology, 1993, 14, pp 163-192.

[4]   Munier R. and Sandstedt H. SR-95: Hypothetical layout using site data from Äspö, Swedish Nuclear Fuel and Waste Management Co., SKB Progress Report 94-53, 1994, Stockholm.

[5]   Narasimhan T.N. and Witherspoon P.A. An integrated finite difference method for analyzing fluid flow in porous media, Water Resources Research, 1976, 12, 57.

[6]   Robinson P.C. Connectivity, flow and transport in network models of fractured media, Ph. D. Thesis, St. Catherine's College, Oxford University, May 1984, Ref. TP 1072.

[7]   Moreno L., Tsang Y.W., Tsang C.F., Hale F.V., and Neretnieks I. Flow and tracer transport in a single fracture. A stochastic model and its relation to some field observations, Water Resources Research, 1988, 24, pp 2033-3048.

[8]   Yamashita R., Kimura H., Particle-tracking technique for nuclide decay chain transport in fractured porous media, Journal of Nuclear Science and Technology, 1990, 27, pp 1041-1049.

[9]   Moreno L., Gylling B., and Neretnieks I. Solute transport in fractured media - the important mechanisms for performance assessment, Journal of Contaminant Hydrology, 1997, 25, pp. 163-192.

[10] Boghammar A., Kemakta Konsult AB, Sweden, 1996, (personal communication).

Modelling Transport in Media with Correlated Spatially Variable Physical and Geochemical Properties

M D Impey, C C Sellar, R C Brown, K J Clark
QuantiSci, UK

K Hatanaka, Y Iriji, M Uchida
PNC, Japan

Abstract

Preferential pathways for the release of radionuclides through heterogeneous media are of concern in performance assessment. There is anecdotal evidence of correlations between spatially variable fields of physical properties such as permeability and of geochemical properties such as sorption. Such correlations could lead to enhanced channelling of radionuclide transport. We wished to evaluate this possibility. This study formed the first stage of the MACRO-II joint experimental-modelling programme.

MACRO-II continues the synthetic media approach to studying spatial variability used successfully in the first MACRO experiment. A synthetic medium has the advantage that it is possible to characterise the medium completely. Having gained an understanding of the complete system it is then possible to understand how partial knowledge effects flow and transport predictions. Eventually this knowledge should be transferred to in-situ experiments on real rock.

For the synthetic experiments to be a useful stepping stone to performance assessment, the experiments need to have similar characteristics to real systems. Realistic spatial variability is achieved by assuming a fractal model for the spatial variability of the permeability, and by then correlating a sorption parameter with the permeability. Variable, possibly correlated, two-dimensional fields of permeability and sorption values are hence computed for a grid of cells., on which flow and transport simulations are carried out. The MACRO-AFFINITY code has been enhanced to incorporate spatially variable geochemical and physical properties. The calculated property fields are the theoretical targets which the construction of the experimental apparatus is intended to simulate.

As in the MACRO programme, the apparatus is a two-dimensional bed made up of blocks of beads which correspond to the model grid-cells. The permeability and sorption properties of these blocks are required to match the target fields. The blocks are made up of a mixture of non-sorbing acrylic and sorbing ion-exchange resin beads. Each block contains beads of the same diameter that determine the permeability of the block. The required sorption behaviour can be controlled by the ratio of resin beads and acrylic beads. Steady flow will be established through the bed, and then a NaCl sorbing tracer will be injected and the plume migration monitored. Parallel model simulations will be carried out.

As a first step, batch and column experiments are being carried out to characterise the sorption behaviour of the ion-exchange beads. The results of the experiments will be used to fit and select the most appropriate mathematical model. Preliminary modelling results using data from the early experiments are presented in this poster. The preliminary calculations show that there are significant differences between modelled plumes for which the sorbing parameter varied by only one order of magnitude. Enhanced channelling was observed in those cases in which the permeability and sorption parameters were negatively correlated.

These results confirm some of the initial concerns over the effect of correlations between physical and geochemical properties. The calculations have provided feedback on the experimental and calculational procedure proposed for the MACRO-II experiment, and on the focus of further preliminary experiments.

1.   MOTIVATION

Preferential pathways for the release of radionuclides through heterogeneous media are of concern in performance assessment. The existence of a preferential pathway implies spatial heterogeneity within the medium in which the radionuclide is migrating. Historically, geochemical spatial variability has been neglected in comparison with spatial variability of hydraulic parameters such as permeability or porosity. There is, however, anecdotal evidence of correlations between spatially variable fields of hydraulic properties and of geochemical properties such as sorption. Such correlations could lead to enhanced channelling of radionuclide transport and so impact adversely upon a performance assessment. We wished to evaluate this possibility within the MACRO-II joint experimental-modelling programme.

The MACRO-II programme is being carried out within the ENTRY-2 facility at PNC's Tokai Works. It follows on from the successful MACRO-1 experiment in the first ENTRY facility, and is expected to be a three year programme. MACRO-2 consists of three experiments on flow and transport in heterogeneous porous media with the objective of considering in turn

(1) transport of a sorbing tracer,

(2) density effects on flow,

(3) reactive solute transport with precipitation and dissolution.

In 1996-97 preliminary experimental and computer modelling experiments were carried out in order to provide input to the final design of the first of these three experiments (sorbing tracer) [1]. The subsequent density effects and reactive solute experiments will use the same basic apparatus, adapted to the specific aims of these later experiments.

2.   THE SORBING TRACER EXPERIMENT

It is intended to use a similar experimental apparatus to that used in MACRO-1 to carry out the MACRO-II experiment. A flow bed will be constructed block by block from individual blocks of glass beads to form a two-dimensional grid. In this way the spatially variable transmissivity field can be specified at the resolution of a grid block.

An individual bead block for the MACRO-II sorption experiment is made up of a mixture of sorbing resin and non-sorbing acrylic beads of uniform radius (Figure 1). Blocks made up from larger diameter beads are more transmissive. MACRO-II continues the synthetic media approach to studying spatial variability used successfully in the first MACRO experiment. A synthetic medium has the advantage that it is possible to completely characterise the medium. Having gained an understanding of the complete system it is then possible to understand how partial knowledge effects flow and transport predictions. Eventually this knowledge should be transferred to in-situ experiments on real rock.

● non sorbing acrylic bead

◐ sorbing resin bead

Figure 1. Schematic diagram of individual bead block. Block size is 10 cm by 10 cm by 5 cm

3.   EXPERIMENTAL DESIGN PROCEDURE

A design procedure was developed in which firstly a 'theoretical target' of correlated transmissivity and sorption parameter fields is specified, and secondly the experimental parameters required to implement the desired fields are derived. These parameter values are the radius of the beads and the mixing ratio of sorbing to non-sorbing beads within each block, and the distribution of the blocks. The experimental apparatus can then be constructed block by block using these derived parameter values.

3. 1.   Derivation of theoretical target

We hypothesise that we can characterise the spatial variability of sorption parameters by a statistically self-affine fractal, and that the sorption parameter can be correlated to transmissivity. A heuristic approach was adopted in order to derive probable forms for the correlation between sorption and conductivity. Consider a model of a porous medium as being made up of cylindrical pathways containing infill material. The conductivity is proportional to the square of the pathway radius. The dependence of the sorption upon this length-scale depends upon the model for sorption. For example, sorption may be assumed to depend upon the surface area of the infill material, or upon the surface area of the pathways. The different assumptions lead to a range of power law correlations between a sorption parameter s and transmissivity T to be hypothesised.

The power law relation between s and T means that if $\log_{10}T$ is a self-affine fractal, then $\log_{10}s$ is also a self-affine fractal with the same dimension. The model for the sorption parameter has the general form

$$\log_{10} s = \log_{10}(\alpha(T - \gamma)^{\beta}) + F \tag{1}$$

where α, β γ are correlation parameters and F is a statistically self-affine fractal.

3.2. Derivation of Experimental Parameters

In MACRO-I the conductivity value corresponding to each of the available bead sizes was determined; these values are used again. The relation between the sorption parameter, bead radius and mixing ratio is more complex. The mixing ratio Σ and bead radius R together control the available sorbing surface area. The identity of the sorbing parameter s depends upon the theoretical model of sorption being used. Depending upon the model for sorption, parameters such as sorption capacity (Langmuir model) and sorption rate (linear model) can be expected to vary with Σ and R. The form of the relation between s, Σ and R was estimated by considering those sorption model parameters which depend upon surface area, and has the general form

$$s(\Sigma, R) = \left(\frac{s_0 R_0}{\Sigma_0} \right) \frac{\Sigma}{R} \tag{2}$$

where s_0, R_0 and Σ_0 are a known set of values. The possible sorption values are found to be constrained by the bounds on the mixing ratio, especially for positively correlated cases. It is anticipated that when further batch and column experimental data is available, the functional form of this estimated relation will be refined.

3.3. Design Procedure

The complete design procedure is therefore:-

1. Generate a spatially variable $\log_{10}$ transmissivity field on a grid of cells using a statistically self-affine fractal model.

2. Quantize the fractal into a limited set of transmissivity values and hence determine the required bead radius for each block.

3. Determine an appropriate sorption model and hence fit the parameter values to the results from batch and column experiments.

4. Choose appropriate correlation model parameter values α, β γ, and hence generate a spatially variable sorption parameter field using (1).

5. Derive the required mixing ratio for each block using (2), subject to constraints of a minimum mixing ratio of sorbing beads to total beads of 0.2 and a maximum mixing ratio of 1.0.

4. DEMONSTRATION TWO-DIMENSIONAL SIMULATIONS

The design procedure was demonstrated using the results from preliminary batch and column experiments carried out by JGC Corporation. A Langmuir model for sorption was found to be a reasonable fit for the batch experiments. The Langmuir sorption model, with linear desorption, models the rate of change of sorbed concentration (per unit volume solid material) S (kg m^{-3}) by

$$\frac{\partial S}{\partial t} = k_f C(Q - S) - k_b S, \tag{3}$$

where C is the aqueous concentration (per unit volume pore space) (kg m^{-3}). The fitted parameters are listed in Table 1. The sorption capacity, Q, is the spatially variable sorption parameter for the Langmuir model.

Parameter	Value
sorption rate k_f	3.27×10^{-3} s^{-1}m^{3}kg^{-1}
desorption rate k_b	1.87×10^{-5} s^{-1}
sorption capacity Q	8.63×10^{-1} kg m^{-3}

Table 1. Fitted parameters for Langmuir sorption model

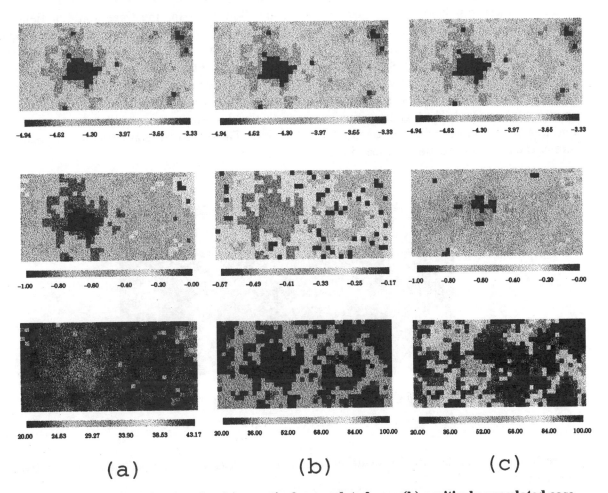

(a) (b) (c)

Figure 2. Theoretical (targets for (a) negatively correlated case (b) positively correlated case (c) uncorrelated case. The fields shown in each case are: top - log10(transmissivity)(m^{2}s^{-1}); middle - log10 sorption capacity) (kg m^3); bottom - mixing ratio (%).

An isotropic transmissivity field with a mean $\log_{10} T$ value of -4.15 m^2s^{-1}, a variance of 0.3 and fractal dimension of 2.7 was generated. Using (1) and constraints of $0.1 < Q < 1.0$, values of the correlation parameters α, γ were derived for the three cases of a positively correlated ($\beta = 0.5$), negatively

correlated ($\beta = -0.5$) and uncorrelated ($\beta = 0$) target sorption field. The constraints on Q were imposed in order to design flow beds with sorption properties consistent with those measured in the batch experiments. Fractal Q fields were then generated using (1), and target mixing ratio fields derived using (2). The theoretical targets which were derived are illustrated in Figure 2.

Flow and transport calculations were carried out using the target transmissivity and sorption fields. A one metre head difference was imposed between the left hand and right hand boundaries, and non-flux conditions set for the upper and lower boundaries. The resulting Darcy flux field is shown in Figure 3. Tracer simulations were carried out on this flow field for the three sorption fields. A constant injection concentration of 0.5 $kg\,m^{-3}$ was used. Figure 4 shows predicted plumes 1000†seconds after tracer injection for (a) the negatively correlated case, and (b) the positively correlated case. The upper part of the figure shows the aqueous tracer concentration ($kg\,m^{-3}$), and the lower part shows the sorbed tracer concentration ($kg\,m^{-3}$) for each case. The increased channelling in the negatively correlated case, can clearly be seen when the plumes in Figure 4a and the flow field in Figure 3 are compared. The high transmissivity regions have low sorption capacity and hence promote rapid transport of radionuclides throughout the flow field. The enhanced retardation in the positively correlated case can, in contrast, be seen in Figure 4b.

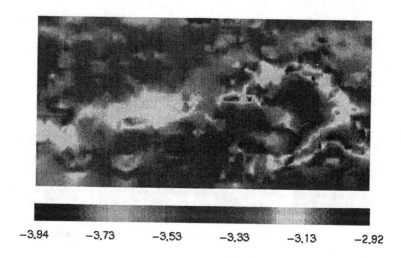

-3.94   -3.73   -3.53   -3.33   -3.13   -2.92

Figure 3. Darcy flux field (ms^{-1}) resulting from 1 m head difference across transmissivity fields from Figure 3.

5. CONCLUSIONS

We have demonstrated the use of simulation in the design and analysis of the MACRO-II sorption experiments We have demonstrated how it is possible to construct a theoretical target for spatial variability and use experimental parameters to construct a physical analogue. The initial simulations have shown the channelling behaviour resulting from correlated physical and geochemical properties. It is recommended to proceed with more batch and column experiments to better determine the appropriate sorption model, and to quantify the errors within the experiments.

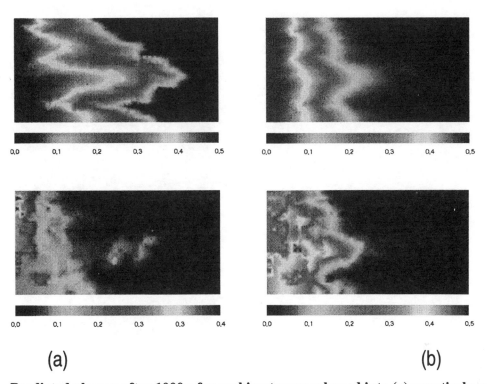

(a)   (b)

Figure 4. Predicted plumes after 1000 s for sorbing tracer released into (a) negatively correlated flow and sorption fields, and (b) positively correlated flow and sorption fields.

6. REFERENCES

[1]   Impey, M.D., Humm J.P., Sellar C.C. and Takase H., Numerical Simulation of the MACRO-II Experiment on the Effects of Heterogeneous Flow and Mineralogy on Sorption Solute Transport. Preliminary Two-Dimensional Numerical Simulations, QuantiSci Report ID5013B-4 Version 1, 1997

Development of a New Transport Model (PICNIC) which Accounts for the Large- and Small-Scale Heterogeneity

A. Jakob
Paul Scherrer Institute PSI, Switzerland

Abstract

Often, transport models have to make use of strong simplifications in order to account for the normal heterogeneity of the host rock surrounding a radioactive waste repository. When using such concepts for safety assessment purposes, these simplifications are considered from a conservative point of view, so the radiological consequences due to such simplifications are never underestimated. In an extreme case, the safety of a repository could perhaps be solely based on the performance of the engineered barriers, the components of which are believed to be homogeneous, geometrically simple and well characterised. However, it is obvious that due importance should be given to the safety provided by the natural barriers of the host rock. This requires an adequate treatment of the heterogeneity in a model. There is already a huge amount of site-specific data available on groundwater flow paths and the host rock surrounding the repository. These data should be used in the transport models. On the technical side, fast and efficient computational tools are currently available which allow the treatment of natural heterogeneity in a model without unnecessary simplifications.

Typically, potential repository host rocks show heterogeneity over a wide range of length scales:

- Radionuclide migration takes place in a network of interconnected water-conducting fractures, veins, fissures, etc., which cause the large-scale heterogeneity.
- Within a water-conducting feature, water may flow in individual channels which are in turn surrounded by a geometrically complex matrix. The variable chemical and physical properties of the matrix, into which the radionuclides can diffuse, cause the small scale heterogeneity.

No exploration programme is able to locate and characterise all relevant water-conducting features within a potential repository domain without destroying the investigated domain; hence, the large-scale heterogeneity is normally only poorly characterised. A modelling study based on a fracture network transport model should be based on the examination of a few carefully selected cases designed to cover the whole range of possible and realistic fracture network configurations and be consistent with real world observations.

The small scale heterogeneity may be much better-characterised, because it can be investigated in the laboratory and is already partially considered in today's transport model parameters. However, because of the rather simple treatment of matrix diffusion, reality still has to be simplified, resulting in a loss of geosphere performance. It is very common that matrix diffusion is considered to be a one-dimensional process within a matrix of limited extension, having homogeneous (and isotropic) properties. Hence, large parts of the matrix, which would in reality be accessed by diffusing radionuclides, are disregarded. Sorption of nuclides in these disregarded parts of the rock matrix could significantly enhance nuclide retardation and consequently reduce the release to a high-conductive feature and to the biosphere.

The new transport code PICNIC (PSI/QuantiSci Interactive Code for Networks of Interconnected Channels) [1], [2] which is going to be used as the principal tool for NAGRA's[1] performance assessments, has the following features:

- The code computes the groundwater-mediated transport of radionuclides together with their radioactive production and decay (whole radioactive decay chains are considered). Transport by advection and micro-dispersion through a network of channels, either of planar or tubular geometry, within a larger-scale network of approximately planar water-conducting features, is taken into account. The channels – each representing a section of the geosphere - may vary widely in their spatial dimensions and frequency; the water-conducting features may have a range of transmissivities constituting the large-scale heterogeneity. Such a network allows a much more accurate representation of large scale heterogeneity than in older models. In addition, the model accounts for two-dimensional molecular diffusion into the porous rock zone surrounding the channels. Each zone may consist of a larger number of sub-regions with a range of properties. Hence, PICNIC is based on the multi-porosity medium concept and - when compared to former models - more completely accounts for the small-scale heterogeneity. Linear equilibrium sorption processes on the walls of the water conducting features and onto inner surfaces of the connected pore spaces retard the migration of the radionuclides.
- The new code allows quantifying the potential consequences of multiple pathways from the near field to a high-conductive fracture and the biosphere. The characteristics of such multiple pathways may vary from pathway to pathway but also along the pathways themselves. In reality, a small number of fast pathways may dominate the maximum release from a repository. The degree to which pathways are interconnected may also be of importance. With the help of PICNIC these questions can be investigated thoroughly and in a systematic way by parameter variations.
- All the parameters within PICNIC which describe geometrical structure and physical properties are time-invariant. An accurate and efficient method to solve the system of transport equations utilises the method of Laplace transforms. First, linear response functions are obtained in the Laplace-domain and subsequently, a numerical inversion is performed with the help of Talbot's algorithm [3]. Hence, it is not possible to account for non-linear processes such as non-linear sorption or complex chemical reactions.
- Presently PICNIC is still being extensively tested, but it has already been used in a first application [4], based on the reference cases and corresponding data of NAGRA's Kristallin-I safety assessment [5]. Code verification is a very important issue and the verification for PICNIC is also based on the invariance properties [6] of the model. Even if certain transport parameters are scaled by an arbitrary factor A, according to the scheme below, the (numerical) solution of the transport problem has to be invariant.
- In the near future it is expected to provide an appropriate interface with existing groundwater flow models, such as NAPSAC [7] (a code currently used by NAGRA), together with a graphical user interface (GUI) for network definition and parameter input.

---

[1]   National Cooperative for the Disposal of Radioactive Waste, Switzerland.

q_f	D_f	ε_f	L	R_f	d	R_p	D_p	ε_p	δ_f
A		A						A	
A	A			A					A
		A	A	A^{-2}				A^{-1}	
					A		A^{2}		A^{-1}
						A	A		A^{-1}

Invariant matrix for PICNIC. Multiplication of some of the transport parameters by an arbitrary factor A, according to this scheme, does not change the solution of the transport problem. A blank field indicates a factor of 1.

q_f [m/s] is the specific discharge, D_f [m^2/s] is the dispersion coefficient, ε_f [-] is the flow porosity, L [m] is the migration distance, R_f [-] is the retardation factor in the fracture, d [m] is the maximum penetration depth for matrix diffusion, R_p [-] is the retardation factor in the matrix, D_p [m^2/s] is the diffusion coefficient, ε_p [-] is the porosity in the matrix, and δ_f [m^{-1}] is the surface to volume ratio of the water conducting feature.

References

[1]   Barten, W., Input File and Fundamentals for the First Phase of PICNIC, PSI-TM-44-95-01, Würenlingen und Villigen 1995

[2]   Barten, W., Linear response concept combining advection and limited matrix diffusion in a fracture network transport model, Water Resources Research, 1996, 32, 3285 - 3296.

[3]   Talbot, A., The Accurate Numerical Inversion of Laplace Transforms, Journal of the Institute of mathematics and its applications, 1979, 23, 97-120

[4]   Schneider, J., Zuidema, P., Smith, P., Gribi., P and Niemeyer, M., Recent developments in the safety assessment of a repository for high-level radioactive waste sited in the crystalline basement of northern Switzerland, International Conference on Deep Geological Disposal of Radioactive Waste, Winnipeg, Manitoba, Canada, September 16-19, 1996.

[5]   NAGRA, Kristallin-I Safety Assessment Report, NAGRA Technical Report NTB 93-22, Wettingen 1994.

[6]   Barten, W., PICNIC-I Test Cases: Fracture Case, PSI-TM-44-95-12, Würenlingen und Villigen 1996.

[7]   Grindrod, P., Herbert, A., Roberts, D. and Robinson, P., NAPSAC Technical Document, SKB Technical Report (Stripa Project) 91-31, Stockholm 1991.

Scale-Dependent Heterogeneities in the Gorleben Area and Implications for Model Studies

Rüdiger Ludwig, Klaus Schelkes
Bundesanstalt für Geowissenschaften und Rohstoffe, Hanover, Germany

Klaus-Jürgen Röhlig
Gesellschaft für Anlagen- und Reaktorsicherheit (GRS) mbH, Cologne, Germany

Jürgen Wollrath
Bundesamt für Strahlenschutz, Salzgitter, Germany

Abstract

The Gorleben salt dome is located near the community of Gorleben in the northeastern part of Lower Saxony. Hydrogeological investigations are being conducted in an area of more than 300 km^2 around the salt dome to study the aquifer system in the sediments above it. During the field investigations and data evaluation, the existence of heterogeneities on various scales (mm to km) has been verified. These heterogeneities predominantly reflect the differences in the lithology of the sediments. The spatial distribution of the hydrogeological properties of the sediments reflects that of the smallest sedimentological units (mm to m) up to lithofacies (tens of m to km). The density of the groundwater, which is a function of salt concentration, also varies spatially. The horizontal variability of the groundwater density is much less than its vertical variability, where density changes significantly within a few meters to tens of meters. The change in groundwater density has to be considered together with spatially variable hydrogeological parameters because it has considerable influence on groundwater movement and contaminant transport.

Spatial connectivity models based on information from borehole logs were developed using geostatistical methods. Different approaches for using soft information as input for these methods will be evaluated. Two-dimensional analyses of stratigraphic information were performed to identify general trends and anisotropy. Three-dimensional, porous medium, freshwater models represent the overall hydrogeological structure and neglect the smaller scale heterogeneities. These models are used to calculate regional groundwater movement and the characteristic trends of the main flow paths under freshwater conditions. Two- and three-dimensional, porous medium, salt/fresh water models are used to calculate changes in the flow field, density distribution, flow paths and travel times, and have demonstrated the sensitivity of the salt/fresh water system to major changes in the hydrogeological structure, as well as other important factors like initial density distribution and period of time being modelled. A code which is currently being developed will soon make it possible to perform more advanced three-dimensional sensitivity studies taking into account the spatial variability of the hydrogeological parameters together with the variability in groundwater density. Deterministic and probabilistic uncertainty and sensitivity analyses have been performed based on groundwater, flow paths, and nuclide transport calculations using a two-dimensional freshwater model. Stochastic methods are tested using distribution functions for hydrogeological parameters of different hydrostratigraphic units and for the presence or absence of low conductivity materials at specific locations.

1. Introduction

The Gorleben salt dome is located in the rural district of Lüchow-Dannenberg near the community of Gorleben in the northeastern part of Lower Saxony. It is about 14 km long and up to 4 km wide (Fig. 1) and extends from a depth of about 3500 m up to 260 m below the surface. The planned repository will be located at a depth of about 850 m, which is about 600 m below the top of the salt dome. Hydrogeological investigations are being conducted in an area of more than 300 km^2 around the salt dome to study the aquifer system in the sediments above the salt dome. This aquifer system consists mainly of slightly consolidated to unconsolidated Quaternary and Tertiary sediments, such as sand, silt, clay, boulder clay and gravel. The caprock at the top of the salt dome can be considered a karstic to fractured aquifer in which most of the fissures have been filled with clay and sand.

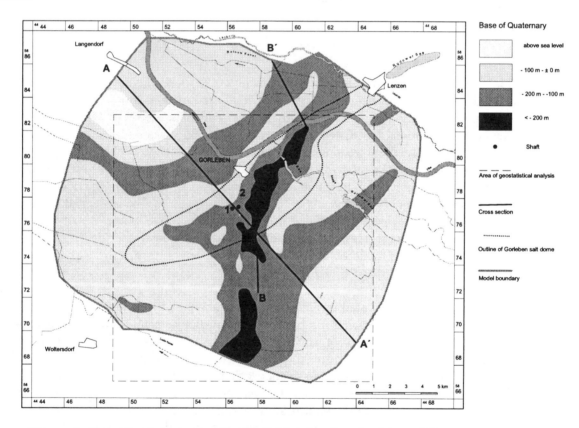

Figure 1. Investigation area at the Gorleben site showing base of Quaternary, model boundaries and positions of cross sections

2.   Heterogeneities in the Gorleben area

2.1   Hydrogeological heterogeneities

During the field investigations and data evaluation, the existence of hydrogeological heterogeneities on various scales (mm to km) has been verified. These heterogeneities predominantly reflect differences in the lithology of the sediments. The spatial distribution of the hydrogeological properties of the sediments reflects that of the smallest sedimentological units (mm to m) up to lithofacies (tens of m to km). The sedimentological sequence has been partly modified by glaciotectonics, permafrost and subrosion (underground solution).

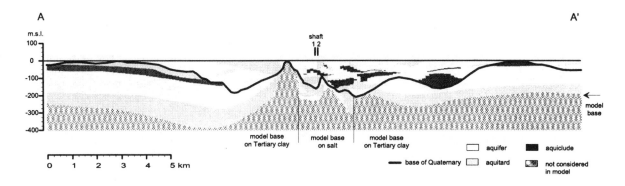

Figure 2. Cross section through the hydrogeological structure model

The hydrogeological cross section in Figure 2 gives an example of hydrogeological heterogeneities on a scale ranging from decameters to kilometers. The Gorleben site is characterised at this scale range by the predominance of aquifers with intercalated aquitards and aquicludes. Above the base of the Quaternary the hydrogeological units tend to be smaller than below.

Figure 3 demonstrates hydrogeological heterogeneities from millimeter to decameter size. The shaft wall map (Fig. 3d) shows hydrogeological "units" ranging in scale from a meter to several decameters [3]. It is possible to subdivide the hydrogeological heterogeneities into different types according to shape:

1.   Diffuse heterogeneities: Diffuse and partly interconnected material is intercalated in a matrix of other material. (Example 1 (Figs. 3c and 3d at -220 m) sand- to clay-filled solution cavities in the caprock; example 2 (Fig. 3d -85 to -120 m) sandy till intercalated in till).

2.   Stratification: Subhorizontal layers with different hydrogeological properties a few meters to a few decameters thick and a lateral extent of several decameters to a few kilometers (Figs. 2 and 3d). The extent of homogeneous bodies is related to the frequency and extent of facies changes. In the Quaternary homogeneous bodies tend to be smaller than in the Tertiary.

3. Lamination: Very rapid changes in hydrogeological properties. In the horizontal direction the extent of the laminations varies between decimeters and decameters and in the vertical direction between millimeters and decimeters. A typical example of this type of hydrogeological heterogeneity is cross-stratification (Fig. 3a, lower part).The original structure is frequently modified by post-dispositional processes (e.g. permafrost, diagenesis). These processes normally reduce the extent of hydrogeologically "homogeneous" areas (Fig. 3a top and center).

4. Heterogeneities caused by glaciotectonics or subrosion: These geological processes normally reduce the extent of hydrogeologically "homogeneous" areas to about a meter to 100 meters (Fig. 3d at -15 to -25 m).

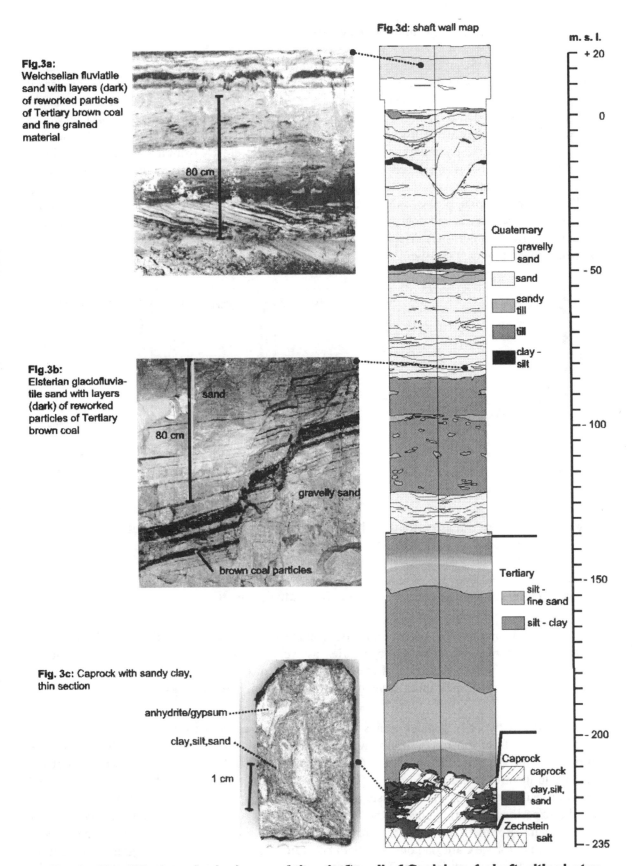

Fig.3a:
Weichselian fluviatile sand with layers (dark) of reworked particles of Tertiary brown coal and fine grained material

80 cm

Fig.3d: shaft wall map

m. s. l.
+ 20

0

- 50

Quaternary
gravelly sand
sand
sandy till
till
clay - silt

Fig.3b:
Elsterian glaciofluviatile sand with layers (dark) of reworked particles of Tertiary brown coal

sand
80 cm
gravelly sand
brown coal particles

- 100

Tertiary
silt - fine sand
silt - clay

- 150

Fig. 3c: Caprock with sandy clay, thin section

anhydrite/gypsum
clay,silt,sand
1 cm

- 200

Caprock
caprock
clay,silt, sand

Zechstein salt

- 235

Fig. 3. Simplified geological map of the shaft wall of Gorleben 1 shaft with photos

2.2　Spatial variability in groundwater density

The density of the groundwater, which is a function of salt concentration, also varies spatially. The horizontal variability in the density is much less than the vertical variability, where the density changes significantly within a few meters to tens of meters. Fig. 4 shows the density of water samples from the Gorleben subglacial channel (Fig. 1, profile B-B´) as a function of depth (Fig. 4A), as well as three examples of groundwater density logs derived from geophysical borehole logs (Fig. 4B, see also Fig. 7) [4]. Generally, the groundwater mineralization increases slowly from the groundwater table to a depth of 160-170 m below mean sea level. Below 170 m there is a very steep increase over a small transition zone to highly saline water. In some parts of the area, a relatively high salinity is observed at shallow depth. At some places, saline water overlies fresh water. These changes in groundwater density have to be considered together with spatially variable hydrogeological parameters (e.g., permeability, effective porosity, dispersivity) because of their significant influence on groundwater movement and contaminant transport.

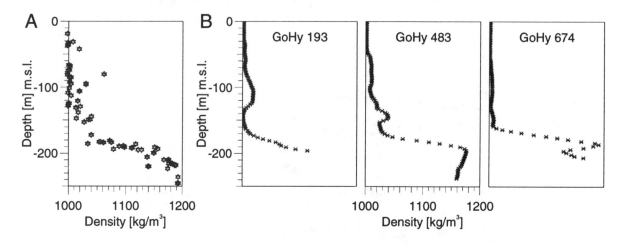

Figure 4. Groundwater density measured on water samples and three examples of groundwater density logs

3.　Spatial hydrogeological heterogeneity: Geostatistical methods

Ongoing work concerns uncertainties of groundwater flow and especially groundwater travel times arising from regional-scale heterogeneities as well as from a lack of knowledge about the hydrogeological structure. Geostatistical methods are used in order to establish spatial connectivity models for several petrographic units based on information from borehole logs.

In order to identify general trends and anisotropies, as well as a first step towards later consideration of soft information, initially two-dimensional analyses of stratigraphic information were performed. As an example, Figures 5 and 6 show results obtained for the depth of the base of the Quaternary in the area marked in Figure 1. The analyses were carried out for both (a) Cartesian and (b) transformed co-ordinates using a system following the shape of the erosion channel. Figure 5 shows the experimental variograms of the depth of the base of the Quaternary using the two co-ordinate systems. For the transformed co-ordinate system, the shape of the anisotropy is much better recognizable than for the Cartesian one. For both co-ordinate systems, several variogram models were used as a basis for ordinary kriging. The ordinary kriging results for the depth of the base of the

Quaternary shown in Figure 6 were chosen as typical examples representing a broad variety of results obtained using the different variogram models.

Of course, the most prominent differences between Figures 6a and 6b occur in areas with little borehole information, especially around the Gartow channel in the southeastern part. Generally, the ordinary kriging results for the transformed co-ordinate system show a higher continuity than those for the Cartesian system. This applies especially to the direction transverse to the erosion channel.

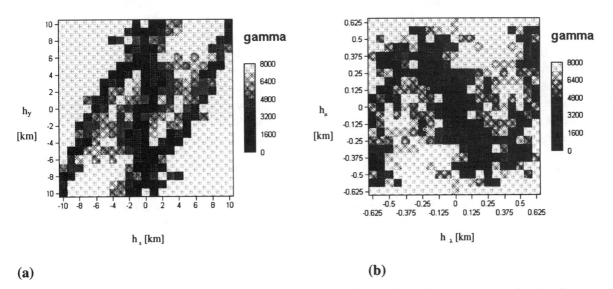

(a)　　　　　　　　　　　　　　　　　　　　(b)

Figure 5.　Experimental variograms of the depth of the base of the Quaternary using both Cartesian (a) and transformed (b) co-ordinates.

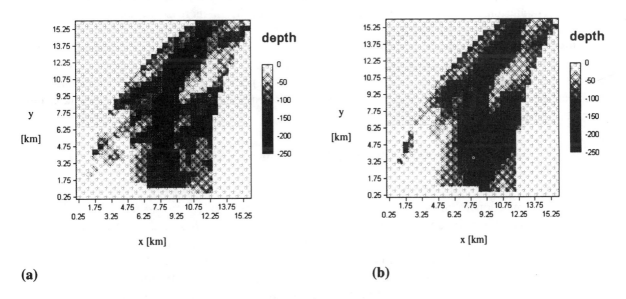

(a)　　　　　　　　　　　　　　　　　　　　(b)

Figure 6.　Examples for ordinary kriging results for the depth of the base of Quaternary [m] using (a) Cartesian and (b) transformed co-ordinates (the latter after inverse transformation).

Within the ongoing working programme, different approaches for the consideration of soft information will be evaluated. Using the results of stochastic simulations based on the connectivity models mentioned above, uncertainty analyses for the groundwater travel times will be performed.

The uncertainty analyses will be carried out using two-dimensional freshwater models in order to make a large number of simulation runs feasable. Nevertheless, these two-dimensional models will be based on a fully three-dimensional geostatistical analysis.

4. Three-dimensional, porous-medium, freshwater models (large scale: geology)

Three-dimensional, porous-medium, freshwater models represent the overall hydrogeological structures (scale: tens of m vertically and hundreds of m horizontally) and neglect the smaller scale heterogeneities. These models are used to calculate the regional groundwater movement and the characteristic trends of the main flow paths under freshwater conditions. Starting with an approach based on expert judgement of the overall hydrogeological structure, and using mean values of the hydrogeological parameters, different model approaches based on our increasing knowledge of the hydrogeology have been investigated over the past years. A large number of sensitivity calculations were carried out to determine the sensitivity of the model to changes in the permeability distribution and in the details of the hydrogeological structures. The most sophisticated model is based on a 100 m•100 m grid in the horizontal direction and grid spacing of metres to tens of metres vertically. The models were calibrated by checking the differences between observed and calculated hydraulic heads in that part of the model where fresh or slightly saline water is present and by comparing recharge and discharge across the upper model boundary with results of measurements.

The results of the model calculations demonstrate the influence of changes on flow-path distribution and provide a good understanding of the flow system in that part of the groundwater system where fresh or slightly saline water is present [2]. Furthermore, the effect of the caprock on the flow field was determined. The variation in hydraulic conductivity and anisotropy of different hydrostratigraphic units in the sensitivity calculations results in an increase in anisotropy of those hydrostratigraphic units where the small scale vertical heterogeneity is not represented in the model. The clustering of increased positive or negative differences in observed and calculated hydraulic heads in distinct parts of the model (e.g. at the southern end of the Gorleben subglacial channel) indicate that these differences cannot be reduced because the geometry of the hydrostratigraphic units cannot be sufficiently resolved and that therefore smaller scale heterogeneities should be taken into account in those parts of the model.

5. Two-dimensional, porous-medium, salt/fresh water models (large to intermediate scale: geology and density)

Two-dimensional, porous-medium, salt/fresh water models as cross-section models of a distinct part of the three-dimensional model initially provide only a simplified representation of the hydrogeology (scale: tens of m vertically and hundreds of m horizontally) and a more realistic representation of the hydrogeology (scale: m to tens of m vertically and tens of m to hundreds of m horizontally) during later modeling phases. Changes in the flow field, density distribution, flow paths and travel times demonstrated the sensitivity of the salt/fresh water system to major changes in the hydrogeological structure, as well as other important factors like the initial density distribution and modelling time. [5,6,7].

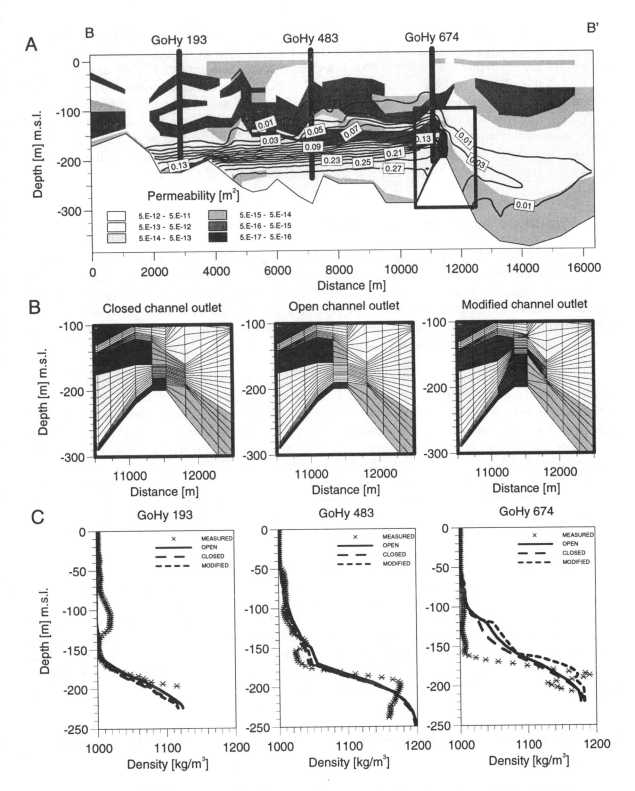

Fig. 7: A) Concentration isolines after 10000 years for the modified
           channel outlet
        B) Different shapes of the channel outlet
        C) Comparison of measured (✽) and calculated (−) density
           logs for the open, closed and modified channel outlet.

311

As one example, Figure 7A shows a realistic representation of the hydrogeological structure of the Gorleben subglacial channel on profile B-B′ of Figure 1. One of the important heterogeneities which influence the flow field on this section is a change in the structure of the outlet of the channel. The area of this channel outlet is marked in Figure 7A, the different configurations in the shape and the size of the channel outlet are zoomed in Figure 7B. As one of the model results, Figure 7A shows the density distribution of one of the model variations (modified outlet) for a specially assumed initial density distribution after 10,000 years of model time. Figure 7C gives a comparison of calculated density distributions for the same initial density distribution and model time with the measured density logs (see Fig. 4) which were used as a measure of the quality of the fit.[6]

6. Spatial hydrogeological heterogeneity: Stochastic methods

One of the goals of the European Community project EVEREST [1] was to demonstrate how deterministic and probabilistic techniques are used for uncertainty and sensitivity analyses in the performance assessment. Deterministic and probabilistic calculations have been carried out for several scenarios. Near-field, geosphere, and biosphere calculations have been performed, each of them based on the nuclide releases calculated for the preceeding compartment. Both parameter and model uncertainties were taken into account.

A „parameter variation and discrete zones" approach was used in the groundwater and nuclide transport calculations, which were performed in order to evaluate the consequences resulting from parameter uncertainties. The analyses were carried out using two-dimensional freshwater models in order to make a large number of simulation runs feasable. Figure 8a shows flow paths calculated for the reference case (groundwater travel times from 2500 to 3200 years). Several paths with travel times between 650 and 91,000 years were calculated for several Monte Carlo realizations of the parameter vector, most of them similar to the paths shown on Figure 8a. Besides several near-field parameters, the sorption coefficients were identified as sensitive to the release of highly sorbing nuclides from the geosphere into the biosphere. For less sorbing nuclides, in a few cases high sensitivity measures were calculated for the conductivity values.

Additionally, uncertainties about the lithology arising at specific locations in the northern part of the channel (right-hand part of Fig. 8) were considered using deterministic and probabilistic approaches. For the probabilistic calculations, randomly chosen models were combined with Monte-Carlo simulations of the parameter vector. Several paths with travel times between 650 and 770,000 years have been calculated. Figure 8b shows the flow paths calculated for one simulation. Again, sorption values have been calculated as the geosphere parameters most sensitive to the release of highly sorbing nuclides. Neither the conductivity values nor the parameters characterizing the choice of the model showed remarkable high sensitivity measures.

7. Three-dimensional, porous-medium, salt/fresh water models (large scale: geology and density)

A simplified three-dimensional, porous-medium, salt/fresh water model of a distinct part (the Gorleben erosion channel, see Fig.1) of the three-dimensional freshwater model represents on a large scale only the main hydrogeological structures (scale: tens of m vertically and hundreds of m horizontally). This is the first attempt to couple flow and salt transport calculations in three dimensions. The results will demonstrate connections between the large scale spatial heterogeneities of the hydrogeological parameters and the spatially variable density distribution, as well as the influence of both on groundwater movement and contaminant transport in three dimensions.

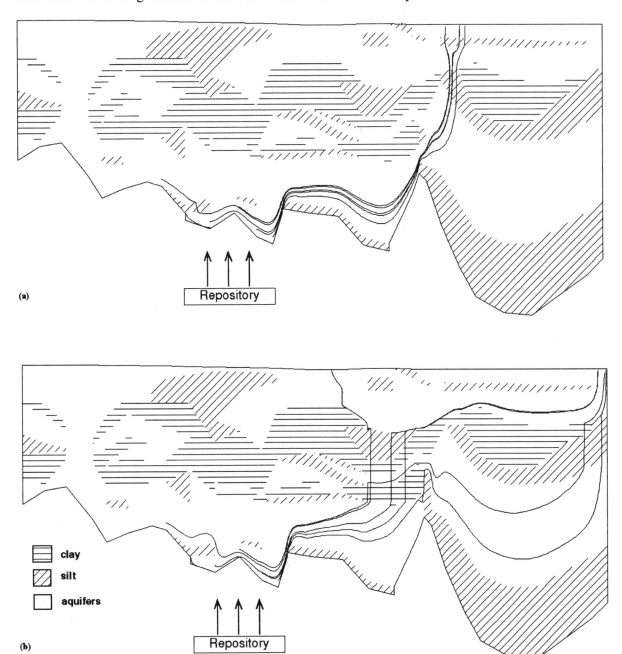

Figure 8. Calculated flow paths for (a) the reference case and (b) one Monte Carlo simulation

More advanced three-dimensional, porous-medium, salt/fresh water models (scale: m to tens of m vertically and tens of m to hundreds of m horizontally) will be used as soon as new model code development has been finished. This new code will make it possible to perform three-dimensional sensitivity studies taking into account the spatial variability of the hydrogeological parameters

313

together with the variability in groundwater density using a more realistic three-dimensional model. Furthermore, dispersion effects arising from heterogeneities on various scales will be treated using a combined Fickian and stochastic dispersion model.

8.   Summary and conclusions

Field investigations and data evaluation succeeded in verifying the existence of spatial heterogeneities of the hydrogeological parameters at various scales (mm to km), as well as the heterogeneity in the density distribution, in the Gorleben area. The influence of these heterogeneities on groundwater flow and contaminant transport is being investigated in a step-by-step study including geostatistical methods, calculation of groundwater movement in three dimensions and coupled groundwater movement and salt transport in two and three dimensions using deterministic as well as stochastic methods. All these investigations increased our understanding of the system behaviour substantially and demonstrated the influence of heterogeneities on groundwater movement and contaminant transport. We are confident that use of different model concepts and numerous different calculations will enable us to understand the system behaviour and that it will provide a sound basis for safety analyses for a radioactive waste repository in the Gorleben salt dome.

9.   Acknowledgements

This work was carried out in research projects which were financed by the Federal Ministry for the Environment, Nature Conservation and Reactor Safety (BMU) and the Federal Ministry for Education, Science, Research and Technology (BMBF) as well as in the site investigation program, which is being conducted by the Federal Office for Radiation Protection. Many scientists have worked in these investigations. The authors wish to thank the official organizations for their support in this research and our colleagues in the different organizations for permission to use their data and results for this publication. The recent three-dimensional freshwater and the preliminary three-dimensional salt/fresh water calculations were performed by Colenco Power Engineering Ltd., Baden, Switzerland on behalf of the Federal Office for Radiation Protection. The authors wish to thank M. Genter, W. Klemenz, A. Rivera and M. Schindler of Colenco for their good co-operation.

10.   References

[1]   Becker, A.; Fischer, H.; Hofer, E.; Kloos, M.; Krzykacz, B.; Martens, K.-H.; Röhig, K.-J.: EValuation of Elements Responsible for the effective Engaged dose rates associated with the final STorage of radioactive waste: EVEREST project. Volume 3a: Salt formation, site in Germany. 1997, European commission, nuclear science and technology, Contract No. FI2W-CT90-0017, Final report, EUR17449/3a EN, 429 p.

[2]   Genter, M.; Klemenz, W.; Ludwig, R.; Wollrath, J.: Simulation of Steady State Groundwater Flow in the Gorleben Area, Germany. Proc. 21st Int. Symp. Scientific Basis for Nuclear Waste Management MRS '97, Davos, Sept. 28. to Oct. 3. 1997.

[3]   Ludwig, R.R.: Die Quartärprofile der Schächte Gorleben 1 und 2. Zeitschrift für angewandte Geologie, 1993, Band 39, Heft 2, 78-83.

[4]   Ochmann, N. and Fielitz, K.: Estimation of horizontal and vertical groundwater flow from well-logging and pressure data in groundwater of variable density above a salt dome. In: Study and Modelling of Salt Water Intrusion into Aquifers (ed. by E Custodio and A. Galofre), 1993,CIMNE, 23-34.

[5]   Schelkes, K. and Vogel, P.: Paleohydrogeological information as an important tool for groundwater modeling of the Gorleben site. In: Paleohydrogeological Methods and their applications, 1993, OECD, 237-250.

[6]   Vogel, P. and Schelkes, K.: Influence of initial conditions and hydrogeological setting on variable density flow in an aquifer above a salt dome. In: Calibration and Reliability in Groundwater Modelling, 1996, IAHS Publ. no. 237, 373-381.

[7]   Vogel, P., Schelkes, K., Giesel, W.: Modeling of variable-density flow in an aquifer crossing a salt dome – first results. In: Study and Modelling of Salt Water Intrusion into Aquifers (ed. by E Custodio and A. Galofre), 1993,CIMNE, 359-369.

A Case Study on the Influence of Sorption Inhomogeneities on the Migration of Contaminants

L. Lührmann, E. Fein

Gesellschaft für Anlagen- und Reaktorsicherheit (mbH), Germany

P. Knabner

Institut für Angewandte Mathematik, Universität Erlangen, Germany

Abstract

A two-dimensional numerical test case study was performed in order to investigate the influence of sorption inhomogeneities on the migration behaviour of contaminants. The sorption processes are modelled with a nonlinear kinetic-Freundlich reaction. Thereby a kinetic approach of first order is supposed. In a first test case homogeneous sorption parameters are used, and in a second case the transport path is divided into regions of low and high sorption. For the latter the numerical solution requires a finer discretization in the region of the sorption inhomogeneities. The nonlinearities caused by the kinetic-Freundlich reaction are resolved using Newton's method. Besides the well known retardation of the contaminants caused by sorption processes the simulations yield the following results: The distribution of the concentration of the sorbed contaminants is discontinuous in the region of the sorption inhomogeneities. The nonlinear behaviour of the kinetic-Freundlich reaction counteracts the spreading effects of the contaminant caused by the diffusive and dispersive transport.

Introduction

In this paper we perform a two-dimensional test case in order to investigate the influence of sorption inhomogenities on the migration behaviour of contaminants. A second purpose of this exercise is to verify our transport code TRAPIC. For that, a test case presented by *Knabner* [4] is chosen in order to compare his numerical results calculated by the finite element code SOTRA with the results of our finite volume code TRAPIC.

The solute transport through porous media with kinetically controlled sorption processes can be described by the equations (e.g. [2 , 4]):

$$\frac{\partial}{\partial t}(\Theta C)(\vec{x}, t) + \rho(\vec{x})\frac{\partial}{\partial t}S(\vec{x}, t) = \nabla \cdot [-(\vec{q}C) + \Theta \mathbf{D}\nabla C](\vec{x}, t),$$

(1)

$$\frac{\partial}{\partial t}S(\vec{x}, t) = k[f(\vec{x}, t, C(\vec{x}, t)) - S(\vec{x}, t)]$$

(2)

with the solute concentration C [M/ L^3], the sorbed concentration S [M/M], the sum of the molecular diffusion and mechanical dispersion $\mathbf{D}$ [L^2/T], the Darcy-velocity $\vec{q}$ [L/T], the porosity Θ [-] and the bulk density ρ [M/ L^3]. The parameters t [T] and $\vec{x} = (x_1, \dots, x_N)$ [L] denote time and space, respectively. Source and sink terms are neglected in the equations (1) and (2). The kinetic rate constant k describes the kinetics of the sorption. An equilibrium sorption is formally obtained by setting the parameter $k = \infty$. In equilibrium state the sorbed concentration S is given by the adsorption isotherm f [M/M]:

$$S(\vec{x}, t) = f(\vec{x}, t, C(\vec{x}, t)).$$

(3)

Typically used adsorption isotherms are the Freundlich- and Henry-isotherms, respectively, which are of the form:

$$f(.,.,C) = K_{nl}C^p \qquad \text{Freundlich,}$$
$$f(.,.,C) = K_d C \qquad \text{Henry.}$$

(4)

Here, the unit of the term C^p is determined by [M/ L^3], which leads to the unit [L^3/M] of the Freundlich constant K_{nl}. The parameter p [-] with $0 < p < 1$ denotes the Freundlich-exponent and K_d [L^3/M] the well known sorption distribution coefficient of the Henry-isotherm. By application of a kinetic-Freundlich reaction the ordinary differential equation (2) becomes nonlinear, leading to a nonlinearly coupled system of the differential equations (1) and (2).

Numerical solution

The transport code TRAPIC is implemented as a module in the program package ug [1]. The ug library is a set of very general data structures and procedures to support the numerical solution of partial differential equations. Main emphasis has been put on re-usability of the different software components for a variety of discretization and solution procedures.

Assume a spatial discretization of the underlying domain with nodes $\{\vec{P}_i\}_{i=1,\dots,n}$, and a time stepping $t^0 = 0$, $t^{m+1} = \Delta t + t^m$. Additionally, assume that the approximation of the concentrations $\vec{C}^m, \vec{S}^m$ defined by $\vec{C}^m = [C(\vec{P}_i, t^m)]_{i=1,\dots,n}$ and $\vec{S}^m = [S(\vec{P}_i, t^m)]_{i=1,\dots,n}$ are known at the time level $t = t^m$, which is true for t^0 by evaluation of the initial data. Applying a spatial discretization given by an upwind finite volume method (e.g. [6]) and taking the values of the concentrations implicit in time, then following the work of *Knabner* [4] the set of discrete equations for the values $\vec{C}^{m+1}$ leads to the form:

$$(\mathbf{M}^{m+1} - \Delta t \mathbf{K}^{m+1}) \vec{C}^{m+1} = \mathbf{M}^m \vec{C}^m + \vec{b}^{m+1}$$
$$+ \mathbf{M}_\rho \frac{1}{1/(\Delta t k) + 1} [\vec{S}^m - \vec{f}^{m+1}(\vec{C}^{m+1})].$$

$$(5)$$

Here, $\mathbf{M}^m$ and $\mathbf{K}^m$ are the mass and stiffness matrices at the time level $t = t^m$, respectively, M_ρ is the mass matrix depending on the bulk density ρ. The ith component of the vector $\vec{f}^m$ is given by $[\vec{f}^m(\vec{C}^m)]_i = f(\vec{P}_i, t^m, C(\vec{P}_i, t^m))$. Finally, $\vec{b}^{m+1}$ contains the contribution from a flux boundary condition and possibly an additional source term.

After $\vec{C}^{m+1}$ has been computed by solution of (5) the approximate sorbed concentration $\vec{S}^{m+1}$ can be obtained by the evaluation of the differential equation (2) to the discrete problem:

$$\vec{S}^{m+1} = \frac{1}{1/(\Delta t k) + 1} \vec{f}^{m+1}(\vec{C}^{m+1}) + \frac{1/(\Delta t k)}{1/(\Delta t k) + 1} \vec{S}^m .$$

$$(6)$$

The use of a node-oriented quadrature rule in the evaluation of the mass matrices leads to "mass lumping", that is, a diagonal form of $\mathbf{M}^m$ and $\mathbf{M}_\rho$. Then the set of nonlinear equations (5) has the following structure:

$$\mathbf{A}^{m+1} \vec{C}^{m+1} = \vec{b}^{m+1} + \vec{g}^{m+1}(\vec{C}^{m+1}),$$

$$(7)$$

where the nonlinearity is such that its ith component only depends on the ith component of the unknown $\vec{C}^{m+1}$. For solving the nonlinear equation (7) an inexact Newton method [3] is implemented in the code TRAPIC.

Test case study

For our numerical test cases the transport is considered in the region Ω and based on a simple flow field with

$$\Omega \;=\; (0, L_1) \times (0, L_2),$$

$$\vec{q} \;=\; (q_1, q_2) \;=\; \text{constant} \;\; \text{with} \; q_2 \;=\; 0.$$
$$(8)$$

The boundary conditions are given as follows: For $x_2 = 0$ and $x_2 = L_2$ a no flow boundary is assumed, the outflow boundary for $x_1 = L_1$ is defined by the assumption of no dispersive/diffusive flux across this boundary. The mass inflow in the normal direction at the inflow boundary ($x_1 = 0$) is given by:

$$F_{\text{in}}(x_2, t) \;=\; q_1 \begin{cases} C_f & x_2 \in [H_1, H_2], t \in [0, T_1], \\ 0 & \text{otherwise.} \end{cases}$$
$$(9)$$

For the initial conditions no concentrations of sorbed and dissolved contaminants are assumed:

$$C_0(\vec{x}) \;=\; S_0(\vec{x}) \;=\; 0, \qquad \vec{x} \in \Omega.$$
$$(10)$$

The concentrations and sorption parameters have the dimensions:

$$C : [100 \, \mu g \, / \, cm^3], \qquad S : [100 \, \mu g \, / \, g],$$
$$K_d : [cm^3 \, / \, g], \qquad K_{nl} : [cm^3 \, / \, g].$$
$$(11)$$

The matrix of the diffusion/dispersion term is taken in the form of Scheidegger's approach:

$$\Theta \mathbf{D} \;=\; \Theta D_m \begin{bmatrix} 1 & 0 \\ 0 & 1 \end{bmatrix} + q_1 \begin{bmatrix} \alpha_L & 0 \\ 0 & \alpha_T \end{bmatrix}$$
$$(12)$$

320

with the constant coefficient of molecular diffusion D_m and constant longitudinal and transversal dispersion lengths α_L and α_T .

The parameters, which are the same in all simulations are listed in table 1. The concentrations will be displayed as concentration profiles. The values of the profiles for the dissolved and sorbed concentrations increase exponentially each by a factor of two starting from $C = 10^{-3}$ up to $C = 0.512$ and from $S = 5.33 \cdot 10^{-3}$ up to $S = 5.46$, respectively.

symbol	value	unit
α_L	0.595	[cm]
α_T	0.0595	[cm]
C_f	1.0	$[100\ \mu g\ /\ cm^3]$
D_m	0.025	$[\,cm^2\ /\ d\,]$
(H_1, H_2)	(1.875, 3.125)	[cm]
k	0.5	$[1\ /\ d\,]$
L_1	10.0	[cm]
L_2	5.0	[cm]
Θ	0.5	[-]
$\vec{q}$	(2.5,0.0)	$[\,cm\ /\ d\,]$
ρ	1.5	$[\,g\ /\ cm^3]$
T_1	8.0	[d]

Table 1: Common parameters used in all test cases

Homogeneous sorption parameters

In the test case 1a the sorption process is modelled by a kinetic-Henry reaction with homogeneous sorption parameter setting by:

$$K_d = K_d(\vec{x}) = 2 \ .$$

$$(13)$$

Figure 1 shows the distribution of the dissolved concentration. The spreading of the concentrations leads to a positive concentration in the entire domain for small times, too. This spreading is caused by the diffusive and dispersive flow. The direction of the flow field and the ellipsoidal shape of the contour lines show the dominance of the advection. Additionally, the differences between the transversal and longitudinal dispersion intensify these effects.

In the test case 1b a kinetic-Freundlich reaction with homogeneous sorption parameters is used, where

$$K_{nl} = K_{nl}(\vec{x}) = 1 \qquad [\,\vec{x} \in \overline{\Omega}\,]$$

(14)

and $p = 0.5$. As the Freundlich-isotherm is not continuously differentiable at $C = 0$, the Newton method cannot be applied directly. Therefore, we use a regularisation of the Freundlich-isotherm given by:

$$f_\varepsilon(\vec{x},t,C) = \begin{cases} f(\vec{x},t,C) & C \geq \varepsilon, \\ [f(\vec{x},t,\varepsilon)/\varepsilon]C & \text{otherwise.} \end{cases}$$

(15)

Figure 2 shows the spatial distribution of the dissolved concentration at $t = 1\,[d]$. In contrast to the first test case the speed of the propagation is finite. This leads to a concentration front which forms the boundary of a downstream region with zero concentration. The effect of finite speed of propagation is theoretically investigated for transport problems in one-dimension by *van Duijn* and *Knabner* [5].

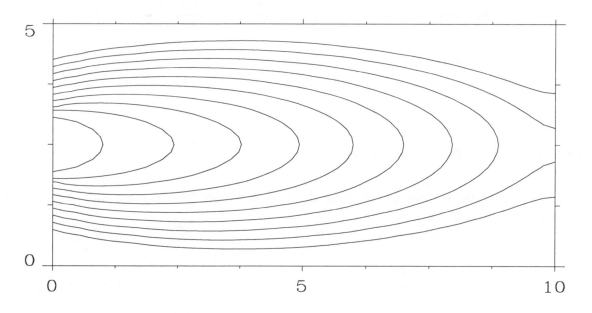

Figure 1: Homogeneous, kinetic-Henry reaction: dissolved phase $t = 1$

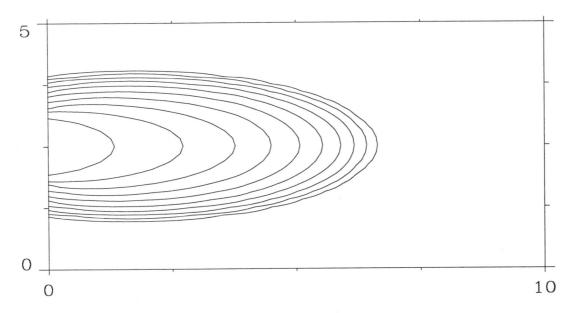

Figure 2: Homogeneous, kinetic-Freundlich reaction: dissolved phase $t = 1$

Inhomogenous sorption parameters

For the second test case a kinetic-Freundlich reaction with spatially inhomogeneous sorption parameters is used, where

$$K_{nl}(\vec{x}) = \begin{cases} 10 & x_1 \in [4, 6], \\ 1 & \text{otherwise} \end{cases}$$

$$(16)$$

and $p = 0.5$.

Figures 3 to 6 show the dissolved and sorbed concentrations for $t = 20$ and 40 [d]. The concentration profiles obtained with the code TRAPIC agree in a sufficient way with the simulations performed by *Knabner* [4] with the finite element code SOTRA.

The layer of strong adsorption acts as an obstacle for the mass flow. Correspondingly the maximum of the sorbed concentration is located there and discontinuities will be obtained at the boundaries between the regions with different sorption properties. After the cut-off of the mass inflow ($t = 8$) the layer of high adsorbed concentrations acts as source for the dissolved concentration resulting from the desorption process.

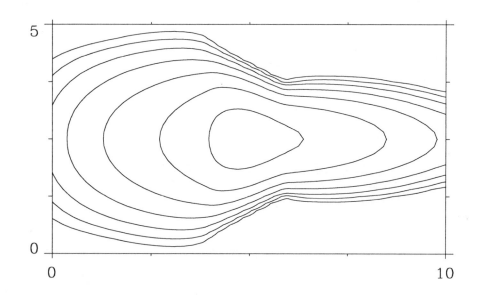

Figure 3: Inhomogeneous sorption: dissolved phase $t = 20$

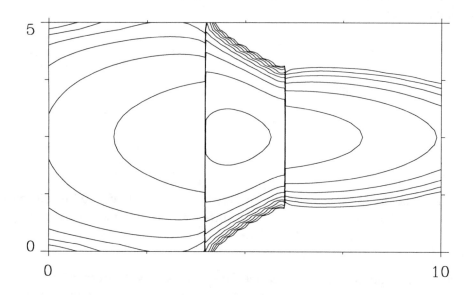

Figure 4: Inhomogeneous adsorption: sorbed phase $t = 20$

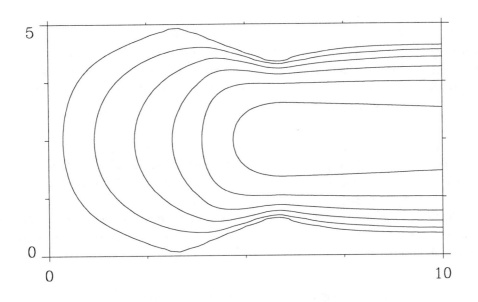

Figure 5: Inhomogeneous sorption: dissolved phase $t = 40$

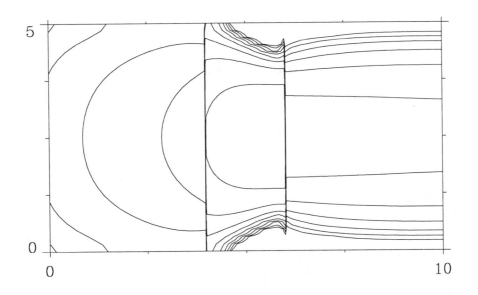

Figure 6: Inhomogeneous adsorption: sorbed phase $t = 40$

Results and Conclusions

We performed a two-dimensional numerical test case study in order to investigate the influence of sorption inhomogeneities on the migration behaviour of contaminants. The sorption processes are modelled with a linear kinetic-Henry and nonlinear kinetic-Freundlich reaction, respectively. Thereby a kinetic approach of first order is supposed. The nonlinearities caused by the kinetic-Freundlich reaction are resolved using inexact Newton's method.

For the first test case homogeneous sorption parameters are used. Besides the well known retardation of the contaminants caused by sorption processes the simulations yield the following results: The nonlinear behaviour of the kinetic-Freundlich reaction counteracts the spreading effects of the contaminant caused by the diffusive and dispersive transport. This leads to a finite speed of propagation, that is, a concentration front which forms a downstream region with zero concentration. In contrast to this the use of a kinetic-Henry reaction leads to positive concentration in the whole domain also for small times.

For the second case the transport path is divided into regions of low and high sorption. The sorption processes are modelled by a kinetic-Freundlich reaction. The distribution of the concentration of the sorbed contaminants is discontinuous in the region of the sorption inhomogeneities. For this the numerical solution requires a finer discretization in the region of the sorption inhomogeneities.

References

[1]   P. Bastian: ug 2.0, Ein Programmbaukasten zur effizienten Lösung von Stömungsproblemen, Preprint IWR, IWR Universität Heidelberg, 1992.

[2]   J. Bear, Y. Bachmat, Introduction to modeling of transport phenomena in porous media. Kluwer Academic Publishers, Dordrecht, 1991.

[3]   C. T. Kelley, Iterative Methods for Linear and Nonlinear Equations. Frontiers in Applied Mathematics, SIAM, 1995.

[4]   P. Knabner, Finite element approximation of solute transport in porous media with general adsorption processes, in Xiao, S. T. (ed.), Flow and transport in porous media, World Scientific Publishing, Singapore, 1992.

[5]   C. J. Van Duijn, P. Knabner, Travelling waves in the transport of reactive solutes through porous media: Adsorption and binary ion exchange, Part 1. Transp. Porous Media, 1992, 8, 167-194.

[6]   C. B. Vreugdenhil and B. Koren, Eds., Numerical Methods for Advection – Diffusion Problems, Notes on Numerisal Fluid Mechanics, Vol. 45, Vieweg, 1993.

Single Well Harmonic Pulse Testing – A Novel Technique for In-Situ Measurement of the Hydraulic Heterogeneity of a Fracture

A.P.S. Selvadurai
McGill University, Canada

P.A. Flavelle
Atomic Energy Control Board, Canada

Abstract

This paper describes and demonstrates a novel hydraulic testing procedure that is intended to interrogate successively larger regions of rock fracture around a single borehole. It is postulated that the radius of influence of a sinusoidal injection/withdrawal test will be controlled by the amplitude and frequency of the applied flow rate. Hence, the test parameters can be varied between tests to increase the radius of influence of the tests. Changes in hydrogeologic characteristics inferred from the tests would provide a direct measure of the possible effects of radial heterogeneity of the flow system.

An existing test facility developed to investigate coupled thermal-hydrological-mechanical responses of fractured and unfractured large granite samples has been modified to apply a controlled sinusoidal flow rate to a sample containing a fracture. Pressure in the internal cavity of the granitic cylinder intersecting the fracture was recorded for various sinusoidal flow amplitudes and frequencies. The low flow / high frequency tests required by the sample size exhibited significant instrument noise, requiring correction.

The present paper discusses trends observed in preliminary experiments. The monitored pressure in the sample cavity, however, compares well with numerical calculations using a finite difference simulator. This simulator will be used to inverse model the tests to determine the radius of influence, and to investigate the effects of fracture heterogeneity for future tests. As a methodology for verifying the validity of the overall approach, a formal analytical solution for the flow equation with radial symmetry has been derived to within an inversion of a Laplace transform containing Bessel functions.

Introduction

The in situ characterization of fracture flow properties is of vital importance for making performance predictions that will be used to site and design a bedrock disposal facility for radioactive waste. The conventional techniques of borehole packer testing involve either steady flow rates, transient pressure pulses, rate of loss of tracer from a test well or breakthrough curves of tracers at a monitoring well. These tests are useful to measure average flow characteristics around a borehole in single-well tests and between boreholes in multiple-well tests.

Heterogeneity of the hydrogeologic characteristics of fractures is important in predictions of groundwater flow that are used in support of safety arguments. Measurements of spatial heterogeneity of flow characteristics for licensing a disposal site could be problematic with current testing methodologies. Typically, flow characteristics are assumed to be uniform and homogeneous over the tested volume (within the radius of influence in single-well tests, and between pairs of wells in multiple-well tests). Detailed measurements of heterogeneity at a potential disposal site would require the installation of a dense array of boreholes, which themselves could impair the integrity of the geosphere barrier. Consequently, alternate techniques to measure the spatial characteristics of the hydraulic conductivity of fractures are considered to be of practical value.

Early in the 1980s a cross-hole sinusoidal testing technique was developed and demonstrated (Black and Kipp [1]; Black and Barker [2]). This technique uses the amplitude attenuation and phase shift of a sinusoidal pressure perturbation between an activation well a monitoring well. Its use has been demonstrated in the Stripa Project (Black and Holmes [3]; Black et al [4]), and more recently by Motojima et al [5] in connection with the study of bedrock. Theoretical development of the cross-hole sinusoidal testing technique has been included in the generalized radial flow models of Barker [6] and Geier et al [7]. This technique, however, also requires the use of monitoring wells, and so suffers from the same need for assuming homogeneity between the wells.

We propose applying the sinusoidal testing technique to a single well but with varying the amplitude and the frequency of the injection/withdrawal flow rate [8]. This would activate only regions of the fracture consistent with the frequency of the pressure pulses. With such a technique, it is anticipated that high frequency pulses should hydraulically excite only regions close to the borehole and low frequency pulses are expected to hydraulically excite larger regions of the fracture. This should allow direct measurement of the radial heterogeneity of a fracture's flow characteristics. In addition to the potential to measure radial heterogeneity from a single well, monitoring wells would allow measurement of directional heterogeneity.

Models

We adopt the simplest case of a borehole intersecting a fracture at right angles for our conceptual model. The test procedure comprises a sinusoidal injection/withdrawal in the borehole, so that there is no net water movement. Flow into and out of the borehole through the fracture will attenuate and delay the sinusoidal pressure in the borehole, compared to the pressure response if there was no fracture present.

The flow loss (gain) from (to) the borehole will depend on the pressure increase (decrease) in the borehole and the length of time the system has to respond to the perturbation. The amplitude of the pressure perturbation is governed by the amplitude of the applied flow rate (Q_0), and the response time is controlled by the radial frequency (w) of the applied flow. The radius of influence of the test is a function of the amplitude and frequency of the injection/withdrawal and the transmissivity (T_f) and storativity (S_f) of that part of the fracture system within the radius of influence.

Invoking axial symmetry, the flow equation can be solved in radial coordinates by the application of a Laplace transform technique. For an applied flow rate of $Q_0 \sin(wt)$ and assuming the pressure at $r=\infty$ is always zero, the pressure in the fracture intersecting a borehole of radius r_a is given by:

$$p(r,t) = \frac{Q_0 \, \mu_w \, \omega \sqrt{T_f}}{2 \, \pi \, (2b) \, r_a \, K_f \sqrt{S_f}} L^{-1} \left\{ \frac{K_0 \left(r \frac{\sqrt{sS_f}}{\sqrt{T_f}} \right)}{\sqrt{s} \, (s^2 + \omega^2) \, K_1 \left(r_a \frac{\sqrt{sS_f}}{\sqrt{T_f}} \right)} \right\}$$

where m_w is the viscosity of water, $(2b)$ is the hydraulic aperture of the fracture, K_f is the fracture permeability, L^{-1} is the inverse Laplace transform, s is the Laplace variable and K_0, K_1 are the zeroth and first order modified Bessel functions of the second kind. This equation can be solved only by numerical inversion.

To determine the hydraulic properties of the fracture within the radius of influence of the test, the equation has to be solved twice: once at $r = r_a$ to give the borehole pressure history (leading to the evaluation of T_f and S_f), and a second time at $p(r,t) \approx 0$, to find r at which the amplitude of the pressure perturbation falls below some measurement limit (to evaluate the test's radius of influence over which T_f and S_f are assumed homogeneous). Hence, the problem of hydraulic heterogeneity can be posed as an inverse problem.

The objective of the experimental testing is to produce the data needed to demonstrate the frequency and amplitude dependence of single well sinusoidal testing of a single (assumed homogeneous) fracture. Testing of heterogeneity effects is reserved for future developments. Initial derivation of the data analysis procedure, based on simplifications of the Bessel functions and convolution theory are approximations which can be further improved.

A finite difference simulator of the conceptual model has been constructed using an object-oriented programming package. Numerical experiments were performed to simulate the experimental procedures. The simulator allows a broader range of interrogation of the system by means of numerical experiments than is possible with the physical samples. Heterogeneity within the fracture can be simulated, to explore the responses that may be observed in field tests.

Laboratory Experimental Investigations

A test facility designed and constructed to examine the coupled thermal-hydrological-mechanical response of fractured and unfractured large granite samples [9] was modified to demonstrate the test procedure (Figure 1). A cylindrical sample of Barre granite (approximately 50 cm long and 47 cm diameter) with a 5.7 cm axial cavity and a single fracture normal to the axis of symmetry was subjected to isothermal testing. Axial stress normal to the plane of the fracture was used to control its aperture, which was measured using a steady-state injection test.

329

Water flow was provided for the harmonic pulse tests by computer control of a constant-rate withdrawal pump and a sinusoidal-rate injection pump. Amplitude and frequency of the sinusoidal flow rate were varied, and the flow rate and cavity pressure were monitored. Interpretation of the measurements to estimate the hydraulic diffusivity was based on the cavity pressure amplitude as a function of injection rate amplitude and frequency.

The test procedure was conducted first on an unfractured rock sample to determine the equipment and background (cavity and rock matrix) response. Then a fractured sample was tested under various values of confining stress to provide a range of fracture hydraulic diffusivity. The testing also included constant rate injection to determine the steady-state hydraulic parameters of the fracture.

The size of the fractured sample required low flow / high frequency tests to maintain the radius of influence within the sample. Flow amplitudes varying from 0.5 to 3.3 ml/min and frequencies from 1 to 4 cycles/min were used. In these ranges, the pumps' noise interfered with the measurements, and correction and smoothing of the data were necessary. An example of the corrected, smoothed data from one test is shown in Figure 2. These same data, normalized for flow amplitude, are shown in the phase plane (pressure vs. flow rate) in Figure 3.

Concluding remarks

The analytical solution of the homogeneous fracture in the Laplace transform domain can be numerically inverted for specific choices of K_f to examine the zone of influence of the frequency and flow amplitude dependent fluid injection-withdrawal rate.

In the absence of an existing solution, inverse modelling using the finite difference simulator was initially done. Those calculations gave results consistent with and exhibiting the same trends as the measurements for similar (but not identical) test parameters. Recalculation of the numerical experiments using actual test parameters is in progress. The initial numerical experiments match the physical experiments sufficiently well to justify using the simulator to investigate the effects of fracture heterogeneity. Under the assumption of a uniform (parallel plate) fracture, both the transmissivity and storativity are regular functions of the radius of influence of the tests. Experimental deviations from the response expected for a uniform, homogeneous fracture can be interpreted as a measure of heterogeneity of the fracture's flow characteristics. This will be the focus of future numerical and laboratory experiments..

This work begins the demonstration of a potentially very powerful test procedure that can interrogate successively larger volumes of a flow system from a single well. Combined with monitoring wells to determine flow system anisotropy, this single-well test permits the measurement of the heterogeneity of flow characteristics between the wells, eliminating the need for assuming uniform conditions. The only other methods of determining this heterogeneity is by inference from cross-hole geophysical measurements or excavation of the fracture system.

References

[1]   Black, J.H. and Kipp, K.L. Jr., Determination of Hydrogeological Parameters Using Sinusoidal Pressure Testing: A Theoretical Appraisal, Water Resources Research, 1981, 17(3), 686-692.

[2]   Black, J.H. and Barker, J.A., Application of the Sinusoidal Pressure Test to the Measurement of Hydraulic Parameters, OECD/NEA Workshop on Geological Disposal of Radioactive Wastes – In Situ Experiments in Granite, Stockholm 1982, 121-130.

[3]   Black, J.H. and Holmes, D.C., Hydraulic Testing Within the Cross-Hole Investigation Programme at Stripa, CEC/OECD/NEA Workshop on the Design and Instrumentation of In Situ Experiments in Underground Laboratories for Radioactive Waste Disposal, Brussels 1984, 116-127.

[4]   Black, John H., Barker, John A., and Noy, David J., Crosshole Investigations – The method, Theory and Analysis of Crosshole Sinusoidal Pressure Tests in Fissured Rock, Stripa Project Internal Report 86-03, SKB, Stockholm, June 1986, 53 pp.

[5]   Motojima, Isao, Kono, Iichiro, and Nishigaki, Makoto, Crosshole Permeability Testing Method in Bedrock, Soils and Foundation, 1993, 33(4), 108-120.

[6]   Barker, J.A., A Generalized Radial Flow Model for Hydraulic Tests in Fractured Rock, Water Resources Research, 1988, 24(10), 1796-1804.

[7]   Geier, J.E., Doe, T.W., Benabderrahman, A., and Hässler, L., Generalized Radial Flow Interpretation of Well Tests for the SITE-94 Project, SKI Report 96:4, SKI, Stockholm, December 1996, 148 pp.

[8]   Selvadurai, A.P.S., Sinusoidal Testing of Fractures to Measure Hydraulic Heterogeneity, AECB Report RSP-0038[2], Atomic Energy Control Board, Ottawa, August 1997.

[9]   Selvadurai, A.P.S., Thermal Consolidation Effects Around a High Level Repository, AECB Report RSP-0029[1], Atomic Energy Control Board, Ottawa, March 1997.

---

1.   Unpublished contractor's report, available from Office of Public Information, Atomic Energy Control Board, P.O. Box 1046 Station B, Ottawa, Canada K1P 5S9.

Experimental Configuration For a Cyclic Flow Rate Test on a Fractured Granite Cylinder

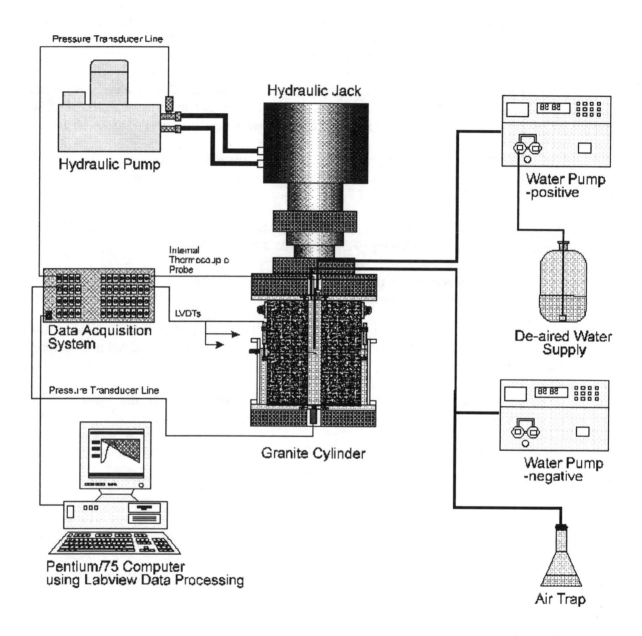

Figure 1. Schematic experimental configuration for conducting cyclic flow rate tests on fractured granite cylinders.

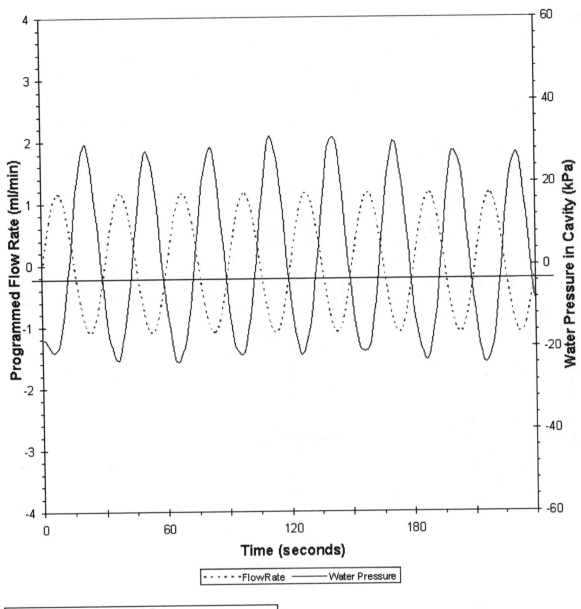

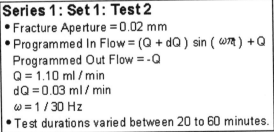

Series 1: Set 1: Test 2
- Fracture Aperture = 0.02 mm
- Programmed In Flow = (Q + dQ) sin ($\omega \pi t$) + Q
  Programmed Out Flow = -Q
  Q = 1.10 ml / min
  dQ = 0.03 ml / min
  ω = 1 / 30 Hz
- Test durations varied between 20 to 60 minutes.

Figure 2. Time histories of the fluid flow rate and the cavity fluid pressure.

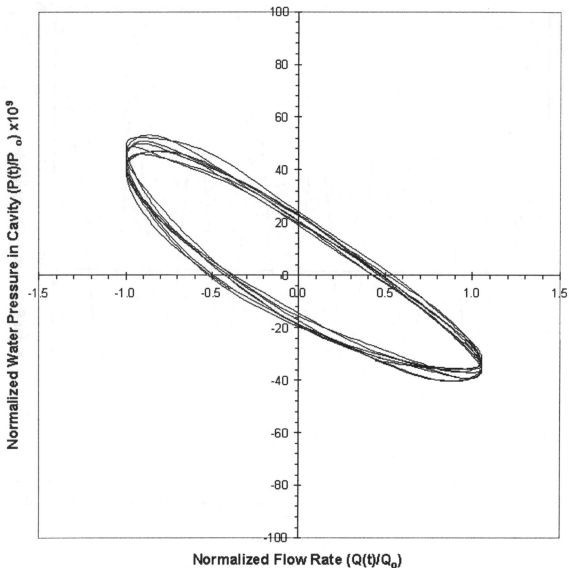

Normalized Flow Rate (Q(t)/Q$_o$)

Series 1: Set 1: Test 2

- Fracture Aperture = 0.02 mm
- Programmed In Flow = (Q + dQ) sin ($\omega\pi t$) + Q
  Programmed Out Flow = -Q
  Q = 1.10 ml / min
  dQ = 0.03 ml / min
  ω = 1 / 30 Hz
- Test durations varied between 20 to 60 minutes.

Figure 3. Experimental results for the normalized fluid pressure in the cavity vs. the normalized flow rate.

Effect of Sorption Kinetics on the Migration of Radionuclides in Fractured Rock

Anders Wörman

Institute of Earth Sciences, Uppsala University, Sweden

Shulan Xu

Institute of Earth Sciences, Uppsala University, Sweden

Abstract

A theoretical model has been developed for transport of radionuclides in fractured crystalline rock which among various other mechanisms also represents the effect of sorption kinetics. Batch tests on the sorption of ^{137}Cs on crushed rock facilitated the establishment of relationships for the sorption kinetics that could be translated to the conditions of intact rock. The relationships are consistent with a Langmuir adsorption isotherm for the equilibrium state. A computer-based simulation study indicates that the effect of sorption kinetics for Cs on crystalline rock can be an essential factor in migration of radionuclides in fractured rock on a length scale commonly used in laboratory experiments. Indirectly, this may have implications for the generalisation of experimental results also for prototype cases and performance assessment analyses.

Introduction

The method most widely considered around the world for the disposal of spent nuclear fuel is isolation in a deep repository in rock. As part of its research programme on the safety of final disposal, the Swedish Nuclear Power Inspectorate (SKI) initiated a project to study how spatial variability in rock chemistry in combination with spatial variability in matrix diffusion affects the radio nuclide migration along single fractures in crystalline rock. The project involves the development of a mathematical framework and a numerical simulation package, batch tests to determine the geochemical parameters and migration experiments. The purpose of this paper is to through light on the possible effects of sorption kinetics on migration of radionuclides in a single fracture. The information of geochemistry derived from experimental works is used as a basis for a simulation study of a hypothetical migration experiment.

Neretnieks [1] proposed a one-dimensional formulation for radio nuclide migration that includes the effect of matrix diffusion and instantaneous matrix sorption. This concept has been developed in two-dimensional formulations ([2] and [3]). The model framework developed in this project [4] also includes first-order sorption kinetics in the rock matrix.

As basis for the simulation study, the most salient parameters were determined for the Äspö-diorite collected at the Äspö Hard Rock laboratory located in south-eastern Sweden. Porosity, effective diffusivity, sorption equilibrium coefficient and sorption rate coefficient should be determined for Caesium-137. In this way, the results from the simulations will be representative of the specific rock types of interest and of an isotope of interest to the performance assessment analysis for the repository. To estimate all parameters under the same conditions could be important due to the influence of parameter correlations.

Model of migration process

The model of solute transport in a single fracture represents the main fracture plane as a rectangle with open upstream and downstream boundaries according to Fig. 1. Micro fissures are connected to the main fracture which allows the radionuclides to diffuse into the rock matrix lateral to the main flow direction. The micro fissures are mutually isolated and closed except at their connection to the main fracture plane. With respect to the factors mentioned so far, the part of the model concept that describes solute transport is identical to the one by Kunstman et al. [3]. The current model, however, also includes non equilibium adsorption. Furthermore, the current transport formulation incorporates variability in aperture and physical-chemical parameters as well as shear dispersion.

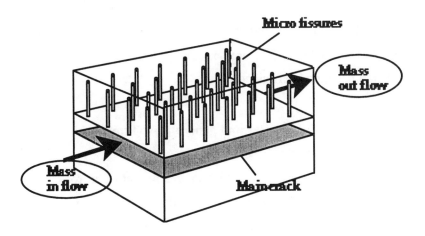

Fig. 1 Schematic illustration of the model framework used to analyse the effects of matrix diffusion and adsorption kinetics on radionuclide migration.

For incompressible fluids, we may write the depth-averaged form of the A/D equation as [4]

$$\frac{\partial \langle c \rangle}{\partial t} + \langle u_k \rangle \frac{\partial \langle c \rangle}{\partial x_k} - \frac{1}{h} \frac{\partial h}{\partial x_k} \left[\langle E_{kj} + D_{kj} \rangle \frac{\partial \langle c \rangle}{\partial x_j} \right] - \frac{\partial}{\partial x_k} \left[\langle E_{kj} + D_{kj} \rangle \frac{\partial \langle c \rangle}{\partial x_j} \right]$$

$$+ \gamma \langle c \rangle - 2 \frac{n D_p}{h} \frac{\partial c_m}{\partial z} \Big|_{z=0} = 0 \tag{1}$$

in which c is the concentration of solute (mass/activity per bulk volume of water), D_{kj} is molecular (ionic) diffusivity tensor in units [m^2/s], indices k = j = 1,2, u is advective velocity in units [m/s] (for inert particles the advective velocity equals the fluid flow velocity), λ is rate of radioactive decay in units [1/s], t is time, x_k is the Cartesian co-ordinate vector, D_p is pore diffusivity ($D_p = D\delta_D/\tau^2$ where δ_D is constrictivity and τ is tortuosity), n is porosity of rock matrix, h is the fracture aperture and E is dispersion coefficient. Brackets < > denote a depth average.

A first-order transfer function for the adsorption process between dissolved and adsorbed phases in the micro-fissures is assumed. For constant porosity, diffusion and adsorption in the rock matrix, the concentrations of solute in pore water and solid phase are stated as

$$\frac{\partial c_m}{\partial t} - D_p \frac{\partial^2 c_m}{\partial z^2} + \gamma c_m + k_r \frac{\phi}{n} (K_D c_m - c_w) = 0 \tag{2}$$

$$\frac{\partial c_w}{\partial t} + \gamma c_w - k_r (K_D c_m - c_w) = 0 \tag{3}$$

337

in which c_w is adsorbed mass phase per unit solid area [kg/m^2], c_m is dissolved mass phase per unit volume of water [kg/ m^3], ϕ is specific surface (solid surface area per unit volume of rock) [m^{-1}] and k_r is rate coefficient [s^{-1}], K_D _ [c_w / c_m]$_e$ a partition coefficient.

Strategy to measure sorption kinetics on particulate matter and generalise to intact rock

A main problem to estimate the effect of sorption kinetics of radionuclides migrating in rock is the difficulty to measure the sorption rate coefficients for intact rock. There does not exist any reliable method to measure instantaneously the state of sorption in intact rock pieces, whereas similar measurements can be readily done with crushed rock, i.e. particulate forms of the rock.

The strategy applied in this study was first to establish relationships between sorption rate as well as equilibrium state of the sorption process (partitioning) and specific inner surface (surface per unit volume of rock) for particulate rock. Further, these relationships can be used for a known equilibrium state of the sorption state in the intact rock to estimate 1) the inner surface of the intact rock characterising the sorption process and, hence, 2) to estimate the sorption rate coefficient for intact rock.

Consider a suspension of particulate granite in a solution of a certain sorbing substance. The rate of adsorption of the solute on the particles can be expressed in terms of entities from surface chemistry as [5]

$$G \equiv \frac{dc_a}{dt} = G_{ad} - G_{des}$$

(4)

in which the adsorption rate

$$G_{ad} = \frac{U}{\sqrt{3} \, 2} \, s \, \phi \, c_d$$

(5)

the desorption rate

$$G_{des} = a \, c_a$$

(6)

U is the squared mean speed of molecules, s is sticking probability, a is rate of departure, c_d is the concentration of the dissolved phase and c_a is the concentration of the adsorbed phase (both concentrations are defined as mass of solute per unit volume of the medium including particles). Eq.

(5) is based on the kinetic theory for molecules moving independently of and rapidly in comparison with the particles.

The form of (4) adopted herein is based on three additional assumptions of which the first is that the sticking probability $s = s_0 (1 - c_a/c_{as})$, the second that the saturation concentration $c_{as} = \alpha \phi$, and the third that the rate of departure $a = \beta \phi$ in which s_0, α and β are constants. Hence, (4) can be written as

$$G \equiv \frac{dc_a}{dt} = k_r \ (k_d c_d - c_a) \tag{7}$$

in which the rate coefficient

$$k_r = \phi \ (\beta + \gamma c_d/(\alpha \phi)) \tag{8}$$

the partition coefficient

$$k_d = \frac{1}{\beta/\gamma + c_d/(\alpha \phi)} \tag{9}$$

and $\gamma = [U/(2 \_3)] \ s_0$. The constants a, α and β should be determined from experiments with particulate matter. The translation of the partition coefficient to intact rock is given by $k_d = \phi \ K_D$. Equation (9) is given a kind of Langmuir adsorption isotherm.

Experiments to determine sorption kinetics

Drill cores having a diameter of 20 cm and a length of 50 -60 cm were used for laboratory experiments to determine various geochemical parameters. One rock type was Äspö-diorite; a medium-grained, porphyritic monzo/grano-diorite. A thin-sections were prepared from the drill cores and studied in transmissive light and the mineral compositions were determined by point counting on the thin-section. The dominating minerals are plagioclase, K-feldspar, quartz and biotite.

To provide a basis for the simulation study, several parameters were determined by first sectioning the drill core in a certain pattern. Porosity was measured by the leaching method [6], effective diffusivity by the through-diffusion technique [7], surface sorption selectivity by means of a microprobe analysis and sorption kinetics in a way subsequently described.

Slabs from the Äspö-diorite were crushed and dry sieved into eight size fractions. Particles with a size larger than 0.65 mm were washed several times with distilled water and dried at 80 °C for

24 hours. The mean diameters of the particle size fractions were determined using a Galai CIS-1 instrument, (0.0065, 0.011, 0.0154, 0.0227, 0.104 mm) except for 0.65, 1.198 and 2.397 mm which were determined according to the sieved size. ^{137}Cs was added in dissolved form into a synthetic ground water according to Allard and Beall [8] with a suspension of crushed granite. Samples were withdrawn from the gently shaken batch solution and immediately separated (by centrifugation) into their dissolved and particulate components.

The partition coefficient between the solid and the liquid phase, k_d, was obtained by measuring the concentration decrease in the solution, using a gamma-counter. The initial concentration of Cs was $1.3*10^{-5}$ mol/l. The particle concentrations were 40 g/l, 60 g/l and 150 g/l for the particle size of 0.0154 mm, and 40 g/l for all the other particle sizes. The batch volume was 200 ml. The experiment was performed at room temperature and under oxic condition.

The particle surface was determined as the BET (Brunauer, Emmet, and Tekller)-surface area using nitrogen gas and a FLOWSORB II 2300 analyser.

Experimental results

The porosity of the slabs decreases with slab thickness. For the largest thickness used, 15 mm, the porosity of the Äspö-diorite was found to be 0.4 % with a variance of 0.003 and a skewness of +0.48. The expected value of the effective diffusivity was $12.75 \ 10^{-13} \ m^2/s$ with a coefficient of variance of 0.017 and a coefficient of skewness of +1.42.

The rate coefficient k_r was obtained by taking it as a constant in each individual adsorption experiment and fitting the theory (solution to (7)) to data by means of a non-linear regression. The constants α, β and γ in (8) and (9) were determined using also a least square regression technique. In Figs. 2 and 3 the results are plotted in terms of $1/k_d$ versus c_d/ϕ and k_r/ϕ versus c_d/ϕ.

The partition ratio for Cs at equilibrium depends on the particle size or the ratio between mantle surface area and inner surfaces of the grains. As the grains become larger than about 1 mm, the inner surface will dominate. Hence, the specific inner surface can be taken as those found for the largest particles (1.198 and 2.397 mm); $\phi = 4500 \ m^2/m^3$. Experimental results of K_D for adsorption of Cs on intact crystalline rock indicate that $K_D \_ 0.05 \ m^3/kg$ [9] which means that $k_d \_ 130$. Hence, Fig. 2 indicates that, for the experimental conditions, the c_d/ϕ-ratio is practically zero and from Fig. 3 one obtains $k_r/\phi \_ 2.46 \ 10^{-4}$. Since, the BET-surface is about $4500 \ m^2/m^3$, one has $k_r \_ 1 \ [d^{-1}]$. These values ($k_d \_ 130$ and $k_r \_ 1 \ [d^{-1}]$) will be used as references for the simulations.

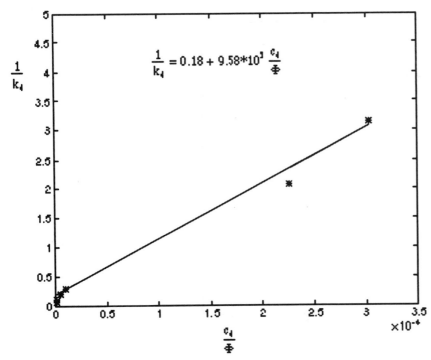

$$\frac{1}{k_d} = 0.18 + 9.58*10^3 \frac{c_d}{\Phi}$$

Fig. 2 $1/k_d$ vs. c_d/ϕ according to experimental results (*) and theory (solid line).

Simulation results of the migration of Cs in a single fracture

The governing system of equations, (1) - (3), can be rewritten in a dimensionless form and the following dimensionless parameters can be identified [4]; $U = u/u_c$, $E^* = E/(L_c \, u_c)$, $D_p^* = D_p/(L_c \, u_c)$, $X = x/L_c$, $K_R = (k_r \, L_c)/u_c$, in which L_c is length scale and u_c is velocity scale.

The scaling technique facilitates a generalisation of the simulation results to a continuous set of conditions and enables us to derive insight of the conditions for which sorption kinetics is an essential factor.

Six simulations were conducted with different values of K_R whereas the other dimensionless parameters were kept as constants. Fig. 4 shows the simulation results in terms of $c_d(t)$ at $X = 10$ for the various parameter set ups of defined in table 1, for a Dirac pulse defined at the upstream boundary and a semi-infinite domain (in the downstream direction). To translate the results in a dimensional form, the scaling variables have to be chosen in a way that is consistent with essential (necessary) constraints of the problem, namely that $Dp = 0.006$ m^2/y and $E = 0.06$ m^2/y . This was ensured by keeping $L_c \, u_c = 1$. An essential constraint specific to sorption of ^{137}Cs on Äspö-diorite is that $k_r = 1$ d^{-1} which leave only one possibility in each simulation to choose the scaling variable L_c. These are the values included in Table 1.

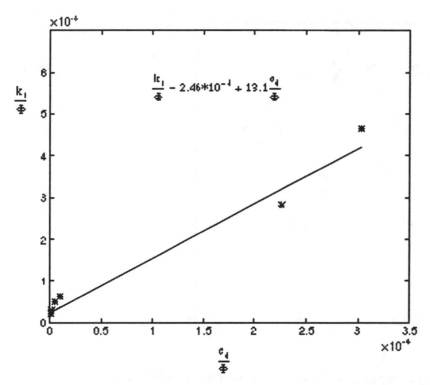

Fig. 3 k_r/ϕ vs. c_d/ϕ according to experimental results (*) and theory (solid line).

The kinetics of the adsorption process during the uptake phase can be interpreted as a decrease in the effective partition coefficient. In contrast, adsorption kinetics during the release phase can be interpreted as a net increase of the partition coefficient. This, in turn, causes a prolonged tail in the breakthrough curve. The simulation results in Fig. 4, for U = 20, indicate that the effect of sorption kinetics increases the height of the peak in the breakthrough by more than 10% if K_R is less than about 0.35. In a dimensional form this implies that if u = 645 m/y, the observation distance should be larger than 0.31 m for kinetics to be an insignificant factor (cf. Table 1). As u decreases to 270 m/y, this observation distance increases to 0.37 m (cf. Table 1).

Table 1 Definition of simulation cases.

Dimensionless variables							Dimensional variables				
U_k	E^*	D_p^*	X^*	K_R^*	L_c	$L_c^*U_c$	u (m/y)	E (m²/y)	D_p (m²/y)	x (m)	k_r (d⁻¹)
10	0,06	0,006	10	5	0,117	1	85,44	0,06	0,006	1,170	1
10	0,06	0,006	10	0,5	0,037	1	270,19	0,06	0,006	0,370	1
10	0,06	0,006	10	0,05	0,0117	1	854,4	0,06	0,006	0,117	1
20	0,06	0,006	10	3,5	0,0979	1	204,24	0,06	0,006	0,979	1
20	0,06	0,006	10	0,35	0,031	1	645,87	0,06	0,006	0,309	1
20	0,06	0,006	10	0,035	0,0098	1	2042,4	0,06	0,006	0,097	1

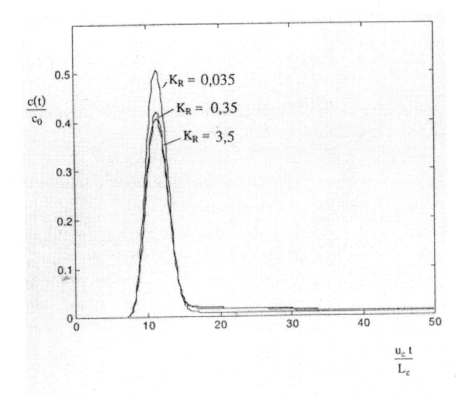

Fig. 4 Breakthrough at a distance X = 10 due to a Dirac pulse of Cs defined at the upstream boundary.

Conclusions

Sorption kinetics can be an important factor in the interpretation of experiments on the migration of Cs in fractured crystalline rock. Results from sorption experiments in combination with theoretical analyses of the migration process indicate that, if the flow velocity is 645 m/y in a fracture that conducts a solution of caesium, the distance of observation should be larger than 0.31 m for sorption kinetics to be an insignificant factor. As the velocity decreases, this distance increases.

Because sorption kinetics may be an essential factor in the interpretation of various laboratory experiments, it is a factor to be taken into account in the build-up of our understanding of the different mechanisms affecting the migration of radionuclides in crystalline rock. A proper understanding of theses mechanisms is essential to the generalisation to a prototype case and performance assessment analyses.

Acknowledgements

This study was financed by the Swedish Nuclear Power Inspectorate (SKI) under the grant number 14.9 940826-96076. Dr. Björn Dverstorp initiated the project and contributed with valuable comments.

References

[1]   Neretnieks, I, 1980. Diffusion in the Rock Matrix: An important factor in radionuclide retardation?, Journal of Geophysical Research, 85, 4379, 1980.

[2]   Moreno, L., Tsang, Y. W., Tsang, C. F., Hale, F. V. and Neretnieks, I., 1988. Flow and Tracer Transport in a Single Fracture: A Stochastic Model and Its Relation to Some Field Observations. Water Resources Research, Vol. 24. No. 12, 2033-2048.

[3]   Kunstman, H., Kinzelbach, W., Marschall, P. and Li, G., 1997. Joint Inversion of Tracer Tests Using reversed Flow Fields. *Journal of Contaminant Hydrology*. Vol. 26, Nos. 1-4.

[4]   Wörman, A. and Xu, S., 1996. Simulation of radio nuclide migration in crystalline rock under influence of matrix diffusion and adsorption   kinetics: Code   development   and   pre-assessment of migration experiment. SKI Report 96:22. Swedish Nuclear Power Inspectorate, Stockholm.

[5]   Attkins, P. W. , 1983. **Physical Chemistry**, *sec. ed., Oxford University Press*.

[6]   Skagius, A.-C. K., 1986. Diffusion of Dissolved Species in Matrix of Some Swedish Crystalline Rocks. *Ph. D Thesis. Dep. Chemical Engineering, Royal Institute of Technology, Stockholm, Sweden.*

[7]   Johansson, H, Byegård, J., Skarnemark, G. and Skålberg, M., 1997. Matrix diffusion of some alkeli- and alkaline earth-metals in granitic rock., Scientific Basis for Nuclear Waste Management XX, edited by W.J. gray and I.R. Tray, Mat. Res. Soc. Symp. Proc. Vol. 465, pp. 871-878.

[8]   Allard, B. and Bell, G. W. (1979). Adsorption of Americium on geologic media. *J. Environ. Sci. Health*, 6, 507.

[9]   Brandberg, F. and Skagius, K. (1991) Porosity, sorption and diffusivity data compiled for the SKB 91 study. SKB-TR-91-16.

LIST OF PARTICIPANTS

Canada

Cliff DAVISON
AECL
Whiteshell Laboratories
Pinawa, Manitoba ROE 1L0

Tel: +1 (204) 753 2311 [ext: 2299]
Fax: +1 (204) 753 2703
davisonc@aecl.ca

Peter FLAVELLE
AECB
280 Slater Street
Ottawa, Ontario KIP 5SN

Tel: +1 (613) 995 3816
Fax: +1 (613) 995 5086
flavelle.p@atomcon.gc.ca

Finland

Aimo HAUTOJÄRVI
POSIVA
Mikonkatu 15A
00100 Helsinki

Tel: +358 (9) 2280 3747
Fax: +358 (9) 2280 3719
aimo.hautojarvi@posiva.fi

Kai JAKOBSSON
STUK
PO Box 14
00881 Helsinki

Tel: +358 (9) 7598 8308
Fax: +358 (9) 7598 8382
kai.jakobsson@stuk.fi

Antti POTERI
VTT Energy
PO Box 1604
02044 VTT

Tel: +358 (9) 456 5059
Fax: +358 (9) 456 5000
antti.poteri@vtt.fi

France

Bernard BONIN
IPSN
CEN-FAR BP No 6
92265 Fontenay-aux-Roses Cedex

Tel: +33 (01) 46 54 73 96
Fax: +33 (01) 47 35 14 23
bonin@basilic.cea.fr

Lionel DEWIERE
ANDRA
Parc de la Croix Blanche
1-7 rue Jean Monnet
92298 Chatenay-Malabry Cedex

Tel: +33 (01) 46 11 80 38
Fax: +33 (01) 46 11 82 08
lionel.dewiere@andra.fr

Isabelle FOREST
ANDRA
Parc de la Croix Blanche
1-7 rue Jean Monnet
92298 Chatenay-Malabry Cedex

Tel: +33 (01) 46 11 81 20
Fax: +33 (01) 46 11 82 68
isabelle.forest@andra.fr

Emmanuel LEDOUX
Centre d'Informatique Géologique
Ecole Nationale Supérieure des Mines de Paris
35 rue Saint-Honoré
77305 Fontainebleau Cedex

Tel: +33 1 64 69 47 02
Fax: +33 1 64 69 47 03
ledoux@cig.ensmp.fr

Ghislain de MARSILY
Université Paris VI
Géologie Appliquée B123
75252 Paris Cedex 05

Tel: +33 (1) 4427 5126
Fax: +33 (1) 4427 5125
gdm@ccr.jussieu.fr

Emmanuel MOUCHE
CEA
DMT/SEMT/TTMF, C.E de Saclay
91191 Gif-sur-Yvette Cedex

Tel: +33 (01) 69 08 66 99
Fax: +33 (01) 69 08 82 29
manu@semt2.smts.cea.fr

Germany

Eckhard FEIN
GRS
Theodor-Heuss-Strasse 4
38122 Braunschweig

Tel: +49 (531) 8012 292
Fax: +49 (531) 8012 200
fei@grs.de

Klaus-Jürgen RÖHLIG
GRS
Schwertnergasse 1
50667 Köln

Tel: +49 (221) 2068 796
Fax: +49 (221) 2068 888
rkj@grs.de

Klaus SCHELKES
BGR
Stilleweg 2
30655 Hannover

Tel: +49 (511) 643 2616
Fax: +49 (511) 643 2304
k.schelkes@bgr.de

Jurgen WOLLRATH
BfS
P.O. Box 10 01 49
38201 Salzgitter

Tel: +49 (531) 592 7704
Fax: +49 (531) 592 7614
jwollrath@bfs.de

Korea

Chul-Hyung KANG
KAERI
PO Box 105, Yusong
Taejon, 305-600

Tel: +82 (42) 868 2632
Fax: +82 (42) 868 8850
chkang@kaeri.re.kr

Spain

Julio ASTUDILLO
ENRESA
Emilio Vargas 7
28043 Madrid

Tel: +349 (1) 566 8120
Fax: +349 (1) 566 8165
jasp@enresa.es

Jesus CARRERA
School of Civil Engineering
UPC
Campus Nord D-2
08034 Barcelona

Tel: +349 (3) 401 6890
Fax: +349 (3) 401 6504
carrera@etseccpb.upc.es

Jaime GÓMEZ-HERNÁNDEZ
Dep. de Ingeniería Hidráulica
UPV
46071 Valencia

Tel: +349 (6) 3879 614
Fax: +349 (6) 3877 618
jaime@dihma.upv.es

Juan. C. MAYOR
ENRESA
R&D Coordination Department
Emilio Vargas, 7
28043 Madrid

Tel: +349 (1) 566 8217
Fax: +349 (1) 566 81 65
nvid@enresa.es

Sweden

Björn DVERSTORP
SKI
Klarabergsviadukten 90
106 58 Stockholm

Tel: +46 (8) 698 8486
Fax: +46 (8) 661 9086
bjornd@ski.se

Allan EMRÉN
Department of Nuclear Chemistry
Chalmers University of Technology
41296 Göteborg

Tel: +31 772 2801
Fax: +31 772 2931
allan@nc.chalmers.se

Gunnar GUSTAFSON
Chalmers University of Technology
Deptartment of Geology
41296 Göteborg

Tel: +46 (31) 772 2050
Fax: +46 (31) 773 2070
g2@geo.chalmers.se

Ivars NERETNIEKS
Royal Institure of Technology
Dept. of Chemical Engineering and Technology
10044 Stockholm

Tel: +46 (8) 790 8229
Fax: +46 (8) 790 6416/105 228
niquel@ket.kth.se

Jan-Olof SELROOS
SKB
Box 5864
102 40 Stockholm

Tel: +46 (8) 459 3425
Fax: +46 (8) 661 5719
skbjos@skb.se

Anders WÖRMAN
Department of Earth Sciences
Uppsala University
Norbyvägen 18B
75236 Uppsala

Tel: +46 (0) 18 471 25 23
Fax: +46 (0) 18 471 27 37
anders.worman@geo.uu.se

Switzerland

Erik FRANK
HSK
5232 Villigen - HSK

Tel: +41 (56) 310 39 45
Fax: +41 (56) 310 3907
frank@hsk.psi.ch

Andreas GAUTSCHI
NAGRA
Hardstrasse 73
5430 Wettingen

Tel: +41 (56) 437 12 38
Fax: +41 (56) 437 13 17
gautschi@nagra.ch

Jörg HADERMANN
PSI
5232 Villigen

Tel: +41 (56) 310 2415
Fax: +41 (56) 310 2821
hadermann@psi.ch

Andreas JAKOB
PSI
5232 Villigen PSI

Tel: +41 (56) 310 2420
Fax: +41 (56) 310 2821
andreas.jakob@psi.ch

Olivier JAQUET
Colenco Power Engineering AG
Mellingerstrasse 207
5405 Baden

Tel: +41 (56) 483 15 76
Fax: +41 (56) 493 73 57
jaq@colenco.ch

Martin MAZUREK
Rock/Water Interaction Group
Institutes of Geology and of Minerology and Petrology
University of Bern,
Baltzer-Str.1,
3012 Bern

Tel: +41 (31) 631 8781
Fax: +41 (31) 631 4843
mazurek@mpi.unibe.ch

United Kingdom

Kathryn CLARK (on behalf of PNC, Japan)
QuantiSci Ltd.
Chiltern House
45 Station Road
Henley-on-Thames
Oxfordshire, RG9 1AT

Tel: +44 (1491) 410 474
Fax: +44 (1491) 57 6916
kclark@quantisci.co.uk

C. Peter JACKSON
AEA Technology
424.4 Harwell, Didcot
Oxfordshire, OX11 ORA

Tel: +44 (1235) 433 005
Fax: +44 (1235) 436 579
c.peter.jackson@aeat.co.uk

Rae MACKAY
School of Earth Sciences
Birmingham University
Birmingham B15 2TT

Tel: +44 (121) 414 6142
Fax: +44 (121) 414 4942
r.mackay@bham.ac.uk

Simon NORRIS
UK Nirex Ltd.
Curie Avenue
Harwell
Didcot, Oxon, OX11 ORH

Tel: +44 (1235) 825 310
Fax: +44 (1235) 820 560
simon.norris@nirex.co.uk

John PORTER
Hydrogeology Department
AEA Technology
Building 7
Windscale, Seascale
Cumbria, CA20 1PF

Tel: +44 (19467) 72332
Fax: +44 (19467) 72996
john.porter@aeat.co.uk

Paul SMITH
Safety Assessment Management Limited
Sladen Stables
Jaggers Lane
Hathersage
Hope Valley
Derbyshire S32 1AZ

Tel: +44 (1433) 651 599
Fax: +44 (1433) 659 008
paul@samltd.demon.co.uk

United States

Gilles BUSSOD
Los Alamos National Lab.
EES-13, Mail Stop J-521
Los Alamos, NM 87545

Tel: +1 (505) 667 7220
Fax: +1 (505) 667 1934
gbussod@lanl.gov

Tom CLEMO
Environmental Evaluation Group
7007 Wyoming Blvd NE
Albuquerque, NM 87109

Tel: +1 (505) 828 1003
Fax: +1 (505) 828 1062
tclemo@eeg.org

Joel GEIER
Clearwater Hardrock Consulting
Oregon State University
14505 Corvallis Road
Monmouth, Oregon 97361

Tel: +1 (541) 928 1829
Fax: +1 (541) 928 2815
jgeier@ibm.net

Lucy MEIGS
Sandia National Laboratories
PO BOX 5800, Mail Stop 1324
Organization 6115
Albuquerque, NM 87185-1324

Tel: +1 (505) 848 0507
Fax: +1 (505) 848 0605
lcmeigs@nwer.sandia.gov

NEA

Philippe LALIEUX
OECD Nuclear Energy Agency
Le Seine St-Germain
12 boulevard des Îles
92130 Issy-les-Moulineaux
France

Tel: +33 (1) 45 24 10 47
Fax: +33 (1) 45 24 11 10
lalieux@nea.fr

Claudio PESCATORE
OECD Nuclear Energy Agency
Le Seine St-Germain
12 boulevard des Îles
92130 Issy-les-Moulineaux
France

Tel: +33 (1) 45 24 10 48
Fax: +33 (1) 45 24 11 10
pescatore@nea.fr

OECD PUBLICATIONS, 2, rue André-Pascal, 75775 PARIS CEDEX 16
PRINTED IN FRANCE
(66 98 09 1 P) ISBN 92-64-16099-X – No. 50203 1998